D1123159

THE MIDDLE EAST WATER QUESTION

Also published by I.B.Tauris:

Water, Peace and the Middle East edited by J.A. Allan

THE MIDDLE EAST WATER QUESTION

Hydropolitics and the Global Economy

Tony Allan

I.B.Tauris *Publishers*
LONDON • NEW YORK

DOUGLAS COLLEGE LIBRARY

Paperback edition published in 2002 by I.B.Tauris & Co Ltd
6 Salem Road, London W2 4BU
175 Fifth Avenue, New York NY 10010
www.ibtauris.com

In the United States of America and in Canada distributed by
St Martins Press, 175 Fifth Avenue, New York NY 10010

First published in 2001 by I.B.Tauris & Co Ltd

Copyright © J.A. Allan, 2001, 2002

The right of J.A. Allan to be identified as the author of this work has been asserted by the author in accordance with the Copyright, Designs and Patents Act 1988.

All rights reserved. Except for brief quotations in a review, this book, or any part thereof, may not be reproduced, stored in or introduced into a retrieval system, or transmitted, in any form or by any means, electronic, mechanical, photocopying, recording or otherwise, without the prior written permission of the publisher.

ISBN 1 86064 813 4

A full CIP record for this book is available from the British Library
A full CIP record for this book is available from the Library of Congress

Library of Congress catalog card: available

Typeset in Baskerville 11/12pt by The Midlands Book Typesetting Co, Loughborough
Printed and bound in Great Britain by MPG Books Ltd, Bodmin

Contents

Figures and Tables

Tables

Chapter 3

Figures

Chapter 6
Figures

Tables

Chapter 7
Tables

Chapter 8
Figures

Acknowledgements

Many minds and individuals interacting in projects, conferences and workshops have made it possible to handle the subject of Middle Eastern and North African (MENA) water in the eclectic way attempted in this book. Hydrologists such as John Sutcliffe and his many colleagues in BGS, CEH (was IoH) and HRWallingford, all located in Wallingford, that haven of world class water science, have enriched the material. John Rodda, Roger Calow, and particularly Alan Hall come to mind. Alan made possible access to the vast international community of soulmates in the world of water through cooperation on the Global Water Partnerships *Framework for Action* activity of 1999–2000. I first met David Grey in his hydro-geological identity while he was at BGS. At that stage he had a UK brief but managed to fit in a simultaneous almost full-time commitment to capacity building in Palestine. On his return to the World Bank David devoted his vision and exceptional energy and diverse entrepreneurship to making a difference to water management and water institutions in Africa. He has helped the understanding of Africa's water and its problems by many including that of the author.

Hydrologists understand water and the environment. Engineers are equally intimate with water. Their advice has figured prominently in the research for this book. Water is engineered locally and as a result engineers get closer to water using communities than most social scientists. A period associated with Martin Adams, a remarkable interdisciplinarian, of Hunting Technical Services in Egypt in the early 1980s was a particular privilege. It deepened my respect for the contribution of the engineering team working with him and of their sensitivity to social issues as well as their concern for concrete structures. The staff of the MacDonald's, Gibb and Partners and Binnie in Egypt and of Brown and Root in Libya have richly informed the analysis. It has been of great assistance to have had occasional meetings over the past two decades with Bayoumi Attiyah of the Egyptian Ministry of Public Works and Water Resources. His depth of understanding is special and his willingness to inform and discuss very generous.

A number of engineer-economists have been especially important in providing insights. Peter Rogers at Harvard, John Briscoe at the World Bank, Chris Perry of IWMI and John Davies formerly of Coopers and Lybrand have shown me what a very strong framework of analysis their chosen disciplinary combination makes.

In the field of economics I am indebted to Dale Whittington who has tutored me a number of times in the past decade on the history of environmental economics and most recently on the theory of the second best. Jo Morris of Cranfield University is amongst my favourite economists. He defaults to evident wisdom rather than to shaky economic theory. I am also very grateful indeed to Steve Merrett for the intellect he has devoted in his association with SOAS to bringing economic theory to bear on water and its allocation and management. Many of us have benefited. Seminars from and conversations with Jeremy Berkoff have provided ideas that have greatly improved the political economy elements of the book which are its most substantial sections. David Storer was also very helpful in providing insights into the controversial field of privatisation. More recently Hilary Sunman has been helpful in drawing attention to the economic value of environmental intangibles. My encounter with David Kinnersley in 1991 was the first of many very pleasurable meetings. His grasp of relevant economics, law and the politics of institutional change is impressive; his capacity to communicate about them extraordinary. He introduced me to the fundamental place of 'political feasibility' in the very political process of water policy reform. Political feasibility is the central inspiration of this book. I also owe a debt of gratitude to the late Paul Howell who enabled me to realise a personal ambition to become an analyst of the waters of the Nile. Working with him on the 1990 SOAS conference on the Nile and the resulting publication was a privilege with many important intellectual and networking spin-offs which have greatly enhanced this book.

The sections on water law and international water law have been immeasurably improved by my acquaintance with Chibli Mallat of the Université de St Joseph in Beirut. For a period when a colleague at SOAS he chose to share my interest in the water resources of the Middle East. He introduced me to a wide range of specialists in the field of water law and ensured that we published productively. Patricia Wouters and Sergei Vinogradov at the University of Dundee have added greatly to my understanding of water law. The impact that they have made in the UK by their research and educational contributions in the field of water law is beyond measure. One can only recommend very highly indeed their on-line databases which made possible a very important section of the chapter on law in this book. They also made it possible to meet Stephen McCaffrey who provided unique insights into the UN ILC process. I am also grateful for the chances to discuss water law issues with Laurence Boisson De Chazournes of the University of Geneva and Ellen Hey of Erasmus University in Rotterdam. My influential colleague Philippe Sands at SOAS has been important in shaping everyone's thinking on international environmental law.

The politics of water have not yet unfortunately been the focus of many prominent political scientists. The ideas of John Waterbury, now at AUB in Beirut, have been very important indeed to all of us involved in MENA water for the past twenty years. Miriam Lowi has also left her mark on the hydropolitical theory of the MENA region. I have had the privilege of observing water politics practised by a number of exceptional individuals. Mahmoud Abu Zeid, Minister in Egypt, has brought intellect, science, leadership and managerial skills to bear in an exceptionally effective way both in Egypt and at the global level. It has been a great privilege to be informed and diverted for many hours by Munther Haddedin, of Jordan. He has been present at some pivotal moments in the MENA region's water negotiations. Ahmed Mango and Aoun Kawshaneh in Amman have over the years provided special insights on the political and legal aspects of water in the contentious Jordan Basin. Meier bin Meier of Israel has also provided me with some otherwise inaccessible insights. Shaul Arlosoroff, ex-Mekorot and ex-World Bank, has a range of experience which enables him to share a rare level of understanding of the water sector. In Turkey one has enjoyed accessing the experience of Olcay Unver, President of the GAP Project in Ankara, and his understanding of innovation and social change associated with major regional projects. Another giant in the field was Zewdie Abate of Ethiopia. His death in 1993 left a big gap in the professional ranks of Ethiopia and left us without a major player in the debates over Nile waters.

It has been my privilege to become acquainted with a group of younger researchers in the field of politics from whom I have derived the political theory which has informed substantial sections of this book. Alan Nicol, of SOAS and ODI, with his deep understanding of Egypt, the Sudan and Ethiopia, helped me grasp a different level of understanding of water in national and regional politics. Jeroen Warner, currently at the Flood Hazard Research Centre of Middlesex University, introduced me to the rich analysis on water and environment of the past decade in the Netherlands, and also to the relevant international relations theory of Buzan. Conversations at the photocopier with Toby Dodge, of the SOAS Politics Department, have been invaluable as were the offprints he troubled to provide. To Charles Tripp of the SOAS Politics Department we are indebted for the friendly concept of 'sanctioned discourse' which by now must be the most cited unpublished comment in scholarship. The term was shared during a brief conversation outside the lift at SOAS in 1996. The relationship with the scientists at Middlesex University is on-going, fruitful and reciprocal and I have appreciated the insights provided by Colin Green, Clare Johnson and Terri Sarch.

At SOAS I have been fortunate to have a group of graduate researchers who have spent long periods in the field producing the fruits that can only come from such intensive and continuous fieldwork. Our understanding of the worst water crisis in the MENA region, that of Ta'iz in Yemen, is uniquely well understood because of the work of Chris Handley. Gerhard Lichtenthaler is produing another exceptional study of another severely water challenged region of Yemen, Sa'adah in the north. Other studies by Arnon Medzini and Ghulam Farouq have also enriched the range of accessible evidence used in the book. I shall always be grateful that Hiroyuki Yoshida defaulted to his preferred philosphy during his thesis research and in the process forced me to come to terms with the insights of mid- and late twentieth century French philosphy. The visit of Tony Turton of the University of Pretoria in the first months of 1999 added immense value to the level of intellectual debate on water issues. He quickly digested what we had to offer and set about incorporating the concepts into a much more comprehensive framework which is having significant impact in South Africa and in the global water community. Major thanks go to him. His introduction to Barbara Schreiner has been of immense value to our understanding in London of the water policy challenges in semi-arid regions with troubled politics. His work with Leif Ohlsson, of the Orebro University in Sweden, is the most cited in the book.

Other groups of scientists and professionals have been very helpful. Involvement in the informal Palestine-Israel workshops in 1992–1997 on the joint management of aquifers was a rare privilege. Marwan Haddad of Nablus University and Jad Isaac of the Palestinian Consultancy Group were just two of many very over-committed Palestinian scientists who shared their knowledge and their judgements. Eran Feitelson and Hillel Shuval, both of the Hebrew University, were prominent on the Israeli team. In Israel it has been a pleasure and a privilege to relate to scientists in the University of Haifa, Arnon Sofer and Nurit Kliot, who both reached the publishers with their important MENA water books in the early 1990s. Uri Shamir at the Haifa Technion was always very generous in his willingness to share his vast knowledge of the water sector. Interactions with David Newman at Beer Sheva were also always very useful.

Other networks, increasingly electronically based, enabled access to individuals with special knowledge, including Andrew Macoun, Christopher Ward and Ines Dombrowsky of the World Bank. At FAO I have appreciated the chance to meet and liaise with Hans Wolter, Bo Appelgren, Wulf Klohn and John Latham. The Erasmus Water Workshops, a European Union Initiative linked kindred spirits in the universities of Europe. I particularly appreciated the leadership of Pierpaolo Faggi in Padua. He made possible productive links with Franca Battigelli in Udine, with Jan Lundquist and Geoffrey Gooch in Linkøping, with Leandro del Moral in Sevilla, and with Thierry Joliveau and Jacques Bethemont in St Etienne. Other friends across the world have been inspirational in different ways. The example of Murray Watson is a demanding one. He constantly demonstrates that understanding only comes with very lengthy acquaintance with communities and political systems. One is aware that there will be too little depth in a book such as this for him.

Work in Yemen led to acquaintance with Dr Chauduri at UNDP and to a very rewarding relationship with Mohamed El Eryani at Sana'a University. The Framework for Action process of the Global Water Partnership led to a rapid expansion of one's information resources. Meeting Ivan Cheret of Lyonnaise des Eaux was as important as the encounter with David Kinnersley. They both know that first steps and last steps are always political if people have water policy reform in mind. To them I make an additional special dedication of this study.

The book would not have existed without Keith McLachlan's long friendship and support and his invitation to contribute, in 1966, to his Joint Libya-London (SOAS) Universities Research Project. The three years of that interdisciplinary study were fundamental to the gaining of any level of understanding of the allocation and management of water in Libya and subsequently in the MENA region more generally. I also benefited greatly from exposure to two economist heavyweights, Edith Penrose and Robert Mabro. Most important was the experience derived from working with Mukhtar Buru and many other staff and students in the University of Libya. One's strong affection for Libya is deeply rooted.

A special level of personal thanks goes to people who have assisted in the production of the book. Yasir Mohileldeen has been a very valuable member of the SOAS community for over three years. His achievements have been numerous and significant. One of his diverse skills is cartographic excellence and I am indebted to him for the production of the figures in the book. His research, including advanced GIS inputs, also underpin a number of the case studies cited in the book. Donald Chambers has assisted in the tedious but essential task of analysing MENA grain imports which are a central feature of the analysis. Ian Griffin's support on what would be an unmanageable water bibliography has been very important. I am indebted to Jonathan Wild for editorial support in editing earlier books on the Nile, which are heavily cited in this book.

A number of friends have looked at parts of the book. In this connection I owe a special debt to Brian and Lynne Chatterton. However, the omissions, mis-emphasis and other errors remain my responsibility. I am also grateful to David Stonestreet of I.B.Tauris for indulging my inability to keep promises and for dealing with production.

Finally I recognise warmly my department at the School of Oriental and African Studies for providing the stream of resources—for most of 34 years in this case—which allow academic staff to write and research. A number of colleagues have been helpful over a long period and I would like to thank Robert Bradnock and Philip Stott for their friendly support these past years. I would also like to thank all the administrative staff of the School for the decades of support and on occasions for their patience. Beyond SOAS one recognises a remote and constructive tension with national government. Without the tax-payer's essential contribution mediated by the Department of Education's unpopular but useful research assessment the book would have taken longer to reach the publisher.

Dedication

This book and much else would not exist without Mary who usefully reminds us that people do not like being written about.

Is there a water problem in the Middle East and North Africa? Another case of the blind people and the elephant?

Accessing global soil moisture as traded virtual water. And on the dangers of scientific analysis by outsiders visiting communities and polities

Is there a problem?
Water is a problem say some
Others say it's not
Water is in short supply say some
Others insist it's not

Like the blind men and the Elephant
It depends on what you touch;
But even more important still
Is who you are and what you vouch

Useful and un-useful science
Water is very short
The hydrologist insists
Short for whom?
Ask the knowing economists?
For they detect virtual water
In food embedded
And wondrous subsidies
For food importers added

Politics
Peoples and governments are blind
Alien scientists insist
As dangerous water fantasies exist
And in beliefs for millennia persist

Wise politicians know that they
Must resist the simple (scientific) truths
And by subscribing to pervasive water lies
Remain in power along with (Their) essential truths
[Allusion to Goleman 1997]

Impaired interpretation
Why not use Our theories say social theorists
They explain all these mysteries
But (alien) scientists are blind fools
Devoted to familiar narrowing tools.
'It's far, far better to be blind
Knowing only what impairment finds.'

The answer

Because the population too fast grows
The answer to the question posed—
is 'Yes there's a regional deficit'
But for the water stressed
A solution also does exist
Through trading to 'entitlement'. [An allusion to Sen. 1981]

Tony Allan, 1997, SOAS, ta1@soas.ac.uk

PART ONE: MENA WATER RESOURCES, ALLOCATION, MANAGEMENT AND PERCEPTION

CHAPTER 1

Water optimists and water pessimists: insiders and outsiders and an analytical framework

It is a paradox that the water pessimists are wrong but their pessimism is a very useful political tool which can help the innovator to shift the eternally interdependent belief systems of the public and their politicians. The water optimists are right but their optimism is dangerous because the notion enables politicians to treat water as a low policy priority and thereby please those who perceive that they are prospering under the old order.

The comparative disadvantage in economic terms of the Middle East and North Africa with respect to water is an extreme and classic case.

Thirty years is a rather short transitional period for the necessary major adjustments in water policies to be developed in response to limited water availability.

Those purveying the economic and environmental facts of life which contradict the deeply held belief systems of whole populations will be ignored if they do not shape their message and pace its delivery to accord with political realities.

Not all waters are equal: some are more evident than others.

The issue of whether there will be enough water for a future global population double its present size is controversial. The answer to the question is of particular importance to the peoples and political leaders of the Middle East and North Africa. The region's economies are already as dependent on global water as they are on the renewable waters of the region. They will be much more dependent on global water in future.

The answer is almost certainly a resounding yes. There will be enough fresh water in the global system. But on the supply side, that is freshwater availability, the science has not yet been done to prove the future capacity of the global freshwater system. A researched estimate of sufficient precision to be useful for politicians and decision makers would cost many billions of dollars. On the demand side, freshwater needs are driven by rising populations. Here the precision of the estimates of future global population vary by over 50 per cent. In this uncertain information domain there is space for numerous pessimists and optimists to spin counter interpretations. Whether we believe the optimists, including the author and most economists (IFPRI 1995 and 1997, Dyson 1994:403), or the pessimists (Brown 1996a and 1996b, Brown and Kane 1995, Postel 1997 and 1999) depends upon the assumptions used by the respective analysts.

Mega questions tend to be ignored; or attract untestable ideological interpretations of religious intensity. Is there enough freshwater for future populations is a mega question and it is not given a fraction of the attention it deserves by scientists. In the 1960s and 1970s there were attempts in the former Soviet Union to review the world's water balance (L'vovich 1969 and 1974). Water availability and use have also been addressed (Shiklamanov 1985, 1986 and 1994, Shiklamanov and Sokolov 1983, Shiklamanov and Markova 1987, Gleick 1994). UNESCO (Shiklamanov 2000) has used the same scientists to refine the estimates.

Meanwhile an agency is prepared to devote five million dollars a year (at least) to providing advice on managing water and on lending much more for water projects without offering a view on the status of global water. Institutions such as the World Bank can produce shelves of reports about allocating and managing water for the interested professional and general reader (World Bank 1993, Serageldin 1994 and 1995) and for the specialist focused on local and regional issues (World Bank 1993, 1994, 1995) without situating the debate in the global resource context. Since 1994 there has been a welcome and responsible attempt by the Food and Agricultural Organization (FAO) to address the subject of water availability for Africa and the Middle East (FAO 1995a and 1997a) and on water use by the major using sector, irrigation (FAO 1995b and 1997b). The publications provide first approximations which should be given prominence so that the reliability of the data can be enhanced by progressive iteration. Meanwhile the water professionals considering the world scale are divided between those that judge that humankind will respond to future resource demands (McCalla 1997) and others who have devoted enormous attention to the need to change the way we respond to the growing water supply challenges (Serageldin 1993 and 1994, World Water Council 2000, GWP 2000).

A balanced perspective on the population, water and food nexus tends to come from the International Food Policy Research Institute

(IFPRI 1995 and 1997) which identifies low, medium and high scenarios for population. These predict respectively 7.7 billion, 9.4 billion and 11.1 billion for the global population by 2200.

> In its 1996 study, the United Nations found that population growth between 1990 and 1995 was 1.48 percent per year, rather than the 1.57 percent projected in 1994. In light of this lower rate, the United Nations revised its projections for population growth in the next century, based on three different assumptions about the fertility of the world's women. The medium fertility model—the one usually considered the most likely—would put the world's population at 9.4 billion by 2050 (half a billion lower than the United Nations 1994 estimates). World population would continue to grow until 2200, according to this model, when it would stabilize at 10.73 billion.
>
> The medium fertility model falls in the middle of a wide range of possible outcomes. Low fertility would result in a world population of 7.7 billion by 2200, whereas high fertility would mean 11.1 billion mouths to feed in 2200. These extremes are by no means unrealistic, and the population debate is far from over. ... Public policy and individual behavior, they say, will ultimately determine the world's population.
>
> (IFPRI 1997)

About Middle East water there is less scientific controversy. The Middle East as a region ran out of water in the 1970s. The news of this important economic fact has been little exposed. In political systems, facts, including those on water, which are judged to have costly political consequences can easily be ignored or de-emphasised. With selection and distortion of information being the norm in political processes it is predictable that discourses on Middle Eastern and North African water will be misleading and confusing.

For political leaders in the region political imperatives are more compelling than scientific facts. On water, these imperatives drive them to assert that their economies have **not** run out of water. An ex-Prime Minister of Egypt (Higazi 1994), and mid-1990s Egyptian ministers of water resources, of agriculture and of planning have vied in public in the vehemence of their assertions that Egypt has sufficient water. (Arab Research Centre 1995) The main reason that there can be such different interpretations is that the politicians do not specify what 'sufficient' means.

By sufficient, scientists mean sufficient 'indigenous' water to meet the total water needs *including water for food production*, for the industrial and municipal water needs of an economy and for household requirements. Politicians do not specify. In practice they mean only part of the 'total' needs. Without making their assumptions clear they imply that 'sufficient freshwater' is that volume necessary to sustain the existing jobs in agriculture and industry and for municipal and industrial needs. In a economy such as Israel this may be only 25 per cent of the water which would be needed for food self-sufficiency. For Egypt it is about 60 per cent of the water needed for food self-sufficiency. For politicians in almost all countries in the region the food gap caused by the insufficiency of water has to be ignored. To draw attention to the water gap and the food gap could be politically suicidal.

Reviewing the water resources across the region it is evident that the water demands of the populations of the Arabian Peninsula and desert Libya had exceeded the capacity of their water resources for food self-sufficiency by the 1950s. Israel and Palestine also ran out of water in the 1950s, Jordan in the 1960s, Egypt in the 1970s. The Maghreb countries have recently entered water deficit. The Tigris-Euphrates riparians have not fully utilised their water resources but will do so in the next decades. The Sudan has some way to go before it fully utilises the share it has agreed with Egypt on the Nile—a share not endorsed by the other upstream riparians.

The major indicator of the scale of the water deficit of an economy is the level of its food imports. The reason food imports are such a strong indicator of water deficit is that the water required to raise food is what an economist would refer to as the dominant consumptive use of water. The use is dominant whether viewed from the point of view of the individual citizen or the national economy. Water used in the agricultural sector exceeds by ten times the water used by the industrial and municipal sectors combined.

An individual needs each year only a cubic metre of drinking water, between 50 and 100 cubic metres for other domestic uses, though a much lower actual use is the norm in many rural communities in countries in other regions, for example in the economies south of the Sahara. By contrast an individual needs each year at least 1000 cubic metres of water, either naturally occurring in soil profiles, or transported to the profiles by irrigation systems, to raise the food needs of that individual. At the national level over 90 per cent of all national water budgets are devoted to the agricultural sector.

In the arid and semi-arid Middle East the dominance of the agricultural water demand is stark. There is little or no naturally occurring soil water even in the winter when parts of the region do receive rainfall. By contrast in the economies located in temperate

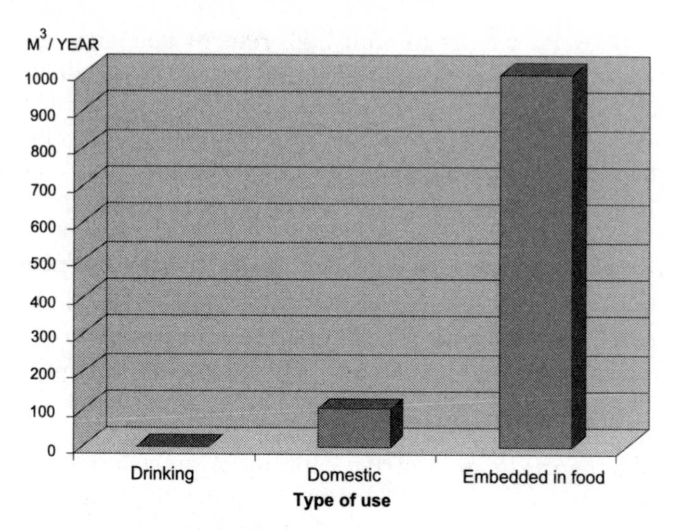

Figure 1.1 Individual water use per year

latitudes, in Europe and the humid tracts of North America for example, the issue of the relative demands of the agricultural and the industrial (including services) sector is scarcely evident. In temperate latitudes crop production is almost totally based on soil water which occurs naturally. Soil water tends to be taken for granted in the economies located in the humid temperate zone. The huge volumes of water utilised by agriculture are not counted as part of the national water budget. Soil water is a free good.

While water is treated as a free good in temperate humid regions, in the semi-arid and arid Middle East and North Africa agricultural water is expensively won because of the costs of storage and distribution. Storage is needed to ensure timely availability and to reduce the loss of water for the economic activities of a political economy. Mobilising such water can in addition be politically stressful both nationally—through environmental impacts, and internationally—through riparian conflict. The comparative disadvantage in economic terms of the Middle East and North Africa with respect to water is an extreme and classic case.

Soil water and the economic efficiency of water use are themes to which we shall return throughout the analysis. The reasons will become clear. First, it is *global soil water* which balances the water budgets of all the economies of the Middle East and North Africa, with the arguable exception of Turkey. Secondly, the *effective allocation of water* between sectors to gain high returns and high levels of employment are fundamental to economic and political stability. The environmental significance of water will not be ignored, but it will be argued that understanding the global availability of water and achieving the economically sound use of water are the major current water policy and management challenges.

The regional water gap and the global water surplus?

The rate at which food imports have been rising in the Middle East and North Africa since the 1970s confirms the scientist's contention that the region has run out of water. The production of every tonne of a food commodity such as wheat requires a water input of about 1000 tonnes (cubic metres). The trend in cereal imports reflects a reasonable approximation of the capacity of an economy to meet its strategic food needs. It will also be shown that the trend is subject to some other factors including particular national food production policies and to some extent to the world prices of commodities such as wheat.

Figure 1.2 illustrates the trend in cereal and flour imports in the Middle East and North Africa since the 1960s. The levelling off in the trend in 1986 reflects changes in particular agricultural and food production policies mentioned above by Egypt and Saudi Arabia, the two biggest players in both cereal and flour production and consumption in the region. Egypt changed its subsidies policies in 1986 which had favoured cotton, so that wheat production became a sound financial option for its farmers. By 1986 Saudi Arabia's irrigation projects had begun to produce sufficient wheat for most

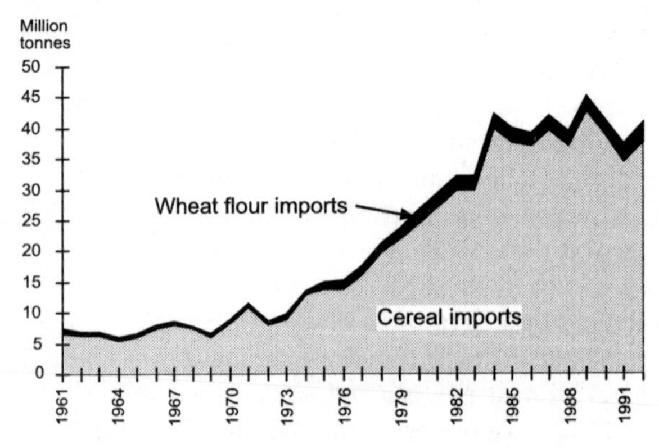

Figure 1.2 Wheat and cereal imports into the Middle East—1961 to 1992

of its needs, and it was about to become a significant wheat exporter in the world market. Saudi has subsequently reduced its wheat production because it was palpably an unrealistic way to use its fossil water.

Scientists can demonstrate that the Middle East is the first region in the history of the world to run out of water. (Rogers 1994, Allan 1994) Meanwhile the peoples and their leaders in the region refuse to recognise these resource and economic realities. Their interpretations of Middle East hydrological and economic contexts are at best underinformed and at worst dangerous; their perception of global hydrological and economic contexts is unsafe.

Table 1.1 summarises the estimated status and utilisation of first, Middle East, and secondly global water resources as researched by international scientists and compares them with the perceptions of these phenomena by politicians and peoples. The table shows that there is scope for dangerous confusion and contradiction at both the

Table 1.1 Middle East and global water uncertainties and the scope for optimism and pessimism

Middle Eastern water for regional self-sufficiency			Global water for global self-sufficiency		
	Evidence (scientific & professional)	Perception (politicians & people)		Evidence (scientific & professional)	Perception (politicians & people)
Availability	Deficit **Pessimism**	No deficit **Optimism [dominant]**	**Availability**	Surplus??? **Optimism & pessimism [contra-dictory]**	No view **Optimism [dominant]**
Utilisation	Productive efficiency possible **Optimism**	Productive efficiency possible **Optimism**	**Utilisation**	Productive efficiency possible **Optimism**	Productive efficiency possible **Optimism**
	Allocative efficiency possible **Optimism**	Allocative efficiency politically impossible **De-emphasis [dominant]**		Allocative efficiency possible **Optimism [contra dictory]**	Allocative efficiency politically impossible **De-emphasis [dominant]**

Source: author

regional and the global levels. Both scales are of vital economic and strategic significance to the region.

Information on future regional and global water supply and demand could be introduced into discourses on water which precede policy making in the region. But political processes facilitate the achievement of the interests of the powerful who in turn want to stay in power. The discourses are in practice 'sanctioned'. The information entering the water discourses is selected, distorted or de-emphasised. Optimistic interpretations and the identification of measures that will be politically safe are the main ideas that contribute to policy. Optimism and de-emphasis are essential short term political tools, and it is these that are deployed by those responsible for the water resource policies in the diverse economies of the region.

Table 1.1 also shows that global environmental knowledge is even more prone to diverse interpretations. In this case scientists cannot provide reliable information on either the supply side—the availability of water, nor on the demand side—the current and future demand for freshwater driven by future populations. Estimates of future population differ by over 50 per cent.

The Middle East's hydrological system is definitely less and less able to meet the rising demands being placed upon it. Food imports are the important indicators of water deficit. (Figure 1.1 and Table 1.2). Meanwhile the global hydrological system is evidently in surplus as it is able to meet the most demanding element of global water demand, the global consumption of food. Assuming a medium water consumption scenario of 1500 cubic metres per person per year (bnm3 ppy), global freshwater needs are about 8.25 billion cubic metres annually. (Table 1.3) A consumption level of 8.25 bn m3 ppy is well within the estimates of global freshwater availability. (Rodda 1996) But there are differing interpretations of the future position vis-à-vis future demands driven by higher world population (Brown 1996, Postel 1997, IFPRI 1996) because of the global freshwater surplus. The region's water deficit is not serious because there are global systems—trade—which balance the Middle East's deficit. (Allan 1994) Note it is economic systems and not hydrological and water engineering systems which achieve water security for the economies of the region.

Figure 1.3 shows that at the regional level there can only be versions of pessimism about the future availability of water, at least for irrigated food production. Optimism exists on the future volumes of MENA freshwater but it is based on non-scientific assumptions.

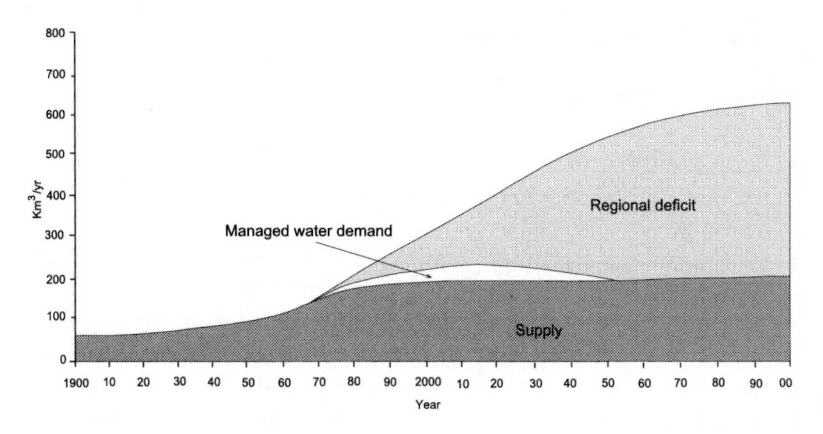

Figure 1.3 The MENA water deficit (Authors estimates)

On the other hand optimism about the capacity to use the region's water more productively is sound. And it is on this aspect of water management that politicians and regional optimists focus, as well as those who have financial and consulting services to contribute from outside the region.

At the global level there is certainty neither about volumes of freshwater available nor about the capacity to use the water effectively. In these circumstances there is evidence to support the arguments of both optimists and pessimists. Because of the mighty 'error bars' on the statistics produced by the as yet inadequate models of global change (Conway 1993:291, Conway and Hulme 1993, Conway, Krol, Alvamo, J. Hulme 1996) very different views emerge concerning the capacity of the world's agricultural systems to raise food, which leads in turn to a very confusing debate. The implications for scientific and political discourses on Middle Eastern and North African water are dire.

The pessimistic global freshwater scenario is persuasive if we assume:

- static patterns of food consumption by individuals
- or worse, patterns even more extravagant than current diets in terms of water consumptive use
- a pessimistic estimate (high scenario) of the demographic transition
- static technology
- totally inflexible political and international institutions

- ineffective trading systems.

If, however, it is assumed that:

- communities can change their food consumption behaviour, just as they can be weaned off tobacco and over indulgence in sun-bathing
- the world's growth in population will level off in the second half of the twenty-first century at about ten billion
- economically sound policies can improve the productivity of water by ensuring that it is allocated to activities and crop production which bring high returns to water
- technology can significantly improve the productivity of water in food production
- the mechanisms of international trade in staple foods continue to operate with proven effectiveness to ameliorate the uneven water endowments of the world's regions

then one can conclude that the world's water and food futures are uncertain but not seriously insecure. Basic to a serious scientific input to the debate is the contribution by the scientific community of best estimates of the status of existing soil water surpluses in addition to the rather more easily estimated future populations of current and future water challenged regions. Unfortunately these numbers are difficult to obtain because the concept of soil water availability is hard to define and agree and even if the concept were generally agreed a method of monitoring comprehensively has not yet been defined.

The insecurity of the Middle East and North Africa arises not just from its poor water resource endowment. The water endowment is a significant factor, but much more important is the capacity of its agricultural sectors, its governments and international institutions to adapt to the resource scarcity and to take measures to find and mobilise substitutes. There is also a knowledge problem. Scientists cannot answer the question posed above about the global freshwater system—is there sufficient water in the world's soil profiles, surface water and groundwaters to meet a notional average water consumption per person for say ten billion people?

The imprecision of current estimates of freshwater demand and potential availability is immense. Will there be seven, ten or twelve billion water users? What is a reasonable estimate of average water consumption in the dynamic political economies of the future?

Should it be 1000 cubic metres per person per year or 2000 cubic metres per year, for all food, domestic, drinking, municipal, industrial and leisure water? How do we factor in environmental water? Some communities in the world are already operating beyond the high end of this consumption pattern, for example in parts of the United States. At the other extreme individuals in the vast semi-arid tracts of Africa south of the Sahara have access to an unreliable and very seasonal water. Most can usually access say 1000 cubic metres of soil moisture in their local soil profiles and a tiny daily use of ten litres per day for domestic and drinking purposes, that is a mere 3.5 cubic metres per year; a total use of about 1000 cubic metres per year.

Current and future water gap uncertainties are not quite of the *Himalayan scale* cited by Thompson and Warburton (1988:2) where they observed a difference between the highest and lowest estimates in a review of soil erosion in the Himalayan region of about 400 times. For Middle East water the estimates of future freshwater demand differ by a factor of at least two. (Table 1.2) The comparatively small range in the estimates should not lead to complacency in that the number of people within the range would be at least four billion.

The two variables affecting the estimate of future freshwater needs are first, growth in population and secondly, the use of water per head. Population growth estimates differ by 40 per cent, with the lowest about 50 per cent higher than the current population. (IFPRI 1997) When this 48 per cent is multiplied by either the low (1000 cubic metres per head) or the high (2000 cubic metres per head) estimates of the freshwater likely to be needed per person the estimates of potential water needs for the Middle East and North Africa move further apart to a high estimate which is 180 per cent higher than the lowest.

Potential water availability globally being unknown, guesses in circulation differ according to assumptions. The lowest is based on the assumption that a limit similar to existing use is imminent. (Brown 1996, Postel 1997) The highest anticipate that water will be able to be mobilised at twice the late twentieth century level.

There should be no anxiety in any region of the world, including MENA, concerning the availability of water for drinking and domestic use and for almost all industrial and service sector uses. Current and future tension will be about water for food production. Food production requires about 90 per cent of a community's water; and should there be insufficient water for local food needs, that is the predicament of the Middle East since 1970, then food water has to be imported. Domestic water may also have to be augmented by desalination.

While the underlying message in this study will be optimistic it will be very strongly emphasised that the types of adjustment anticipated by the optimists cannot be achieved quickly. Thirty years is a rather short transitional period for the necessary major adjustments in water policies to be developed in response to limited water availability. Adjustments such as the changes in the public perception of the value of water take time. The associated political discourses enabling fundamental changes in water policies take at least decades. So deep are the belief systems and so challenging any proposal that beliefs should change that politicians are loathe to contradict them, even though the measures are essential for the stabilisation of the political economy. Case studies from the Middle East will illustrate these processes in chapter 4 (Allan 1996:77–83, Feitelson and Allan 1998).

How is it that the two camps, pessimists and optimists, can exist simultaneously and dispense such confusing contradictory interpretations? There are four main reasons, three on the supply side and one on the demand side.

The first supply side reason is the simple matter of the diverse assumptions adopted by the two groups. The second is that the most important variable of all, global soil water, is ignored by pessimists and is not well understood by the optimists. The third is the imprecision in the estimates of the other important components of the global hydrological system, of the freshwater at the surface and underground. The quality of these freshwaters are also poorly quantified. Finally on the demand side there are so many factors affecting the demand of a community for water that it is possible to extrapolate demands for water which differ by four or five times between the low estimate and a high one. Pessimists tend to assert numbers at the high end of the range and optimists to the low end.

The divergence in assumptions between the water pessimists and the water optimists is so great that they cannot communicate. The arguments of **both** pessimists and optimists neglect the most important water of all in the water budgets quoted whether these be water budgets of the MENA region or of other regions in the world.

This silent unaccounted water is soil water located in the soil profiles of the MENA economies themselves, as well as in profiles elsewhere in the global system. The neglect is especially true of the pessimists (Brown 1997, Postel 1997). Soil water and soil moisture

Table 1.2 Uncertain Middle East & North Africa scenarios: population and water consumption scenarios and interpretations of water demand

		End of 20th century	End of 21st century	Population scenarios		
Population-billions		0.35 bn	0.6 bn Low	0.7 bn Medium	0.8 bn High	
Water consumption scenarios		Bn m3/year	Water use Bn m3/year	Water use Bn m3/year	Water use Bn m3/year	Demand management remedies
Water/person	@1000m3/ppy	350	600 [2]	700	800	Economic, social & political
Water/person	@2000m3/ppy	700	1200	1400	1600 [1]	development;
Water available						reform' food consumption
Surface & groundwater		196	220	230	230	patterns;
Soil water (excludes Sudan)		**60**	**60**	**60**	**60**	**demand management of water**
			Demand management remedies			
Source: population estimates 1996			**Economic, social and political development; family planning**			

Assumptions

[1] **Pessimistic assumptions**
- static patterns of food consumption by individuals
- even more extravagant than current diets in terms of water consumptive use
- a pessimistic estimate of the demographic transition
- static technology
- inflexible national political institutions
- inflexible international institutions
- ineffective trading systems

2 **Optimistic assumptions**
- changed food consumption patterns - less livestock products
- optimistic estimate of the demographic transition
- success of economic demand management policies—by allocating to high returns to water activities
- success of economic demand management policies—by techno logical improvements to high returns to water activities
- effective trade

Table 1.3 Uncertain global scenarios: population and water consumption scenarios:: interpretations of global water and food demand and availability

		End of 20th century	End of 21st century	Population scenarios		
	Population-billions	5.5 bn	7.7 bn Low	9.4 bn Medium	11.1 bn High	
Water consumption scenarios		Bn m3/year	Water use Bn m3/year	Water use Bn m3/year	Water use Bn m3/year	Demand management remedies
Water/person	@1000m3/ppy	5500	7700 [2]	9400	11100	Economic, social & political development;
Water/person	@2000m3/ppy	11000	15700	18800	22200 [1]	reform' food consumption patterns;
Water available						
				Demand management remedies		**demand management of water**
				Economic, social and political development; family planning		

Source: population estimates - IFPRI1996

Assumptions

[1] **Pessimistic assumptions**
- static patterns of food consumption by individuals
- even more extravagant than current diets in terms of water consumptive use
- a pessimistic estimate of the demographic transition
- static technology
- inflexible national political institutions
- inflexible international institutions
- ineffective trading systems

2 **Optimistic assumptions**
- changed food consumption patterns - less livestock products
- optimistic estimate of the demographic transition
- success of economic demand management policies—by allocating to high returns to water activities
- success of economic demand management policies—by techno logical improvements to high returns to water activities
- effective trade

Table 1.4 Population and water in the Middle East and North Africa illustrating the range of future low and high scenarios

				Water scenarios			
End 21st century				**Low (1000 m3/ppy)**		**High (2000 m3/ppy)**	
		[1995 position]	Per cent in 1995				
		5.5 bn	**100%**	**5500 bn**	100%		
Population scenarios	**Low**	[in 2100] **7.7 bn**	140%	**7700 bn**	**140%**		
	High	[in 2100] **11.1 bn**	202%			**22200 bn**	**404%**

do not even appear in the indexes of their books, or if they do they refer to examples in some local irrigated circumstances (Postel 1997:103) and not to the pivotal role of naturally occurring soil water in feeding the world's future populations. In brief if at least half of the water needed to feed the Middle East and North Africa's people in the 1990s lies in the soil profiles of temperate humid environments in North America, South America and Europe it is surely scientifically derelict to ignore that water.

The intent here is not to discredit the inestimably important contributions of Lester Brown and Sandra Postel to international and local discourse on the under-researched and intuitively worrying state of the world's water resources. Their focus in highlighting the urgency of achieving major shifts in perception and policy is extremely important. However, by discussing only surface water and groundwater, a partial perspective on the waters of importance to human populations is presented. In neglecting to address a comprehensive version of global hydrology including global soil water, at least until 1999 (Postel 1999) they deny themselves and their readers a balanced perspective. An analysis of the energy future of hydro-carbon short national economy would be as misleading if it neglected options to import oil and gas, or for someone wanting to get from Cairo to Khartoum to assume that the only travel options were on a boat on the River Nile. Even more neglectful is the de-emphasis of the impressive international trading systems which transfer commodities requiring high water inputs from water surplus areas to water poor ones. In practice more water flows into the Middle East each year in this 'virtual form', embedded in cereal

Table 1.5 Middle East and North Africa—water availability (surface and groundwater)—1995

Country		Water Actual km3/year	Population 000 1995
Middle East			
Arid Middle East			
Bahrein		0.116	564
Cyprus		0.900	742
Iraq	usable	35.420	20449
Iraq	unusable	35.000	
Jordan		0.880	5463
Kuwait		0.020	1547
Lebanon		4.407	3009
Oman		0.099	2163
Qatar		0.053	551
Saudi Arabia		2.400	17880
Syria	usable	10.260	14661
Syria	unusable	16.000	
UAE		0.150	1904
Yemen		4.100	14501
Total		**58.805**	**83434**
Less arid Middle East			
Iran		137.510	67283
Turkey		183.762	61945
Total		**321.272**	**129228**
Total Middle East		**380.077**	**212662**
North Africa			
Arid North Africa			
Algeria		14.300	27939
Egypt		58.300	62931
Libya		0.600	5407
Malta		0.016	366
Mauritania		11.400	2274
Morocco		30.000	27028
Sudan-northern Nile		18.500	28098
Tunisia		4.120	8896
Total		**137.236**	**162939**
Less arid northern Africa			
Sudan—other		60.000	
Total		60.000	
Total North Africa		**197.236**	**375601**
MENA total	**196.041**	**577.313**	**588263**

Source: FA0 1997:14. With a reclassification of the countries by ecological zone and region. With some additional comments on the actual availability of water for economic use: for example Turkey, Syria and the Sudan.

imports, as is used for annual crop production in Egypt. Egypt is by far the most water rich country in the arid part of the Middle East and North Africa.

The main purpose of this chapter and the book as a whole is not to contradict the concern of the eco-pessimists and those afflicted with hydro-paranoia vis-à-vis the Middle East, but to enlighten that concern by refocusing the analysis at a higher and more appropriate level, namely at the level of the international political economy. It is at the global level that explanation is to be found concerning why economies operate as they do and why water policies are as they are in the Middle East. Explanations are not to be found by narrow analysis at the catchment and water budgets at the national level or even at the regional level. Catchment hydrologies are not a complete source of explanation because they are not determining of the options available to those managing a national economy. If national hydrological systems restrict economic options then politicians have to find remedies in systems which do provide solutions.

United States water resource lawyers in Colorado have coined the term, the 'problemshed', to make the idea of the 'catchment' or 'system' in which solutions can be found more accessible. National economies operate in international political economic systems—in problemsheds—and not just in hydrological systems—or watersheds. The political economy of the global trade in staple cereals is the relevant 'hydrological' catchment for water deficit economies. Access to, and management of, the problemshed is not just via an environmental system but via an extremely powerful and flexible economic one.

The shared and limiting catchment hydrologies with all their troublesome international relations are dangerously misleading frameworks of analysis. The existence of solutions in 'problemsheds' enable politicians to avoid the stresses of inter-riparian relations in the 'watershed'. In brief if a quarter of the water needed to feed the Middle East and North Africa's people in the 1990s lies in the soil profiles of temperate humid environments in North America, South America and Europe it would be scientifically obtuse to ignore that water. It would also be neglectful of scientists in general and the governments of water short economies in particular to ignore predictable future competition for this water. The scientific neglect is just as serious as the predictable devious and selective hydropolitics engaged in by political interests.

The study will further emphasise that the coming fifty year transformation of the perceptions of the value of water (Allan and

Radwan 1998) will be a tough environment for those looking for quick solutions. There is no reliable check list nor a handy box of tools with which to change the perceptions of large groups of people quickly. And people 'convinced against their will [tend] to be of the same opinion still'. Even when perceptions have transformed and shifted, the implementation of new water policies, especially water re-allocation policies, has to be gradual. These processes will all be slow; sometimes painfully slow for agency officials and for economists and engineers in a hurry because of limited funding and career horizons.

In addition the foundation scientific concepts for economically and environmentally sound water policies require reforms and adjustments which politicians can see will bring high 'political prices' (World Bank 1997a:146). Politicians will defer for as long as possible paying these 'political prices', which can be so extreme that they would involve loss of power. It requires an inhuman level of political courage for a political leader of a country which for five thousand years has enjoyed water security to announce that water resources are no longer adequate. To make the announcement would be political suicide.

Those innovators purveying the economic and environmental facts of life which contradict the deeply held belief systems of whole populations will be ignored if they do not shape their message and pace its delivery to accord with political realities. The authors of the 1997 World Development Report (World Bank 1997:144) put it another way, by quoting Machiavelli who captures the political flavour of innovation:

> 'The innovator makes enemies of all those who prospered under the old order, and only lukewarm support is forthcoming from those who would prosper under the new'. Machiavelli, N. 1513, *The Prince.*

It is a paradox that the water pessimists are wrong but their pessimism is a very useful political tool which can help the innovator to shift the eternally interdependent belief systems of the public and their politicians. The water optimists are right but their optimism is dangerous because the notion enables politicians to treat water as a low policy priority and thereby please those who perceive that they are prospering under the old order. Pessimists also bring more sensational stories to the media. Optimists bring a version of unsensational good news. The good news is also complicated and indigestible as well as unsensational.

A conceptual framework

This preliminary discussion has been eclectic. Reference has been made to hydrological systems, to the environment, to environmental management implying engineering and institutional instruments, to water, to economics, to social theory (belief systems), to law and international law, to politics and to international relations. The scope is challenging but unavoidable if an analysis of any depth is to be achieved. Table 1.2 provides a framework for analysing water endowments, water management and water policy. The framework has developed over a decade of interdisciplinary endeavour. One of the outcomes of that research is a conviction that explanation cannot be found in a single discipline and an even more strongly held conviction that analysis based on a single disciplinary approach is unsafe.

Interdisciplinarity is a demanding 'non-discipline' and especially difficult because anyone adopting the approach is at risk of attracting the very sharp and even destructive analysis of elements of their argument by scientists with more specific disciplinary expertise in a particular episteme. Unfortunately, while there are referees in scientific 'disciplinary' games, albeit narrow, often biased and much questioned, there are no referees at all in 'interdisciplinary' games. Yet firms, governments and other entities which get things done have to address problems which are inherently interdisciplinary. Their 'referees' are market performance and political success, whether sanctioned by democratic processes or not.

In taking the scientifically hazardous interdisciplinary road one must be prepared for the ambushes of the scientifically well armed. The aim of the analysis is to provide a contribution which will be useful and improving. Meanwhile it is hoped that those who read the book from the perspective of an informed specialist will recognise the utility of the broad and comprehensive perspective attempted.

The scope of the approach is summarised in Figure 1.4. The diagram shows seven major areas of scientific endeavour where explanations have been found.

The identification of the areas of science where explanations can be found together with the analytical tools which are integral to such science is an important first stage in framing the analysis. The second is to identify a categorisation of those involved in the discourses over water, both users and providers, as well as those involved in water policy making, in investing in water managing hardware and institutions and in influencing and implementing water allocation and management policies.

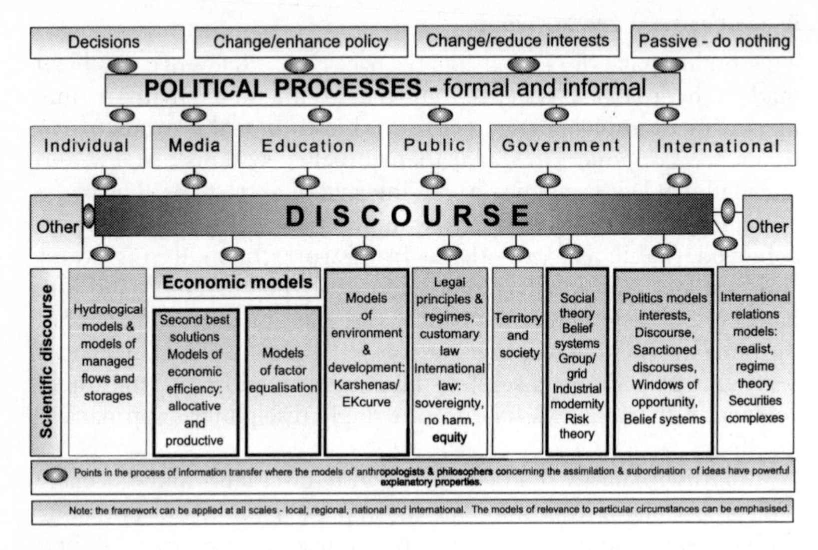

Figure 1.4 A conceptual framework: water resource science and political discourse

Table 1.6 is a first approximation of a categorisation of those involved in water policy development and water use. The numbers participating will vary at the levels of the individual, the sector and the institution according to the perspective of the analyst. The regional actors, shown in the lower part of the diagram are numerous and easy to identify. Despite being numerous and predictably different in interest at regional and national level the underlying perceptions of the actors on many basic issues, such as water not being an economic resource and on the primacy of water rights over other features in a water dispute, are remarkably uniform. This is true for example between riparians of a shared surface water resource, as well as between groups at the sectoral levels within individual economies where farmers and urban water users might be in contention. These perceptions stem from the long history of water surpluses in the tracts where populations have concentrated over the six or so millennia of settled agriculture in the Middle East and North Africa. Features of these uniform perceptions are:

- a de-emphasis of the reason for the growing water deficit, namely the steady and inexorable increase in demand for freshwater which follows the increase in population.

Table 1.6 Categories involved in water policy development and water use: users, scientists and institutions

Scale	**Users and managers**	**Scientists**	**Consultants —engineers & economists etc**	**Institutions —international**	**Institutions —national**
Global		[A diverse category —different assumptions —diverse epistemes]	[Diverse approaches —mainly WIER —consultants listen to clients]	[Economics inspired & well developed epistemic group]	
[Global in perspective; but with deep involvement in the MENA region including, studies investments and business]	de = diverse estimates	**Global change** (Very diverse estimates) **Global hydrology** (de: optimists & pessimists) **Demographers** (Diverse estimates) **Economists on global agriculture** (Generally optimists) **Economists on global trade** (Generally optimists)	International consultants—engineering, economics, trade Professional bodies ICID Engineering Inst's etc	**World Bank**(WIER) **UNDP** (WIER) **FAO** (WIER) **UNEP** (WIER) **ILC/ILA** (WIER?) UNCED/Dublin (WIER) **International Grains Council** **EU** eg MEDA Prog. NGOs/WWC/ (WIER?) WWC/WWP(WIER)	

[**WIER** they contend that Water Is an Economic Resource]

Table 1.6 Categories involved in water policy development and water use: users, scientists and institutions (continued)

Institutions —national	[Category with uniform underlying perceptions] (Believe WINER)	[Category with uniform underlying perceptions] (Believe WINER)	[Category with uniform underlying perceptions] (Believe WINER)	[Category with uniform underlying perceptions] (Believe WINER)	[Category with uniform underlying perceptions] (Believe WINER)
(MENA) and national	***Farmers*** ***Industry*** **Services** ***Municipalities*** ***Families—communities*** ***Environmental use***	*Regional & national environmental and* social scientists	***National and local consultants—*** **engineering &** ***economic etc*** Professional bodies	*Arab League* *AOAD* ECOWAS *ACSAD* *Arab Fund* *Regional bi-lateral Funds (eg Kuwait) and agencies*	***Minstries of:*** ***Water*** ***Environment*** ***Agriculture & Land Reclamation*** ***Power*** ***Planning*** ***Trade***
	[**WINER** - they contend that Water Is **NOT** an Economic Resource]			*Political parties-secular* *Political parties-religious*	***NGOs*** *-Local Int'ern'l*

Table 1.7 Goals and principles for water allocation and management and a framework for the analysis and development of policy and practice

Goals of water using activity	Guiding principles	Policies	Instruments	
			Institutional	Engineering
Facilitation of political circumstances to enable optimum resource use	Minimisation of conflict; promotion of cooperation in the areas of water use at all levels	Conflict resolution; identification of **reciprocal arrangements** to promote economically and socially beneficial water use and the use and the installation of such arrangements	Water sharing arrangements —traditional and new; recognition of water rights & of the ownership of water; consultation between legislators, officials (local, national and international) ('**democratic'** institutions); introduction of **new economic and legal instruments** to shift access to water to the most beneficial users and uses	Earth observation (remote sensing); in situ monitoring & information systems
Water use efficiency **Allocative efficiency** **Productive efficiency** *('Development')*	Economic returns to water, sustainablity of water supplies	Investment in sectors, activities and crops which bring optimum returns. **Demand management** & international subsidies and pricing imply water metering	Water pricing, agricultural subsidies, crop pricing and other intervention. Advanced pricing systems imply water metering. Agreements both local Water efficiency studies and water management programmes	Large and small civil works works for water abstraction, treatment, delivery & distribution, recycling, water metering

Table 1.7 Goals and principles for water allocation and management and a framework for the analysis and development of policy and practice (continued)

Goals of water using activity	**Guiding principles**	**Policies**	**Instruments**	
			Institutional	**Engineering**
Equitable use	Social returns to water Social benefits	Identification of the social benefits and disbenefits of water use and the promotion of beneficial uses	Land reform, water regulation, new legislation, reduction of illegal water use, changes to traditional rights	Water control systems, irrigation management
Safe use	Provision of adequate volumes & quality of water	Identification of approp- riate systems —traditional and new —promoting the safe provision of water use, re-use and disposal	Monitoring, legislation, regulating institutions (traditional and new)	Planning for future demands, water control systems, water treatment, maintenance for reliability
Environmentally sound use *('Conservation' Cultivating the* world *as if you would live forever)*	Sustainable use of landscape and amenity including intangibles	Identification of approp- riate systems—traditional and new—for sustainable water use	Monitoring, legislation, regulating institutions (traditional and new)	Quality monitoring, water treatment, wastewater treatment, waste disposal

- an unwillingness to recognise that there is a current water deficit
- that the solution to any future water deficit will be successfully achieved by introducing improvements in water use efficiency by increasing technical (productive) efficiency.
- that water rights are substantive and intrinsic to any negotiating strategy over shared waters and that there are legal instruments which should assist in achieving water rights.

The following are unwelcome and assumed to be unnecessary:

- the achievement of water use efficiency by allocating water to sectors which achieve a higher return to water. There are signs that the re-allocation of water within sectors to achieve high allocative efficiency are gaining priority amongst policy agenda. (Egypt's justification of the New Valley Project in 1997 when it was argued that the use of water would be more efficient than water allocated to sugar-cane and rice production. Also water policy in Tunisia and Morocco. (Jellali and Jebali 1994))
- the idea that it is 'virtual water' embedded in food imports that secures current economic stability and will do so even more emphatically in future.

A shibboleth which puts specialists and professionals on water on different sides of a great economic divide in the issue of the 'value' of water. Outsider scientists from the international community have adopted the idea that water has an economic value. We refer to this group of scientists which populates the upper part of Table 1.6 as WIERs. WIER is an acronym deriving from the sentence—*Water Is an Economic Resource*. In the MENA region the communities of insider scientists, engineers, government officials, politicians and farmers are much more numerous than visiting scientists and professionals and they contend to great practical effect that *Water Is Not an Economic Resource*—WINERs. The economists of the MENA region have no more success than the international scientists and consultants in gaining a place for the WIER ideology, or so it was at the millennium.

Water in international and regional systems: understanding parallel discourses

The subject of Middle East and North African water generates insider and outsider interpretations of the status of the water resources and of the allocative and management options. Different assumptions underlie the arguments of professionals and most scientists from the region from those on which outsiders build. The different perspectives are highlighted at this stage because they figure in a very big way in interpretations of MENA water resources and their management. It is not possible to understand the parallel water discourses without the help of the insider-outsider model.

Such is water's fundamental place in sustaining life and livelihoods that human societies have devoted much political energy and substantial economic resources to ensuring secure supplies, albeit without ensuring equitable access or freedom from the risk of water resource shortages for everyone. From the late nineteenth century until the 1970s, the Northern industrialised economies which dominate the global perspectives on resource development including water were dominated by the vision and politics of what has been termed the *hydraulic mission* (Swyngedouw 1998, Reisner 1984). This mission was first and most fully implemented in the United States (Reisner 1984) and emulated in a very different polity, the former Soviet Union (Breznhev 1978). A similar mission was integral to the rhetoric and resource allocation politics of many other political economies such as post-imperial Spain in the first decades of the twentieth century. (Del Moral 1996, Swyngedouw 1999a and 1999b)

The hydraulic mission was a feature of modernity, a term used to describe the processes of change in the industrialising North of the late nineteenth century and the twentieth century. The hydraulic mission became a casualty of the environmental movement. After a couple of decades of contentious discourse in the 1960s and the 1970s green priorities were widely adopted in the North. Consideration for the environment and sustainability principles affected the water sector by the mid-1970s in the North. Such changes were a feature of the end of modernity and the transition to post-modernity. Post-modern perspectives, policies and policy outcomes involve different ways of interpreting evidence about the concrete world as well as of the worlds of ideas and aesthetics. Giddens and Beck prefer the notion of 'reflexive modernity' to that of 'post-modernity' (Beck 1992, 1995, 1996a & b, Beck et al. 1996, Giddens 1990). They regard the shifts in perception and approach since the mid-1970s as a further phase of modernity. They show that this concept is useful in explaining the changes in public perceptions of the environment in the North of the last quarter of the twentieth century.

Globalisation and the awareness of risk are features of this last phase of 'industrial modernity'. Giddens and Beck argue that the shifts in awareness and perception of reflexive modernity have occurred as the result of the impact of reflexive processes in the North. Reflexive, in this case, means the responses of individuals, communities and institutions to the rapid dissemination of the awareness of risks to their economies, livelihoods and environmental amenity if status quo water resource allocation and management continues. The concept of reflexive modernity is especially useful in the analysis of socio-politically determined approaches to managing the environment and especially in understanding the way that water resources have been perceived during and since the 1970s in the North. Here we are in the world of political ecology (Stott and Sullivan 2000, Allan 2000) which provides analytical tools to interpret how politically significant communities inherit and construct knowledge about their environments. This constructed knowledge is the basis of environmental, including water, policies which are the result of discursive political processes.

In the Northern economies thirty year discourses characterised the protracted contention over water policy which heralded the transition to industrial modernity (Beck 1995, Carter 1982:76, Hajer 1996, Allan 1996b). President Carter, for example, found participation in the transition particularly rough. What he thought was a solution appeared to be a problem to the US legislature and especially to the very powerful institutions which had been created to take forward almost a century of the hydraulic mission:

> ' Had a rough meeting with about 35 members of the Congress on water projects. They are raising Cain because we took those items out of their 1978 budget, but I am determined to push this item as much as possible. A lot of these projects would be ill-advised if they didn't cost anything, but the total estimated cost of them at this point is more than $5 billion, and my guess is that the final cost would be more than twice this amount.' JC Diary, 770310, in Carter 1982:78

> 'I understood the importance of these long awaited projects to the legislators, but during the years since their initial conception, circumstances had changed, environmental considerations had increased in importance, costs and interest charges had skyrocketed, other priorities had become much more urgent, and any original justification for some of the construction had been lost forever. *Still the inexorable forces toward legislative approval moved on. Other recent Presidents, graduates of the congressional system, had looked on the procedure as inviolate. I did not, and dove I head first to reform it.'* JC Diary, 770310, in Carter 1982:78–80

These quotations capture the notion of 'inherited' and 'constructed knowledge' and their role in supporting the interests of the politically influential. Explanations of the transition, from water policies dominated by the 'old water knowledge' inspired by notions of the hydraulic mission, to the 'new water knowledge' characterised by environmental sensitivity and economic efficiency—are diverse. Social theory and political theory have contributed greatly to explanations concerning why, how and when transformation in water policy and water management practice occurred in the North. Such theory will be very helpful in analysing the uncomfortable and politically hazardous transition which the Middle East and North African political economies are experiencing in their consideration and possible adoption of the new environmentally, economically and outsider inspired water wisdom. The theory shows that for the Northern outsider the process of adoption was long and politically stressful in the privileged circumstances of economic diversity and strength in social adaptive capacity. (Ohlsson 1998) The theory will be even more helpful in showing that the transition will be just as long and possibly longer in the economic, cultural and political circumstances of the MENA region where economic and socio-political adaptation will be more challenging.

After 1950 many political economies in the Middle East and the Mediterranean, including Spain, entered a period of progressively more serious national water deficits—defined as insufficient water to meet all needs including that for self-sufficient food production. The communities experiencing this transition into water deficit were not equipped to deal with the new circumstances. Their ideas were based on perceptions that 'water could be analysed as being a resource that embraces both "material" as well as "symbolic" interests' (Bourdieu 1977:182). In the MENA region water is perceived in a different way by both using communities and by professionals in the institutions which govern and allocate resources. The outsider assumption that water is just a 'material' resource is not readily accepted by communities which perceive water as a 'symbolic' resource. Ideas purveyed by alien scientists may come to be understood by part of the population, usually an elite acquainted with Northern science, but it is unacceptable in the water discourse that determines policy and practice in the region. Even in the rapidly industrialising Israel of the 1960s and 1970s the notion of the importance of high returns to water outside the agricultural sector recognised in 1963 by an outsider (Palmer 1963) was not openly discussed. It was not until 1986 that policies re-allocating water were implemented by the Israeli legislature. In less well found economies without the economic capacity (Karshenas 1994) and the social adaptive social capacity (Ohlsson 1999) the process could take much longer.

In the South, just twenty years on, contention such as that engaged in by President Carter in the United States in the late 1970s, has scarcely begun. Evidence of the equivalent reflexive process in the MENA region to date have, with the exception of Israel (Lonergan and Brooks 1993, Feitelson 1996) only found expression and involvement in Northern international agencies and in the writing of alien environmental scientists and economists addressing water policy reform. (Brooks 1996, Allan 1996a:100, Serageldin 1995, World Bank 1993, 1995, 1997a, 1997b) This alien construction of knowledge accelerated in the post-Cold War circumstances of the 1990s. But with respect to water in the MENA region, the reflexive process is only at the stage where alien advocates of the 'new water knowledge' have communicated principles to technical elites in some of the political economies in the MENA region. This very preliminary stage of awareness has been referred to as the phase of 'mutual knowledge'. (Giddens 1984) The diffusion of awareness of the new 'sound' approaches is in progress. (World Bank 1998b) In the MENA region the 'new water knowledge' has been adopted by only one political economy, that of Israel, and is beginning to shape water policy in four others. Only Jordan, Tunisia, Morocco and Cyprus have started on transformation of water policy. As a consequence water pricing of municipal water has been adopted. And in Tunisia and Morocco schemes to charge for irrigation water are being discussed and introduced. (World Bank 1998b:8–9) Likewise in Cyprus. At the millennium the international prominence of the new 'vision' for water is being amplified by an orchestrated global effort by agencies and governments to make water a millennial priority (GWP 2000).

A politically feasible ameliorative alternative: why the contradictory Northern and Southern discourses co-exist

Years of immersion by the author in the integrated river basin management literature and working in the company of professionals in that field reinforced a natural inclination to believe that hydrological and environmental principles could be prime in reaching sound conclusions on how to manage water resources. This impression was further reinforced by the apparent adoption of watershed principles by groups convened to seek cooperation over water resource management in major river basins such as the Nile.

The series of annual conferences convened under the title of Nile 2002 since 1993 is such an initiative (Nile 2002 1993–2000).

International lawyers labouring in conventions coordinated by the International Law Association and by the International Law Commission of the United Nations have also been drawn to hydrological principles. This emphasis was especially the case in their initial analysis of the non-navigational uses of international waterways (Khassawneh 1995, McCaffrey 1995, 1997a and 1997b). In the protracted three decade contention over legal principle and definition (McCaffrey and Sinjela 1998) it became evident that hydropolitics rather than hydrology were prime. The pattern of adoption of the 1997 ILC Convention on the Non-navigational Uses of International Waterways by the nations of the global community will reveal further the political nature of shared water resources. Upstream riparians will be slow to sign up to the 1997 Convention, and those in the midst of major civil works affecting river flows, such as Turkey, will be especially slow.

The river basin remains a conceptual icon of immense material and symbolic value to environmental scientists. In cleaving too closely to the concept environmental scientists prevent themselves from contributing effectively to the interdisciplinary discourse in which the explanation of water policy lies. In practice the uncompensatable resources in the environmental system of the river basin are often a relatively minor influence on water management policy in both Northern and Southern political economies. And this paradoxically can be especially the case when an economy runs out of water. The reason is that the communities and nations that live in the river basins operate in 'open' economic systems where resource shortages can be compensated. When politicians with communal and national responsibilities encounter water stress in their 'limiting' hydrological systems, or in those parts to which they have legitimate or practical access, they will seek solutions outside their accessible watersheds. They find readily available and stress free solutions in 'problemsheds' via whatever operational system is to hand. They reach beyond local constraints to regional and global markets. Usually it is the global trading system which provides the most effective alternative resources. Regional systems are less likely to provide solutions, at least for basic commodity shortages, because all the national economies in the region tend to endure similar natural resource endowments. Global players, especially in the water intense food sector, can provide solutions to local water resource deficits via virtual water in the water, food and trade nexus (McCalla 1997).

The political advantages of virtual water are substantially greater than their economic ones for peoples and politicians in the MENA region. The reason is that the importation of virtual water is not a political problem provided that attention is not drawn to the water, food and trade nexus. Virtual water has the immense advantage of being non-stressful if it remains as politically invisible in the political system as it is economically invisible in national and international economic systems. Water reform policies on the other hand, such as regulatory regimes, water markets, privatisation and care for the environment, inspired by economic principles and by principles of environmental sustainability, confront overstretched Southern politicians with [political] problems with high associated political costs.

The reason that such secure and easily available virtual water is significant to water short economies is that the politicians in these economies can defer dealing with the impacts of their accumulating water deficits because these reserves of accessible virtual water exist. The availability of virtual water in effect dampens the widespread awareness of the extent of national water deficits. The political significance of virtual water on the pace of water policy reform is immense if incalculable. Politicians inside the MENA region need a politically feasible ameliorative alternative to the political stress of rapid adoption of the new water management knowledge advocated by earnest water professional outsiders. Importing virtual water provides that politically feasible alternative. The second best works (Lipsey and Lancaster 1956).

Technical know-how, cultural perceptions and social processes

The discussion so far has established that water policy is driven by perceptions; by informed perceptions and by uninformed perceptions; as well as by enduring perceptions based on old knowledge that has been in place for millennia. These perceptions support deeply held beliefs which it is politically hazardous to contradict. Perceptions of the technical communities bringing new, outsider knowledge are based on new and locally untested approaches.

> 'So ... risks are at the same time "real" and constituted by social perception and construction; their reality springs from the impact of ongoing industrial and scientific production and research routines. On the other hand their knowledge, quite differently, springs out of the history of symbols and one's culture (the understanding of Nature, for example) and the social fabric of knowledge. This

is one of the reasons why the same risk is perceived and handled politically so differently throughout Europe and other parts of the globe.' Beck 1999:76

These new perceptions can be defined as 'risk statements' constructed by the alien agencies and scientists that purvey them in the fields of hydrology, environmental science and economics. Those managing local circumstances on the other hand find the remedies recommended by alien agencies to be definable as alternative 'risk statements' as they threaten social and economic and livelihood norms. Establishing a perception of risk, whether culturally or technically inspired, is a powerful element in the politics which influence policy, including water policy. Beck makes the useful point about the position when a notion of risk has been established in this case in the globalised and risk message receptive North.

> Established risk definitions are thus a magic wand with which an immovable society can put the fear of God in itself and thereby activate itself in its political centres and become politicised from within. The public (mass media) dramatisation of risk is in this sense an antidote to current narrow-minded 'more-of-the-same' attitudes. ... Only a few want to turn the rudder around. Most want both—they want nothing to happen and they want to complain about that very fact, because then everything is possible: enjoying the bad good life and the threats to it as well. Beck 1999:76

Only Israel has experienced these political and social processes in the MENA region. Israeli water scientists, water professionals and journalists have constructed the idea of the red lines below which water levels should not fall. The notion of the red line is used to gain and reinforce the attention of the national population for periodic problems with the water reserves of Lake Kinneret/Tiberias and the West Bank aquifers (Ha'aretz 2000).

> Even more chilling are the forecasts for next year. In the event of a drought year, Israel will face a shortage of drinking water, even if the reservoirs are pumped beyond red lines, and even if all supply of fresh water for agriculture is cut off. Ha'aretz 14 June 2000

It has been shown in the earlier discussion that water policy in the last three decades of the twentieth century has been subject to the influence of many new risk 'constructs' in the North. Risks to the environment have figured most strongly, possibly because they could be articulated in an arresting cultural register urging the reconstruction of lost natural virtue. The language of 'wild' rivers and 'clean', ecologically diverse and amenity rich wetlands has been commonplace in the North since the late 1960s. The arguments of water use efficiency lagged by a decade and have proved to have been much less culturally acceptable and much more contested in both North and South.

A problem with risk statements is that they are both factual and value statements at the same time (Beck 1999:76). A statement such as there is a risk of armed conflict over water as the MENA region's demographically driven water deficit worsens is something which can be approximately quantified. The water deficit numbers can be approximated. They have certainly worsened. But the armed conflicts have died out. They have not become more numerous. The above risk statement is at the same time a very narrow 'value statement' in that it is based on deterministic environmental assumptions. Out in the interdisciplinary community there are economic geographers and trade economists who would interpret the statement very differently based on their additional awareness of the water readily available in global traded commodities. Substantial contention is generated amongst the groups who believe that hazardous water shortages are imminent and will be serious and those who argue that water shortages can be completely ameliorated now, and *probably* completely ameliorated in future, by virtual water (Allan 1999, Ochet 1999). The existence of virtual water facilitates a classic case of a community wanting 'nothing to happen [despite the perceived risk] and ... to complain about that very fact'.

The peoples of the MENA region do not perceive themselves to be part of global water systems. Nor do most scientists perceive that the water resources of the MENA region are subordinate to the global trade in water intensive commodities such as grain. Very few scientists perceive that such global trade is subordinate to the global hydrological system. Subordinate in this case means that the 'subordinate' water dependent activity, first local food production and secondly global trade in food commodities, can only be secure, that is without risk, if the global freshwater resources and global hydrological systems have sufficient accessible freshwater to sustain the subordinate systems. The differing awareness of risk and remedies has generated a lively discourse. The notion of virtual water is a remedy to alien analysts; to those struggling with water deficits and the challenge of negotiating water rights in the distorted Palestinian-Israeli politics it is not relevant (Isaac 1999).

> 'Perception is always and necessarily contextual and locally constituted. This local contextuality is only extendible in the imagination and with the aid of such technologies as television, computers and the mass media.' Beck 1999:77

The pessimists in the community and the pessimist scientists are not aware of the economic impact of virtual water in ameliorating the resource problem nor are they aware of its political impact in making the risk socially invisible. Beck points out that in the risk society the 'point of impact is not obviously tied to the[ir] point of origin. At the same time their transmissions and movements are often invisible and untrackable to everyday perceptions. This social invisibility means that, unlike many other political issues, [environmental] risks must clearly be brought to consciousness, and only then can it be said that they constitute an actual threat, and this includes cultural values and symbols as well as scientific arguments' (Beck 1999:76).

The MENA region and its water resources are a useful example of the disjuncture between 'knowledge' and 'impact'. Such disjuncture explains why knowledges of such uncertainties as future demography, future water demand, the volume of unaccounted resource transfers such as those of virtual water, and the invisible impacts of such numbers are not part of the information in local currency. Even when the impact of the MENA population increases and the region's limited freshwater resources, as estimated by scientists and professionals, should have become evident, such impacts could be rendered invisible by the availability of global solutions. At the local MENA level the disjuncture between the knowledge of the risk of the acknowledged current water deficit and its impact has been achieved by accessing unacknowledged virtual water. At the global level the processes are fathomable.

Knowledge and power in the MENA water sector: a risk society in waiting

The last three decades of the twentieth century in the North has been a period when perceptions of water resources have been transformed as part of the reflexive response to awareness that water environments were being put at risk (Pearce 1992, McCulley 1996) by policies underpinned by the assumptions of the 'hydraulic mission'. In the 1960–1980 period environmental activists and activist scientists influenced communities, constituencies and politicians to operationalise a different evaluation of water environments. The reflexive response of Northern water science and the communities and polities which manage water has not been taken up by an equivalent suite of activists, activist scientists and persuadable officials and politicians everywhere in the MENA region. Although this position is beginning to be challenged (Brechin 1999), circumstances in the MENA region are paradoxical. The MENA region is the first region to encounter what has been argued should be strategically and economically damaging water deficits but it has been able to cope with the associated challenges without adopting economically or environmentally inspired water management policies.

The complex interplay of unrecognised economic solutions, belief systems and political processes has explained the range of perceptions that determine as yet inflexible approaches to water use and allocation. Operationally effective, and water deficit ameliorating, global trading systems have been shown to exist but awareness of them was of such destabilising potential that the social and political systems were trapped in a 'sanctioned discourse' of non-awareness.

The MENA region is an example of how a significantly water stressed part of the South has reacted in terms of perception and water policy reform to the ideas generated in the transition from industrial to reflexive modernity in the North. Risk society theory can be helpful in providing an overarching interdisciplinary framework providing relevant analytical categories for the technical and the social aspects of knowledge [awareness and non-awareness] and impact. The water stressed MENA communities and political economies have most reason of all the Southern regions to move from water policies of non-reflexive to reflexive water policies. In practice the era of the MENA region's version of industrial modernity is being prolonged by the manipulation of awareness of risk by politicians' natural inclination to remain in harmony with the belief systems of their peoples. Belief systems about the fundamental place of water in livelihoods are best left uncontested.

The MENA region is possibly an exceptional example of how the perception of risk, in this case the risk of water shortages, can be manipulated if socio-political circumstances are enabling. MENA water resources are perceived as much via cultivated non-knowledge as by knowledge. Silent solutions to the water resource risk are de-emphasised. As a result the disjuncture between knowledge and impact is extreme.

The MENA region is a risk society in waiting. It has the major risk of water shortages hanging over it but it does not yet have the will to interrogate the problem. Consequently we face the paradox that at the very time when threats and hazards are seen to become more dangerous and more obvious, they simultaneously slip through the net of proofs, attributions and compensation with which the legal and political systems attempt to capture them (Beck 1999:83).

The MENA case study is especially powerful in exemplifying how risk, in this case water shortages, are socially invisible in the region's current phase of social, political and economic transformation. They are also invisible at the national level. Making societies conscious of the risk requires some scientific argument and a great deal of cultural contestation. 'Thus the politics of risk is intrinsically a *politics of knowledge*, expertise and counter-expertise' (Goldblatt 1996).

A conceptual framework

The conceptual framework (Figure 1.4) is the foundation of the rest of the analysis. It will be used to highlight the key sources of explanation. It will also be used as a check list to ensure that the significance of as many influencing factors as possible are evaluated. It will be shown that politics is the richest source of explanation of why freshwater is managed and allocated as it is. Economic explanations are the most elegant but though highlighted in this study insofar as mere words can, they will for the most part remain silent in the region in terms of explanation and inspiration until they are deemed fit for public discourse.

Social theory will be shown to provide explanation which is extremely powerful especially when allied with models from politics. Models from environmental science are shown to provide only a small part of the explanation of why water policies and water using practices have evolved as they have. It will be suggested that the limited explanatory role of models from environmental science has had a negative impact on the effectiveness of the contribution of international law. Lawyers like engineers tend to assume that access to surface and ground waters in river basins is the fundamental concern of those responsible for water. They do not recognise that water is only one of many, often linked, concerns of politicians and water users. The river basin is a bounded environment for hydrological scientists, engineers and lawyers. Politicians do not just manoeuvre in that bounded environment. Nor should international lawyers. Inter-riparian relations over water have to reach a level of cooperation such that contending parties can negotiate rule making agenda as a preliminary to agreeing rules (Ben Venisti 1997). Progress in inter-state relations must precede discussions of principles of international law. And lawyers would make a more effective contribution if they applied their regulatory endeavours to potentially operational economic transactions over water rather than attempting to breathe life into non-operational principles of water rights.

The attempt to categorise the actors involved in water policy and use (Table 1.6) is offered as a means of simplifying an otherwise impenetrable reality. It is hoped that the simplification has not been so extreme as to make the analysis invalid. The simplification helps the author to refer to the different interests affecting both the discourses on freshwater and the outcomes of them. It is hoped that it will also assist the reader in identifying the global and regional actors and to grasp the different assumptions and experiences which motivate and constrain them in their approach to water.

Water ideologies

The analysis will face in many directions. It will attempt to provide critical comment on any numbers that exist on the supply side—the regional renewable water resources. Non-renewable water will also be discussed but will receive relatively little attention, not least because there has been no comprehensive regional review of groundwater. The nature of regional sectoral demands and the likely directions such demands will take in future will also be addressed.

Since perceptions are more important than well verified estimates, not to speak of the soft estimates which are all that we have, particular attention will be paid to the beliefs and expectations of users and politicians. These beliefs determine attitudes to data but even more important to economic and demographic concepts, which are the basis of innovative demand management approaches being advocated by visiting scientists and consultants.

The study identifies with the innovators. It attempts to set out a comprehensive account of the resources that exist, deploys the analytical tools of political economics, but at the same time draws particular attention to social theory.

The users of MENA water for whom these chapters are mainly written, are enmeshed in the enduring communal traditions and perceptions which to date have been impervious to new interpretations of the 'value' of water. The only regional economy, that of Israel, which has adopted comprehensively the principles of demand management, to which much reference will be made in the coming chapters, has achieved the position in spite of its enduring, if somewhat 'invented' traditions, of a rural and water intensive way of life (Feitelson and Allan 1996). These are still held sacred by a majority of Israelis, and asserted by its Minister of Infrastructure in June 1997 (CNNI 1997). This Sharon media intervention demonstrated that water can be used as a political and international relations tool by the pragmatic negotiator (Shuval 1997). Strategic water, in this case, located beneath the West Bank is a reason for continuing to hold strategic territory (Wolf 1996a).

The Israeli case demonstrates that the development of a diverse and vital post-modern industrial economy can submerge deeply held traditional views on water and much else. Those who claim that an economy can only become strong, diverse and secure if it can access a minimum of 1000 cubic metres of freshwater per head for its population have to confront very powerful contrary evidence in the case of Israel. Israel can only access, inequitably in the eyes of its neighbours, just over 300 cubic metres per head per year of highly disputed water.

Water is of palpable significance to the state, to rural communities and to farming families. But it has been demonstrated that livelihoods are of even more significance (Allan 1996a) and the argument will be rehearsed in the coming chapters. Once secure livelihoods are in place it is also possible to extol, but at the same time do nothing about, the primacy of water and its food securing capacity. The transformation of economies, to provide a sufficiency of livelihoods which are not constrained by the limited national water budget, is the important idea. Water for livelihoods is the key; not water for food producing livelihoods. Once secure livelihoods have been generated at the national scale both the state and the individual family can get on with their lives accommodating their hydroparanoia and their familiar but misguided environmental resource perceptions.

CHAPTER 2

The resource: the supply

More water 'flows' into the Middle East each year as 'virtual water' than flows down the Nile into Egypt for agriculture.

The 'virtual water' embedded in wheat though technically incalculable is nevertheless of inestimable economic value to the MENA economies.

Not all waters are equal: some are more accounted and accessible than others

The purpose of this chapter is to provide a perspective on how much water is available to the peoples of the Middle East and North Africa. Emphasis will be given to water which we shall call *evident* water. Evident water is that water of which users are aware. The water which is used from surface sources such as rivers and streams, and water from the ground which comes from springs or which can be pumped from wells is evident water. (Falkenmark's (1986b) blue (fresh) water.) Non-evident water is water that is just as wet and just as productive but it is only utilised if vegetation and crops make use of it while it is in the soil profile. Falkenmark (1986b and 2000) has referred to this water as brown water. Water resides in a soil profile for a period determined by what are known as the mechanical qualities of the soil, that is its texture, and the intensity of a preceding rainfall event. Fine textured soil tends to hold water. Gravity takes water through coarse textured soil more quickly. It is the roots of vegetation and crops that intercept the soil water on its way downwards to the groundwater store. In semi-arid regions such as the Middle East and North Africa many vegetation species are adapted to respond to rainfall events which follow long dry periods. The major crop species essential for human populations require a relatively constant supply of soil moisture for periods of one hundred days or more. In semi-arid regions rainfall sourced soil water has to be supplemented by supplementary irrigation from surface or groundwater. Non-evident soil water has to be supplemented by evident water. Associated transaction costs are incurred in bringing evident water to the crop.

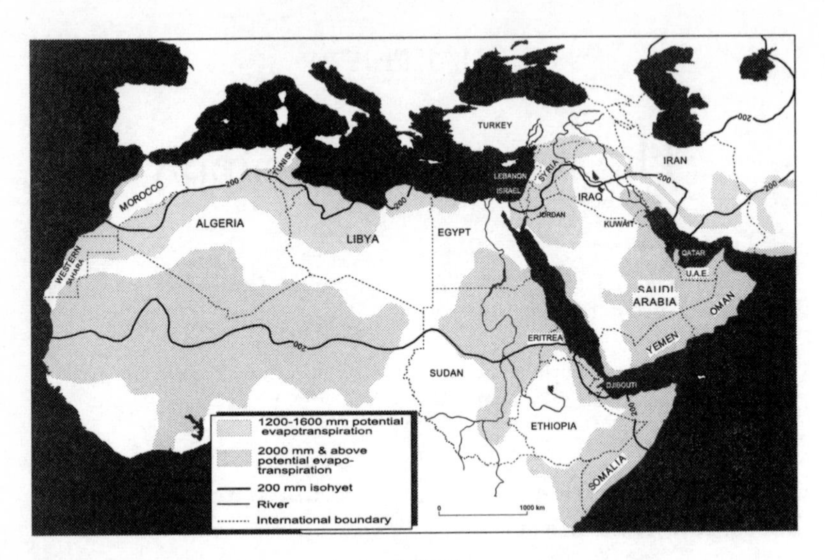

Figure 2.1 Renewable water in the MENA region by sub-region

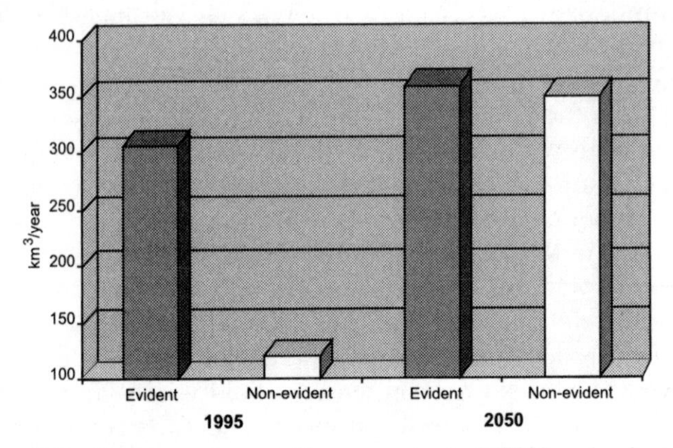

Figure 2.2 Evident and non-evident water in the MENA region by type—1995 and 2050

In the previous chapter it was shown that some water is more *evident* than others. On whether an issue is evident or not an infamous European dictator of the mid-twentieth century concluded that people are prepared to fight for what they care about, care about what they understand and understand what they at least know about. That an issue is evident, that is known about, can make it a matter of personal and communal significance and even of national significance. Water users in the political economies of the Middle

Table 2.1 A typology of Middle East & North African waters

Freshwater in natural hydrological and engineered systems and some information on cost of access and visibility and invisibility

	bn m3/yr	**Cost of access**
Naturally occurring freshwater:		
For all uses depending on costs of delivering		
• precipitation—rain, snow, dew feeding soil water for crops	50 (± 40%)	**Silent**/non-accounted, free
• **surface water** for irrigation	**>200** [1]	Low cost $0.01–0.10/m3
• **groundwater** for irrigation	**100** [1]	Variable cost $0.1–>1.0/m3
• re-used irrigation water for crops (e.g. in the Egyptian Nile system)	20 (+20%)	**Silent**/non-accounted
Other transported water		
Not economical for crop production		
• in **pipelines**		
(e.g. Great Man-Made River(s)—Libya)		2.5 High cost > $0.6./m3
(Israeli Water Carrier)	0.4	High cost-about $0.6/m3
• in ***tankers and bags***	>0.1 (by 2005)	High cost > $1.0/.m3
• 'virtual water' in grain & flour	40 (rising)	No cost; in subsidised commodities
Other produced water		
For irrigation		
• **re-used urban waste-water**	**1.0** rising	High cost > $1.0/m3
For domestic, municipal and industry/services		
• **desalinated water**	**1.0** rising	High cost > $1.0/m3

[1] FAO 1995: 32, Table 18.These are very approximate numbers deduced from the areas estimated by FAO for irrigated cropping. Assuming a one metre depth of water application on average the approximate water use of 200 cubic kilometres per year from surface water and 100 cubic kilometres of water from groundwater can be deduced.

Notes

1 not evident [shown in plain type]—e.g. precipitation
2 evident [shown in bold type]—e.g. **surface water**
3 potentially evident but not yet operational as a major contributor to water consumption [shown in italic bold type]—e.g. ***tankers and bags***

Sources: FAO 1997a and 1997b where shown. Author's estimates

Table 2.2 Sources of supply of evident and non-evident water: estimates of current and future use

Type of water	km3 Evident	km3 Non-evident	km3 Evident	km3 Non-evident
Naturally occurring freshwater				
Precipitation—rain etc		50		50
Surface water—mainly irrigation	200		220	
Groundwater—mainly for irrigation	100		120	
Re-used irrigation water		30		35
Transported water				
Pipelines	3		4	
Tankers and bags	0.1		0.5	
Produced water				
Re-used treated urban waste water	1.2		10	
Desalinated water	1.7		5	
Virtual water				
In cereals and flour	40			100
Other				
Total	306	120	359.5	185
Per cent	72	28	66	34

East and North Africa are afflicted by a deficit of accurate information on water.

What people know about is a social construct. The body of knowledge about water in currency in the region is the result of a centuries long process of selection, emphasis and de-emphasis. The outcome is knowledge which accords with the interests, including the peace of mind and sense of security, of the powerful who use water. The outcome especially accords with the interests of those who lead and manage the political economy as a whole. Another feature of knowledge is that it depends to be rooted in the past. Changes in knowledge systems are resisted. New information which does not fit current assumptions and can easily be rejected by an alliance of water users and the political leaders concerned not to disoblige them. Information about the strategically important virtual water upon which regional economic and political stability depends is especially de-emphasised because it draws attention to the potentially destabilising national water deficits in the region. Water users and

Table 2.3 Middle East and North Africa—water availability (surface & groundwater)

Country	**Population**	**Internal renewable**					**External renewable**				**Total renewable water resources**			
							Surface water		**Groundwater**					
	1	2	3	4	5	6	7	8	9	10	11	12	13	
	000 1995	Surface	Ground-water	Overlap (common to 2 & 3)	Total 2+3–4	IRWR/ inhab-itant	Natural	Actual	Natural	Actual	Natural 5+7+9	Actual 5+8+10	ARWR ppy m3 ppy	
Middle East														
Arid Middle East														***Arid Middle East***
Bahrein	564	0.004	0.000	0.000	0.004	7	0.000	0.000	0.112	0.112	0.116	0.116	206	Bahrein
Cyprus	742	0.830	0.300	0.230	0.900	1213	?	?	?	?	0.900	0.900	1213	Cyprus
Israel	5600	0.600	0.500	0.100	1.000	179	0.500	0.500	0.700	0.700	2.200	2.200	393	Israel
Jordan	5463	0.400	0.500	0.220	0.680	124	0.200	0.200	0.000	0.000	0.880	0.880	161	Jordan
Kuwait	1547	0.000	0.000	0.000	0.000	0	0.000	0.000	0.020	0.020	0.020	0.020	13	Kuwait
Lebanon	3009	4.100	3.200	2.500	4.800	1595	0.037	-0.393	0.000	0.000	4.837	4.407	1465	Lebanon
Oman	2163	0.930	0.955	0.900	0.985	455	0.000	0.000	0.000	0.000	0.985	0.985	455	Oman
Qatar	551	0.001	0.050	0.000	0.051	93	0.000	0.000	0.002	0.002	0.053	0.053	96	Qatar
Saudi Arabia	17880	2.200	2.200	2.000	2.400	134	0.000	0.000	0.000	0.000	2.400	2.400	134	Saudi Arabia
Syria	14661	4.800	4.200	2.000	7.000	477	37.730	17.910	1.350	1.350	46.080	26.260	1791	Syria
UAE	1904	0.150	0.120	0.120	0.150	79	0.000	0.000	0.000	0.000	0.150	0.150	79	UAE
Yemen	14501	4.000	1.500	1.400	4.100	283	0.000	0.000	0.000	0.000	4.100	4.100	283	Yemen

Table 2.3 Middle East and North Africa—water availability (surface & groundwater) (continued)

Country	**Population**	**Internal renewable**					**External renewable**				**Total renewable water resources**			
							Surface water		**Groundwater**					
	1	2	3	4	5	6	7	8	9	10	11	12	13	
	000 1995	Surface	Ground-water	Overlap (common to 2 & 3)	Total 2+3-4	IRWR/ inhab-itant	Natural	Actual	Natural	Actual	Natural 5+7+9	Actual 5+8+10	ARWR ppy m3 ppy	
Middle East - Iraq, Iran, Turkey	**68585**		18.015	13.525	9.470	22.070	**322**	**38.467**	**18.217**	**2.184**	**2.184**	**62.721**	**42.471**	**619**
***Iraq:** its 'humid' north & its major rivers give Iraq a misleadingly favourable water resource status*														
Iraq	20449	34.000	1.200	0.000	35.200	1721	61.220	40.220	0.000	0.000	96.420	75.420	3688	Iraq
Iran	67283	97.300	49.300	18.100	128.500	1910	9.010	9.001			137.510	137.501	2044	Iran
Turkey	61945	192.800	20.000	16.800	196.000	3164	4.700	-12.238	0.000	0.000	200.700	183.762	2967	Turkey
Total Iran & Turkey	**129228**	290.100	69.300	34.900	324.500	**2511**	13.710	-3.237	0.000	0.000	338.210	321.263	**2486**	
Middle East Total	**197813**	**342.115**	**84.025**	**44.370**	**381.770**	**1930**	**113.397**	**55.200**	**2.184**	**2.184**	**497.351**	**439.154**	**2220**	**ME Total**
North Africa														
Arid North Africa														
Algeria	27939	13.200	1.700	1.000	13.900	498	0.400	0.400	0.000	0.000	14.300	14.300	512	Algeria
Egypt	62931	0.500	1.300	0.000	1.800	29	84.000	55.500	1.000	1.000	86.800	58.300	926	Egypt
Libya	5407	0.100	0.500	0.000	0.600	111	0.000	0.000		0.000	0.600	0.600	111	Libya
Malta	366	0.0005	0.015	0.000	0.0155	42	0.000	0.000	0.000	0.000	0.016	0.016	42	Malta
Mauritania	2274	0.100	0.300	0.000	0.400	176	11.000	11.000	0.000	0.000	11.400	11.400	5013	Mauritania
Morocco	27028	23.000	10.000	3.000	30.000	1110	0.000	0.000	0.000	0.000	30.000	30.000	1110	Morocco
Sudan	28098	28.000	7.000	0.000	35.000	1246	119.000	53.500	0.000	0.000	154.000	88.500	3150	Sudan
Tunisia	8896	2.310	1.210	0.000	3.520	396	0.600	0.600	0.000	0.000	4.120	4.120	463	Tunisia
NA Total	**162939**	**67.211**	**22.025**	**4.000**	**85.236**	**523**	**215.000**	**121.000**	**1.000**	**1.000**	**301.236**	**207.236**	**1272**	**NA Total**

Table 2.3 Middle East and North Africa—water availability (surface & groundwater) (continued)

Country	Population	Internal renewable					External renewable				Total renewable water resources			
							Surface water		Groundwater					
	1	2	3	4	5	6	7	8	9	10	11	12	13	
	000 1995	Surface	Ground-water	Overlap (common to 2 & 3)	Total 2+3–4	IRWR/ inhab-itant	Natural	Actual	Natural	Actual	Natural 5+7+9	Actual 5+8+10	ARWR ppy m3 ppy	
Less arid northern Africa														
Sudan - north and south Sudan should be analysed in two parts - arid north and humid south. Not achieved here													Sudan	
MENA total	**360752**	**375**	**105**	**48**	**432**	**1197**	**267**	**136**	**3**	**3**	**799**	**646**	**1792**	

Source: FA0 1997: 14. With a reclassification of the countries by ecological zone and region. With some additional comments on the actual availability of water for economic use: for example Turkey, Syria and the Sudan

Note: Data for *Israel* must have been too controversial for FAO to include; those shown are estimated by the author. *Iraq* shown separately as *Iraq's* data distort those for the region

Summary of water resources in ecologically different regions of the Middle East and North Africa

	1	2	3	4	5	6	7	8	9	10	11	12	13	
ME- *Iq,In,Tur*	360752	375.326	104.850	48.370	431.806	1197	267.177	135.980	3.184	3.184	702.167	570.970	1583	
Iraq	20449	34.000	1.200	0.000	35.200	1721	61.220	40.220	0.000	0.000	96.420	75.420	3688	Iraq
Total Iran & Turkey	381201	409.326	106.050	48.370	467.006	1225	328.397	176.200	3.184	3.184	798.587	646.390	1696	
ME Total	**20449**	**34.000**	**1.200**	**0.000**	**35.200**	**1721**	**61.220**	**55.200**	**0.000**	**0.000**	**96.420**	**90.400**	**4421**	**ME Total**
NA total	**782851**	**852.651**	**22.025**	**96.740**	**969.211**	**1238**	**718.014**	**407.600**	**6.368**	**6.368**	**301.236**	**207.236**	**265**	**NA Total**
MENA total	**386608**	**409**	**107**	**48**	**468**	**1210**	**328**	**176**	**3**	**3**	**799**	**647**	**1674**	

Table 2.4 Summary of renewable water resources by region in the Middle East [ME] and North Africa [NA]

	km3/yr Total internal	km3/yr Total external	km3/yr Total renewable		m3 ppy Total internal	m3 ppy Total average		mn Population
Arid ME	22	20	42	Arid ME	320	600	Arid ME	69
Iraq	35	40	75	Iraq	1721	3700	Iraq	20
Iran & Turkey	325	-3	321	Iran &	2500	2500	Iran & Turkey	129
	Total internal	**Total external**	**Total renewable**	**Total internal**	**Total average**	**Population**		
ME total	380	57	440	ME total	1930	2200	ME total	197
NA total	85	122	207	NA total	523	1270	NA total	163
MENA total	465	180	647	MENA total	1200	1580	MENA total	360

politicians reinforce the water misinformation in currency because it is politically expedient so to do.

Insider knowledge on water is more likely to be determined by politics than by scientific research. Outsiders unencumbered by the politics of water have different perspectives on water and its values. The outsider writing this chapter, for example, will be attempting to quantify the volumes of water available in the region's economies (see Figures 2.4–2.6 and Tables 2.1–2.4). Much rhetoric has been generated by outsiders about the 'evident' water which is in current or imminent deficit. In practice 'non-evident' water, the soil water in the soil profiles of other regions, 'imported' as virtual water embedded in for example grain, solves the immediate and future MENA water deficit. Figure 2.2 indicates how important virtual water is likely to be by 2050. But in the absence of information about the non-evident water it is impossible to mobilise insider understanding or any sort of insider communal will to relate appropriately to the economics of water or the environmental values of water. Politics dominates the discourse. In an information deficit situation the constructed knowledge from the past on water enjoys an ongoing prominence. This prominence reflects the mutual interests of the numerous heavy water users in agriculture and of the politicians who are sensitive to their interests.

In complex industrialised economies the role of water is very important but that role may *not be as visible* as it is in a subsistence system where all inputs are locally accessed: the soil, the water, the livestock, the cultivating systems and the labour. In industrialised economies only a small part of the water used is evident, for example the drinking and domestic water. Some other water is also evident, for example any water laboriously delivered to fields for crop production. On the other hand some even more important water, namely naturally occurring soil water, is less evident, unless there is a drought. In these circumstances the absence of soil water has an economic impact. Such soil water is not even accounted in national water budgets. Water use by industry and services is misunderstood and when considered the volumes involved are usually overestimated. How many people know how much water or how little their firm, or their office or their school uses per year?[1] At the extreme of invisibility, the water embedded in imported commodities such as food, has no profile at all. This last water in practice gives life to the individual and balances and stabilises water deficit economies. The political significance of virtual water is incalculable.

The relevance of the discussion of water being evident or not is that water that is not evident is difficult to relate to and even more difficult to take into account. Such water does not easily gain a political profile. The latter tendency is reinforced if long term users of local water have a vested interest in a major non-evident resource of water remaining economically and politically invisible. Farmers and politicians have already been identified as parties interested in reinforcing the invisibility of virtual water in the political domain. Defining the water resource evident to those participating in an economy is not as straightforward, therefore, as might first appear. Most participants in the insider discourse on water in the MENA region are motivated to reinforce the idea that there is sufficient water in the respective national systems.

Accessibility to water can be defined in environmental as well as economic terms and these two sets of defining factors differ in scale, time, relevance and the extent to which they can be managed. The

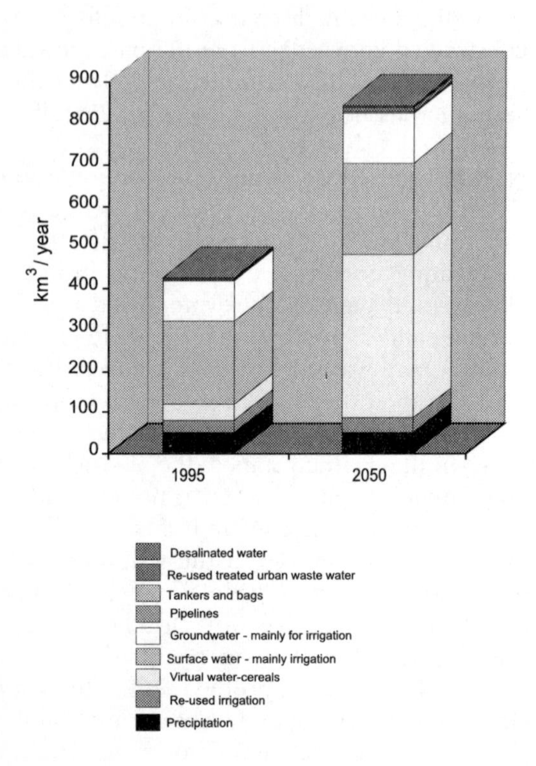

Figure 2.3 Evident and non-evident MENA water by water type, 1995–2050

way these factors differ is relevant to this discussion in that they differ widely in the rates at which they change. Environmental phenomena are relatively stable. Economies on the other hand can be very dynamic over short periods but generally tend to achieve major transformations over two or three decades. In the Middle East there are at least two types of economic change which have been massively transforming. The oil economies were unrecognisable by the 1980s compared with the end of the 1950s; in the same three decades Israel developed from a rural agricultural economy into a post-industrial economy without the help of oil but with unique levels of external support. Other economies, such as that of Egypt, had to transform to cope with their rapidly rising populations, but for most of the period they were only able to match economic growth with population growth rather than exceed it.

The analysis will take a political ecology approach. This approach assumes that the management of the environment, including water, is the management of change. The concept of change refers both to changes imposed by environmental resource users and environmental policy makers as well as by the seasonal and longer term natural dynamic changes which characterise non-equilibrium environments. The non-equilibrium concept refers to the realisation that semi-arid natural environments do not achieve a status of stable equilibrium (DeAngelis and Waterhouse 1987, Benke et al 1993). Rainfall varies within the year from season to season. The MENA region is characterised by the Mediterranean winter rains regime in the north and by a summer monsoon regime in southern Arabia in Yemen and western Oman. Droughts can affect a number of years together. And there is much discussion of the global warming and climate change with possible associated changes in levels of rainfall (Mann 1998).

Coping with these environmental changes is integral to the political ecology of water resources and the peoples who use it and governments which manage them. Peoples and governments have technical, engineering, fiscal, regulatory and institutional tools, albeit not everywhere in equal measure, with which to cope with the natural environmental change. This study is concerned with the political ecologies of water resource management in the MENA region; with understanding the political, economic and technical challenges of coping with environmental change and with rising water demand. Coping measures can be deployed for mere survival in the political ecology of a subsistence mode. Different coping measures are deployed where water contributes to high value economic products and services in complex post-industrial political economies. Both extremes exist in the MENA region. Both extremes are associated with different constructions of the value of water and with what different views on what is reasonable to do with that water. The subsistence mode is associated with a narrow range of water using options. Post-industrial economies are blessed with many options. No single approach will address the challenges faced by groups operating in different political ecologies. Outsiders, coming from circumstances of diverse options, are particularly likely to bring solutions which are seen as problems by those embedded in their local political ecology of water allocation and management, often unchanged for decades.

Within a political ecology both the water environments and the human and social processes which impact and manage them are subject to change. These different change processes occur at differing rates and are subject to different politics. The political processes of adopting water managing technologies and their impact on the water resources of a region or country take place over long periods; perhaps over a century. Water resources can be subject to non-equilibrium natural change as outlined above over relatively short periods of decades. By contrast economies can be transformed within three decades. Extreme environmental events such as droughts and floods can be much more concentrated and generate much more intense politics. Such events are referred to as 'emblematic events' (Hajer 1996). Such events can be especially important in bringing about change. They play a special role in gaining the attention of water users and water policy makers and their occurrence can significantly change water related politics for long enough for changes in policy to be achieved. The different rates of change in the environmental and socio-economic processes are not transparent. Water politics are conducted with very partial and selective stakeholder knowledge.

The following analysis will concentrate on the environmental aspects of water availability and changes in availability with some mention of the economic consequences of availability. The latter will be discussed more fully in chapter 3.

Accessibility of evident and non-evident water

All types of water vary in the degree of their accessibility. Soil water can be seasonal in presence; it also varies in availability from year to year. These are characteristic patterns in all semi-arid environments as rainfall variability increases as the average level of annual rainfall falls. A range of 40 per cent above and below the average is common for 300 mm per year rainfall zones. Surface water can also reflect seasonal rainfall patterns, for example the Blue Nile flood resulting from monsoon rains in Ethiopia in July and August becomes the Nile flood during September and October in northern Sudan and as far as Lake Nasser/Nubia. The Blue Nile flow regime delivers 90 per cent of the water in five months or less.

Mere presence or absence determines availability but so does location. Water can either be on the surface in rivers or lakes, or in the ground at various vertical levels from the soil profile to deep in subsurface strata.

Soil water is subject to downward gravitational and upward capillary forces as well as to the suction of the root systems of natural and cultivated trees and plants. Continuous presence is dependent on the occurrence of levels of rainfall above 1000 mm per year in temperate zones and over 2000 mm per year in high temperature arid zones such as the core Middle East and North Africa. The former comprise less than five per cent of Iran and Turkey and a negligible proportion of the region as a whole. Tracts enjoying 1000 mm of rainfall per year are again mainly in Iran and Turkey and there are minor favoured, but mountainous, tracts in the Lebanon, Yemen and the Maghreb.

A proportion of all the water that falls on the surface returns to the atmosphere. Open water and soil surfaces evaporate and vegetation and crops transpire. The maximum world rates of potential evaporation and transpiration are recorded in the deserts of the central Sahara and Arabia. Open water such as that of Lake Nasser/Nubia at Aswan, soil water at the surface everywhere and some vegetation and crop cover in the deserts of the central Sahara and of central Arabia evaporate and/or transpire at the rate of three metres depth of water per year (See Figure 2.1). Human intervention to build water storage dams and develop irrigated agriculture in desert regions is always associated with actual evapotranspiration. Lake Nasser/Nubia loses between six and thirteen cubic kilometres per year (Stoner 1995) from its 160 cubic kilometres storage depending on the level of the lake. The nominal average flow at Aswan is 65.5 cubic kilometres. The Nile Waters Agreement of 1959 assumed a ten cubic kilometres loss on the basis of an average flow for Egypt of 55.5 cubic kilometres as a result of evaporation and seepage at Lake Nasser/Nubia.

Most of the MENA countries suffer seasonal rainfall levels under 300 mm per year insufficient to support all but a limited range of trees and plants. Naturally occurring species may be suitable for animal feed and field crops such as cereals, wheat and barley, and tree crops such as olives and almonds can exist in regions of low rainfall. Over 80 per cent of the arid Middle East and North Africa receives less than 50 mm per year—the MENA region is without equal in the extent of its deserts.

Vegetation cover is particularly effective in accessing and using the ephemeral soil water. Soil water may reside in a profile for a very short time; perhaps only a few days or less if the original rainfall event brought less than 10 mm.

Groundwater resides in more stable and more secure reservoirs. Provided the groundwater reservoir is not naturally saline or saline through local mismanagement then the groundwater can be usefully augmented by infiltration from rainfall via the soil profile. The reservoir may lie at a metre below the surface from which it may require negligible effort and cost to lift water for use in agriculture and other economic sectors. Groundwater can lie at 1000 or more metres down which means that the costs of lifting the water to the surface may be over one US dollar per cubic metre. At these costs of delivery such water can only be used as an input to high value adding economic activities.

Accessibility may be seasonal with water present for only parts of the year. Accessibility may also be geographically specific with a river or lake located in tracts which have not been heavily settled because the technologies available for water management and, just as important, the actual need to develop the water, were not in the past such as to stimulate development. Examples are areas that remained unused for agriculture for millennia and have only been subject to very intensive water resource development in the twentieth century. Examples are the 1930s Gezira Project in the Sudan (Chesworth 1995) and the GAP project in south-eastern Turkey in the 1990s (Bagis 1989, Gokce 1998, Unver 1999). The main Nile tributaries, the Blue Nile and the Atbara, are still not used in the source region of mountainous Ethiopia. The Nile in Egypt, on the other hand, flows through some of the best soils and some of the most gloriously level irrigable tracts in the world. The Egyptian Nile lowlands were used for five millennia for flood recession irrigation until the nineteenth century. Since then the surface flow has been subject to progressive control until 1970 when the completion of the High Dam at Aswan achieved total regulation (Hurst et al. 1966, Waterbury 1978a, Said 1993, Bethemont 1999).

Accessibility is determined by many terrain variables. Both surface and groundwater may be found at relatively high elevations enabling economical transfer over long distances. Once the fossil water in the Nubian Aquifers of south-eastern Libya has been raised to the surface, albeit at great expense, they can fall by gravity through pipelines to the coast of the Mediterranean (McKenzie and Elsaleh 1995:104). 1.8 cubic kilometres per year was the planned capacity of the eastern pipeline (1986–95), and 0.8 cubic kilometres per year in the western pipeline (1990–97)—the latter requiring some pumping at the Jebel Nafusah. By contrast the water lifted through the earlier (1962–64) but much smaller capacity water carrier in Israel (0.4 cubic kilometres per year) utilised a significant proportion of Israel's total power budget in its early years. The Israeli water carrier water and any water lifting activity are impacted by rapid increases in energy costs. Irrigation in the New Lands to the west of the Nile delta in Egypt can move from cost-effective to cost-ineffective as a result of energy price shocks such as that of 1979/80 (Pacific Consultants 1980:3, Allan 1983a:476).

Variable temporal and geographical accessibility have cost of delivery consequences which will be further discussed in chapter 3. It is sufficient here to note that the supply cost of water has a source component as well as quality and mode of delivery components. Issues relating to these last costs will be developed in more detail in later chapters.

Economic accessibility of non-evident water

Finally 'non-evident' water, as well as real water, can be more or less accessible. In practice 'virtual water' has been fairly readily accessible for thousands of years. Grain was exported to Rome from the Nile delta, and from places which are now Libya and Tunisia, in the period of Roman domination of the Mediterranean. All of these regions have been called 'the granary of Rome', even if only the Nile delta really justified the term.

Wheat, the major traded food commodity by the end of the twentieth century, kept its high rank in the global trade league for millennia. The recent history of the wheat trade in the region has been the most remarkable in all of history, whether measured by volume, economic significance or in the way it has testified to the dominance of the politics over economics in the affairs of the world.

The numerous wheat importing economies of the region have, for almost half a century, enjoyed the progressive fall in the real price of wheat (Dyson 1994). The unprecedented capacity the economic *behomeths*, the US Department of Agriculture (USDA) and the European Union (EU), to subsidise their agricultural sectors has effectively competed down the world price of wheat. There has never been a peace time trading event like it, either in terms of its length or its economic impact. There was another US-European union of interests which enabled the unnatural stability of a commodity price, namely the price of crude oil between 1901 and 1969. It could be argued that the oil event exceeded the wheat event in its unprincipled non-economic excesses. Outsiders advocating market-economy rationales for water users in the MENA region should reflect on their own consistency. They should ponder why it has been so important to have in place the strategically important global wheat and oil non-markets of the twentieth century which their own policy makers, past and present, have done so much to shape. Water is also a globally significant strategic resource. Water is likely to be subject to the non-economic measures that have attended wheat and oil.

For the water short MENA economies the USDA-EU wheat trade folly of the second half of the twentieth century has, however, brought a massive and wonderfully silent water bonus and a very important political bonus.

The economic value of the 'virtual water' embedded in wheat though technically incalculable is nevertheless of inestimable economic value to the MENA economies. Should such water become less accessible then all the economies of the region would be in a vulnerable position. In chapter 1 it was argued that there were no immediate signals that there was insufficient global 'virtual water' to meet regional demands. However, it behoves the governments of the MENA region to evaluate the position with as much vigilance as those who manage another vulnerable economy, that of Japan, study global energy resources and the future availability of key strategic minerals.

Types of water in the Middle East and North Africa

An analysis of the water balance of the Middle East and of its individual economies can be approached in a number of ways. Table 2.1 analyses MENA water according to its natural or engineered provenance, type of use and approximate cost of delivery. The discussion of water as an economic resource in chapter 3 will examine the relationship between demand and cost more fully. It is sufficient at this stage to note that the accessibility of water in terms of the effort and cost required to mobilise it varies greatly.

Those uses which need low cost water such as irrigated agriculture should not access water costing more than a few US cents per cubic metre, at least according to the rules of economics. As subsidised agricultural inputs are the rule world-wide subsidisation of water for agriculture is in practice common and is only one of a number of forms of agricultural subsidy. Water subsidisation is very common in the Middle East and North Africa. These practices are politically legitimised by the well developed WINER ideology which claims that Water Is Not an Economic Resource. This idea is the theme of chapter 4.

The information in Table 2.1 can be analysed to show some revealing features of current regional hydrology and some very important characteristics of the region's future hydrology.

Unequal endowments: a regionalisation of Middle Eastern and North African water

Generalising about Middle Eastern and North African water is difficult and misleading because there are some very distorting elements which seriously affect regional totals and average figures. The MENA region in this study refers to the economies from Morocco in the west to Iran in the east and from Turkey in the north to the Sudan in the south.

At the core of the MENA region are the deserts of North Africa and Arabia. The relatively humid regions are in the north—the mountains of Turkey and the Elburz Mountains of Iran. Precipitation in Turkey and Iran feeds the Tigris and Euphrates rivers. In the south, the mountains, of Ethiopia and East Africa, feed significant, but by no means mighty rivers. The Nile and the Tigris-Euphrates convey less than ten per cent of the water of a river such as the Yangtse in China and less than one per cent of a river such as the Amazon. The Nile and the Tigris-Euphrates are sufficiently large, however, to have met the needs of all the modest populations living along their banks until 1970 in the case of the Nile and until the present and the foreseeable future in the case of the Tigris-Euphrates.

The surface water data for Turkey, Iran and Iraq severely distort any hydrological averages for the region. The southern Sudan does the same in the south but to a much lesser extent.

Table 2.3 provides an overview based on the very useful estimates of the water resources of the region completed by FAO for water resources in general (FAO 1997a:14) and for irrigation water use (FAO 1997b:32). These data have been compiled according to sound hydrological concepts, except for the neglect of naturally occurring

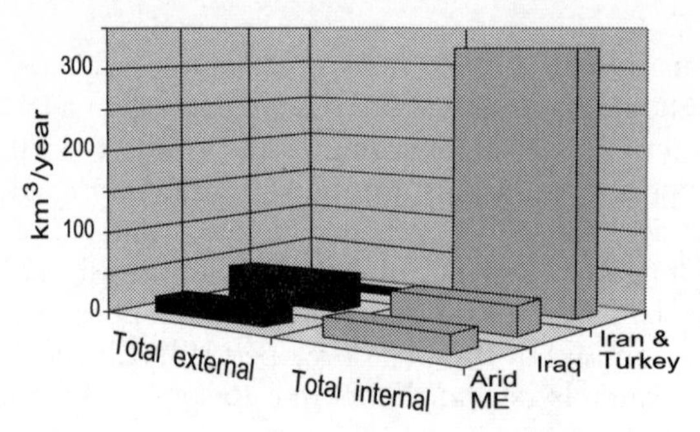

Figure 2.4 Renewable water per head in the Middle East and in North Africa—m3 ppy

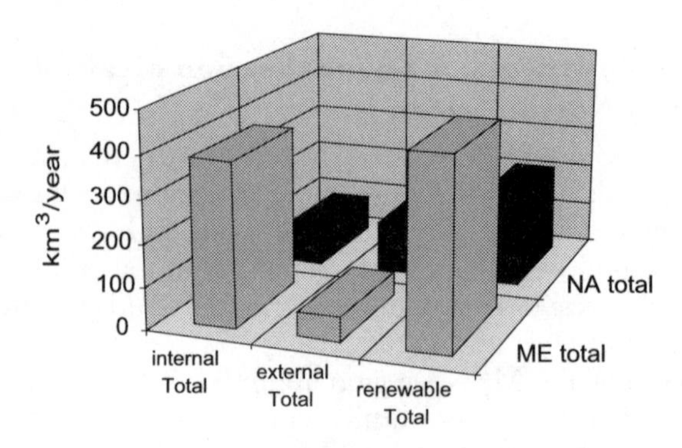

Figure 2.5 Renewable water per head in Middle East and in North Africa—m3 ppy

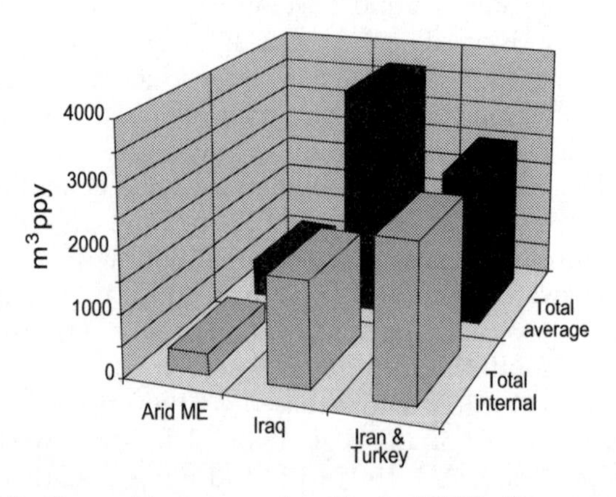

Figure 2.6 Renewable water per head in the Middle East countries—m3 ppy

soil water, and are admirably comprehensive. The absence of Israel from the list is an indicator of the unavoidable political pressure on any attempt to review strategic resources which are judged to be valuable and for which there are predictable counter claims. The data included by the author for Israel to augment the numbers provided by FAO will be controversial and disputed as they were at the time of writing subject to negotiation in the Middle East Peace Talks (see Oslo Accord in appendix of Allan 1996a and Allan 1999). The data for Iraq and the Euphrates generally are also unreliable for similar reasons.

There are a number of predictable biases in the interpretation of shared water resources. Upstream riparians tend to understate the volume of water in the natural system. The bias is found in estimates for both surface water resources and groundwater. Upstream riparians, assuming they have control over river flows, want to establish a low figure for the volume of the 'natural' resource so that any negotiated outcome based on a share of the flow would nominate a volume which is smaller than actually available.

The upstream riparian anticipates that it would then be able to utilise higher volumes than agreed, either derived from the surface resource or pumped from a groundwater resource because effective monitoring would be impossible. Downstream riparians, on the other hand, tend to overstate the 'natural flows' in the hope of being able to gain an agreed 'entitlement' as high as possible in any negotiation of a share of a natural flow. Those who have tried to understand the data on the Tigris and, especially, the Euphrates, rivers have encountered a very serious version of this type of hydropolitics (Kolars and Mitchell 1991, Medzini 2000). The Nile Waters Agreement of 1959 has turned out to be favourable to the downstream user, at least thus far, because an average flow was assumed on the basis of data from 1900–1955, which has not been achieved in the decades since (Evans 1995).

Apart from these biases there are other dangers arising from regional averaging. Table 2.3 and the three figures (2.4, 2.5, 2.6) derived from these data illustrate the problems. Figure 2.3 shows how very different are Turkey and Iran from the rest of the region. Also that the water resources of Iraq can be shown to be higher than for other economies in the MENA region, excluding Turkey and Iran. Once the Turkish, Iranian and Iraqi numbers are incorporated into the Middle East regions data they give the impression of a relatively water rich sub-region; certainly very much richer in water than North Africa.

The arid core of the MENA region is a much safer unit about which to generalise. This is especially the case when water availability is related to levels of population. Figure 2.5 shows the levels of water per head endured by the economies of the core arid Middle East are 600 cubic metres per person per year (m3 ppy) compared with those of Iraq—3600 m3 ppy, and of Turkey and Iran—2300 m3 ppy. The figure for Iraq also shows how dependent it is on external water, that is water flowing into the country in the major rivers and tributaries of the Tigris and Euphrates systems. The high figures for Iraq assume that the flow of the Euphrates has not been reduced by about 50 per cent from an unreliable average flow of over 30 cubic kilometres per year at the Turkish border with Syria in the 1960s to a reliable 15.8 cubic kilometres (equivalent to a 500 cusecs) by the dam building and irrigation developments in south-eastern Turkey in the 1970s, 1980s and 1990s.

Figure 2.4 is especially misleading as the data for Turkey, Iran and Iraq have been included with those of the poorly endowed Middle Eastern (Mashrak) countries, making it appear that all were sufficiently endowed with water to meet their water needs.

The terms internal and external water have been coined by FAO to identify that water which is sourced within a national economy and that which comes from outside its borders. Internal water included the springs and rivers occurring within national boundaries fed and enlarged by local rainfall, together with the groundwater within the national boundaries. External water is water that crosses into a national entity as surface flow or groundwater flow. Some economies are almost totally dependent on internal water, for example that of Saudi Arabia on its groundwater. Egypt by contrast is almost totally dependent on external water in the form of the Nile and mainly on the Blue Nile and other tributaries originating in Ethiopia (Evans 1995).

Surface water: the Nile, the Tigris-Euphrates, the Jordan

Surface water has throughout history been a significant source of water in the MENA region and by the end of the nineteenth century it had become the dominant water resource outstripping the rainfed water in soil profiles. Surface water in the arid and semi-arid MENA region is concentrated in two major rivers, the Nile and the Tigris-Euphrates and a number of small river systems. Of these only the Jordan is of major significance. This importance stems from the basin's political rather than its hydrological scale. The Jordan basin is significant because of its location at the heart of the region's strategic security sub-system (Buzan et al. 1998 and 2000)—the Palestine-Israel conflict and peace process. Water is one of the five major issues in dispute along with Jerusalem, settlements, borders and refugees. Water has been shown to be the issue most amenable to compromise. In the Jordan-Israel Peace Agreement of 1994 Jordan accepted terms which were less than generous. The pragmatic process of a multi-factored negotiation always reduces the relative salience of disputed water.

The Nile

The Nile is an important Middle Eastern river although all its waters come from tropical and equatorial Africa. The provenance of the water means that an understanding of the past, current and future water resources of Egypt and the northern Sudan require that the hydrology of the southern, water generating part of the system, needs to be explained. The Nile is a long river but not a big river in terms of the water which flows in the system. By the time its major tributaries join at Khartoum the flow is about 84 cubic kilometres (billion cubic metres) per year.

The ten riparians of the Nile comprise about 29 per cent of the surface area of the African continent. The parts of these riparians lying within the Nile catchment cover approximately 12 per cent of the African continent. The tropical and equatorial south of the catchment comprises over 50 per cent of the catchment, including the water-tower of the Ethiopian highlands. The humid segment of the catchment enjoys average annual precipitation of over 1000 mm per year. The arid half of the basin receives no useful rainfall as potential evaporation ranges from two to three metres depth per year. The maximum potential evapotranspiration occurs in the area of Lake Nasser/Nubia in the northern Sudan and southern Egypt. Throughout Africa the environmental and economic effectiveness of rainfall is low because potential evaporation and transpiration rates are high at over two metres depth per year. All these environmental factors conspire to make the Nile a long but low water volume river.

The surface waters of the MENA region were not much engineered until the early nineteenth century. The lower Nile was only significantly controlled by the new works at Aswan in 1906 which were augmented in the first two decades of the twentieth century. These works allowed secure access to a minimum of about 30 cubic kilometres per year by the 1930s enabling about 80 per cent of Egypt's irrigated area to be double cropped. Works in the Sudan at Sennar in 1925 on the Blue Nile commanded water sufficient to irrigate the

Gezira Scheme just south of Khartoum—initially one million feddans (400,000 hectares) extended in the 1960s to 800,000 hectares. Not until 1970, with the building of the Aswan High Dam, was the utilisation of the surface flow of the Nile boosted to an average of about 40 cubic kilometres per year—plus the unaccounted re-use in irrigation. The additional water from the new storage at Aswan was sufficient to enable double cropping throughout the six million irrigated feddans (2.5 million hectares) of Egypt by the early 1970s.

The anticipation of the new water from Lake Nasser/Nubia also stimulated the most ambitious land reclamation programme in Egypt since the reign of Mohamed Ali at the beginning of the nineteenth century (Allan 1983a:472). In the 1960s 1.2 million feddans (500,000 hectares) were subject to land reclamation activity, but only 300,000 feddans (125,000 hectares) of the reclaimed land was operating beyond a break even economic performance by the late 1970s (Hunting Technical Surveys 1980). Two million feddans (800,000 hectares) have been a perpetual land reclamation target since the early 1980s.

The flows of the Nile systems are amongst the best understood and monitored in the world; the records are certainly the longest for any river (Hurst et al. 1966, Said 1993, Evans 1995:42–43). The records are for the most part reassuring in that they indicate that although there have been periods of low and high flow these have not been livelihood or economy destroying for the levels of population being supported in previous centuries. And for the optimists of today, who base their optimism on the demonstrated capacities of economies to respond to environmental events, even the threat of lower flows through climate change would be disturbing but not economically determining. Meanwhile current climate change models as frequently predict higher flows as they do lower ones for the Nile (Conway & Hulme 1996:290, Conway et al. 1996) because rainfall futures are much harder to model than temperature futures.

The map of the Nile and its tributaries (Figure 2.7) indicates the length of the system and the large number of riparians, ten in all. Table 2.5 and Figure 2.8 show the key features of Nile hydrology. (Sutcliffe and Lazenby 1995:173–179) The important characteristics of the river and its political fragmentation are:

Ethiopian tributaries—Blue Nile (Abay), Atbara

- over 80 per cent of the water comes from the Ethiopian highands. The Eritrean section of these highlands is not a significant source of water.

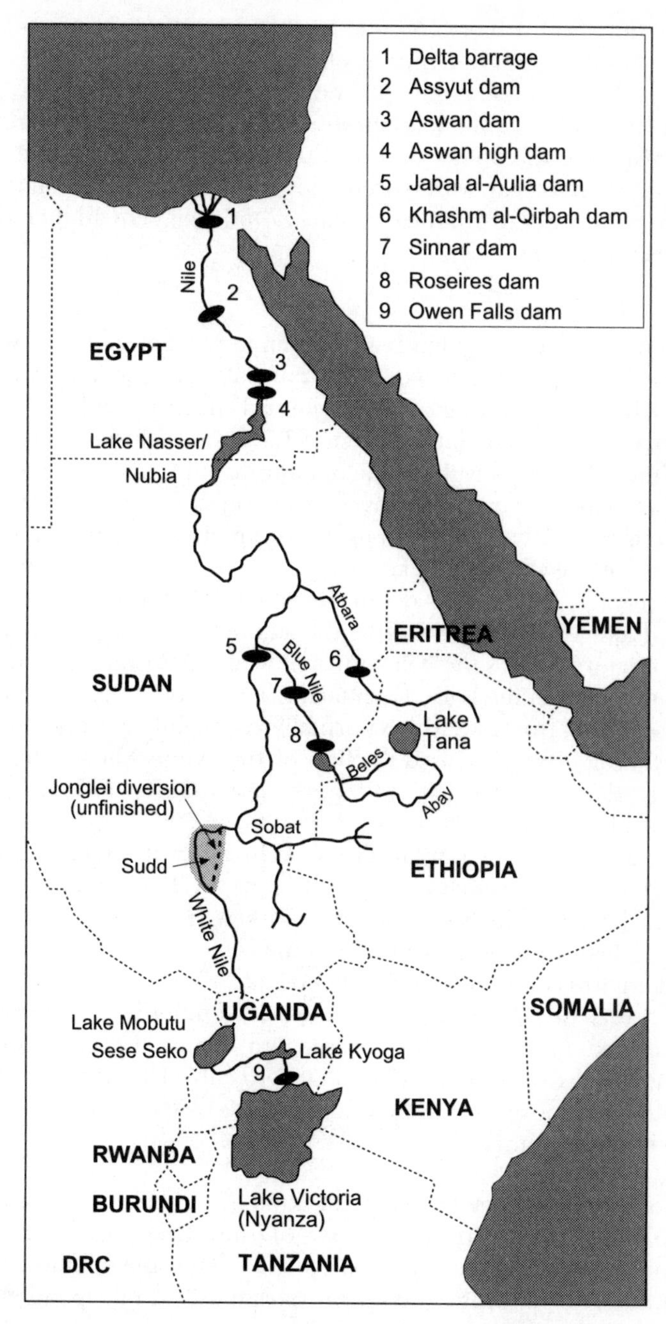

Figure 2.7 Map of the Nile and its tributaries, the riparians and the major civil works

- the flow from Ehiopia is concentrated into the summer months of July-October. A minor element of the Ethiopian flow contributes to the White Nile flow via the Sobat (Baro).
- the Ethiopian tributaries, the Blue Nile (Abay in Ethiopia), the Atbara and the Sobat (Baro in Ethiopia) tributaries vary in flow from year to year with extreme low flows recorded at only half the average and high flows can be 50 per cent higher than average (Evans 1995:52).

White Nile tributaries

- the East African highlands are also an important source of water but this water is regulated, partly naturally in Lake Victoria, and partly by engineering works at Owen Falls in Uganda.
- because of the regulatory effect of the Lake Victoria storage, the White Nile flows without fluctuation through the year and also from year to year. There have been exceptions; the 1962 rainfall event caused Lake Victoria to rise two metres, which for a water body of the extent of Lake Victoria created a significant increase in storage. Two metres depth over the 67,000 square kilometres of Lake Victoria stores 270 cubic kilometres. This is a number to conjure with as the Nile drains only about 25 cubic kilometres annually from the Lake. Even taking into account the evaporative losses from the Lake which normally are slightly less than inflow, the 1962 event increased the flow of the White Nile for the next two decades (Evans 1995:55–57, Sutcliffe and Lazenby 1995:175–179).

  Another major rainfall event occurred in the Lake Victoria region in 1998 which also impacted the level of the Lake and a year later of The Sudd wetlands (Birkett et al. 1999).

  A feature which should be emphasised is that the 25 cubic kilometres contributed annually by Lake Victoria to the Nile flow is a very tiny proportion of the water coming into and leaving the Lake in the Lake Victoria system. Even an evaporation of one metre depth would shift about 67.5 cubic kilometres per year into the atmosphere. Both inflow and the evaporation dwarf the White Nile outflow.

A summary of surface hydrological flows of the basin are shown in Table 2.5 to demonstrate the scale of the surface flows and the evaporative losses in the White Nile system. The vulnerability of the system to changes in the rates of evaporation and precipitation are clear. For example the runoff and evaporative processes in the Lake

Table 2.5 Nile tributaries surface hydrology

River	Station	Start date	mn km2 Area	km3/yr Mean runoff	mm depth	km3 St'd dev'n	% Coef' of vari'n
Nile	Dongola	1870	1610	89.6	55	34.1	38.3
Atbara	Mouth	1903	113	11.7	103	3.7	32.1
Main Nile	Tamaniat	1911	1470	73.3	50	12.2	16.6
Blue Nile	Khartoum	1900	260	49.7	191	11.2	22.5
White Nile							
Sobat	Malakal	1905	1140	29.6	26	5.1	17.3
Sobat	Mouth	1905	232	13.7	59	2.7	19.5
Bahr el Ghazal	Mouth	1938	274	0.3	1	0.2	58.8
Jur	Wau	1930	49	4.5	91	1.6	36.1
Bagr el jebel	Mongalla	1905	483	33.1	69	12.2	36.8
Semliki	Bweramule	1940	30	4.2	140	1.1	25.9
Victoria Nile	Jinja	1900	254	25.0	98	9	36.1
Kagera	Nyakanyasi	1940	56	6.4	114	2	31.4

Source: Sutcliffe and Lazenby 1995:173

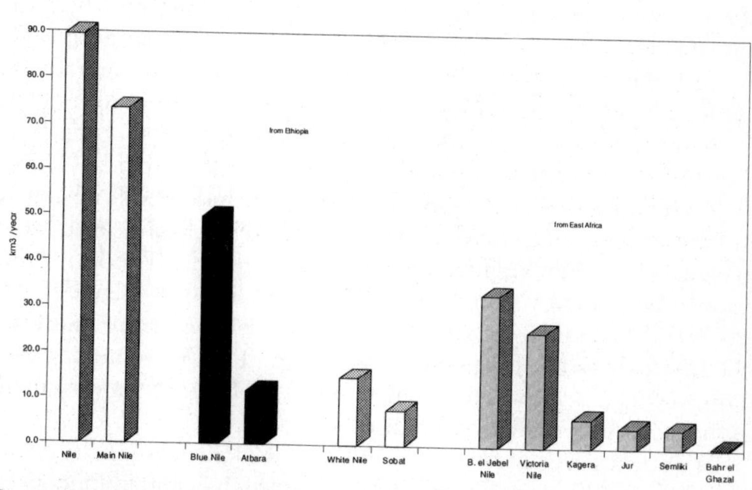

Source: after Sutcliffe and Lazenby 1995:173

Figure 2.8 Nile surface hydrology: summary of flows

Victoria system are more than five times larger than the White Nile flow at Jinja. Rises in temperature which have a predictable impact on evaporation would explain the substantial changes in lake levels in Africa in the past. Unfortunately the impact of higher and lower temperatures on the rainfall regimes of the world, including on those of the Lakes Basin of the East African Highlands are not predictable either globally or regionally.

Figure 2.8 illustrates some very important features of the upper Nile hydrology. The system generates about 120 cubic kilometres of surface flow in Ethiopia and East Africa. However, over 50 per cent of the East African water is subject to evapotranspiration in the Sudd, Machar and other marshes of the southern Sudan (Sutcliffe and Parks, Y. P., 1987, 1989, 1995:252–53, Howell and Lock 1995:245–46). As a result the long term average flow in the northern part of the system—the northern Sudan and Egypt—benefiting the two downstream riparians, was assumed to be 84 cubic kilometres per year on the basis of the six decades of data up to 1955. The flow reflects only 70 per cent of the water which left the Nile Basin's African highlands. Figure 2.8 shows the contributions to the east African tributaries and an indicative level of evaporative losses. Figure 2.9 shows the variation in major lake, wetland and reservoir levels during the 1990s (Birkett 1999).

The Jonglei Project—overtaken by new principles

The phenomenon of the Sudd and its neighbouring marshes was until the 1980s regarded by outsiders as a water resource ripe for recovery. The flow of the White Nile could be boosted annually by first four in a first phase Jonglei Project and by a further four cubic kilometres by a second phase. The channels, accelerating the flow of water would reduce evaporation from the Sudd wetlands (Collins 1990, 1995; Howell et al. 1988). Outsiders included both northern Sudanese, Egyptians and a wide range of colonial and later international agency engineers from across the world. The 1959 Nile Waters Agreement (Nile Waters Agreement 1960, Godana 1985: 186–191) between Egypt and the Sudan made express provision for schemes such as the Jonglei drainage ditch. The Agreement stated that investment in works to enhance the availability of water should be shared. The benefits would also be shared.

1950s and 1970s perspectives on the 260 kilometres Jonglei canal are not universally appreciated. The construction began in June 1978 at a time when draining the second biggest wetland in Africa was not a green issue of global significance. By the time the excavation ceased in 1984 as the result of armed intervention by forces from the peoples of the region world opinion had changed concerning the environmental and economic wisdom of such schemes. By the time the project had been closed down for 15 years world environmental opinion had adopted Californian (Reisner 1993:307–308) green principles. The resumption of the project without the approval of either the local Nilotic communities, mainly the Dinka and the Nuer, or international funding bodies made its resumption improbable.

The New Valley Project—water management reform enables the hydraulic mission one more play in Egypt

With the Jonglei option in abeyance and probably dead the chances of augmenting the waters in Nile of the Sudan and Egypt by hydraulic rather than institutional instruments are limited. The last two years of the twentieth century, however, brought an unprecedented initiative. The Egyptian leadership decided that the Nile could yield one more engineering monument in Egypt's long and monumental hydraulic mission. The difference between this initiative and other barrages and dams was that the water to make it secure had to come from Egypt's existing share agreed with the Sudan in 1959.

The New Valley Project was to bring about five cubic kilometres of water to the Kharga oasis area to the north and west of Lake Nasser/Nubia. The water could come initially from the overfull storage that had characterised the 1990s. However, the experience of the 1980–1987 period showed that accumulated rainfall deficits in Ethiopia could bring Lake Nasser/Nubia to such a low level that Egypt faced a 25 per cent cut in its irrigated area in the 1988 season. The only way that additional water for the New Valley could be assured was if water were to be differently allocated in the existing irrigated area. The Egyptian plan was to reduce the area of sugar in Upper Egypt near Lake Nasser/Nubia and the area of rice in the delta area.

This change of policy was a major departure from normal Eygptian policy. It was a version of demand management; the re-allocation of water to activities which could bring a higher return. The debate rages whether the New Valley agricultural and industrial projects will be the success that their champions predict. The New Valley land reclamation projects of the 1960s were not a success. The low levels of investment were the main reason for this failure. The political and financial commitment of the Egyptian Government were of a different order in 2000 from those of the 1960s. The

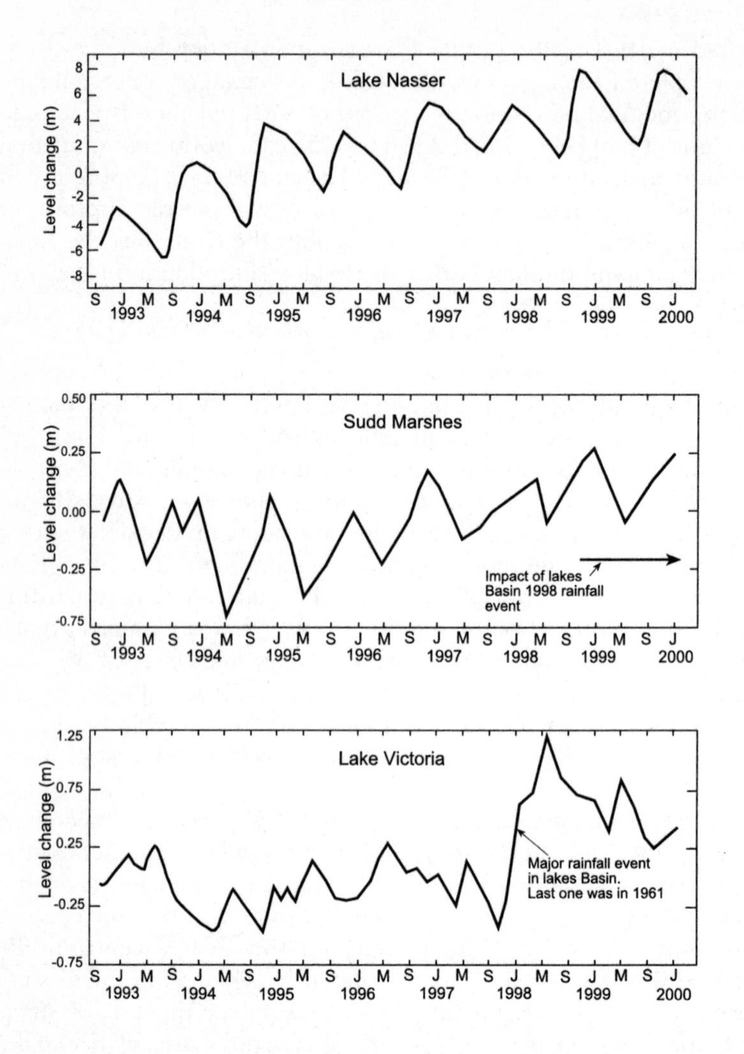

Figure 2.9 Variation in lake and wetland levels of Lake Nasser/Nubia, the Sudd and Lake Victoria 1993–1999 Source: Birkett et al. 1999

involvement of international Arab finance was also very substantial in the new project. The approach which involved the international private sector was also unfamiliar for Egypt. The project has many of the enthusiasms, vigour as well as the environmental and political risk taking of southern California in the 1920s and 1930s evaluated so graphically by Reisner in the book *Cadillac Desert*. Californian advisers are involved in the New Valley project. The jury is out.

For the Egyptian leadership the project addresses many local, national and international priorities. It provides a focus for investment in the south of Egypt which is a neglected and poorly developed region. For the people of the south of Egypt the project is intended to symbolise the commitment of the centre to the economic and social needs of the communities of Upper Egypt. Regionally it is a reminder to the Sudan that Egypt is concerned about its territory and resources which border the Sudan. The investment being directed to this border region confirms that concern. The utilisation of the five cubic kilometres of water in the New Valley was and remains a very important signal to the Sudan and the riparians further south that Egypt intends to enter any negotiations over water with every drop of Egyptian water fully committed and also efficiently committed. More important the water will be being used on water efficient projects that will be being constantly enhanced in the levels of efficient water use achieved.

Additional water from Lake Nasser/Nubia

A possible associated project which could provide some new water could be the management of the regime of Lake Nasser/Nubia. At least five cubic kilometres of water annually could be saved by managing reservoir storage at a low level (Storer 1995). The power generation potential would be reduced if the level of the lake were to be reduced. By 2000 this option was an acceptable trade-off as only a small proportion, less than ten per cent and falling, of Egypt's electric power was by then coming from the hydro-power stations at Aswan. Economic factors will determine whether to take a new approach to the management of Lake Nasser/Nubia.

New water in Ethiopia

The potential of the valleys of the tributaries of the Abay (Nile) in Ethiopia for the production of hydro-power and for consumptive use in irrigation is well understood (US Corps of Engineers 1957, US Bureau of Reclamation 1958). Many studies were also carried out at the turn of the twenty-first century on behalf of the Ethiopian Government. The options for Ethiopia are restricted by its limited economic and institutional capacity. Also by the complex hydro-politics of the Nile basin and the unstable security position in the Horn of Africa. The main potential international sponsor the World Bank is constrained by its own rules and in particular by its Operational Procedure OP 7.50 (World Bank 1992) which prevents the funding of water projects not approved by every potentially affected riparian. The

World Bank was active in the late 1990s in supporting open discussions of professionals via the Nile 2002 process (Nile 2002) and via confidential discussion at the highest levels.

Ethiopia could create structures that would regulate the flow of the Nile tributaries. A number of additional cubic metres per year would be added to the water available to farmers in Ethiopia and downstream. Ethiopia could over a period of two decades or more develop schemes to consume in irrigation volumes of water in excess of ten cubic kilometres per year. This number is alarming to downstream riparians but it should not be. Egypt has already coped with an annual water deficit bigger than this. In twenty years time when its economy will be immensely more diverse and strong and water use efficiency even further improved coping with a reduction of flow of this level will not be a significant economic or political challenge.

Tigris-Euphrates

Both the Euphrates and Tigris Rivers rise in the mountains of eastern Turkey and join the sea at the head of the Persian Gulf. The Euphrates enters Syrian territory at Karkamis, downstream from the Turkish town of Birecik. It is joined by its major tributaries, the Balik and the Khabur, as it flows south-east across the Syrian tablelands before entering Iraqi territory near Qusaybah (see Figure 2.10).

The Tigris flows through Turkey until the border city of Cizre; from there, it forms the border between Turkey and Syria for 32 km, then crosses into Iraq at Faysh Khabur. Within Iraq, it gathers several tributaries (the Greater Zab, the Lesser Zab, the Adhaim, and the Diyala) from the Zagros Mountains to the east. The Tigris and Euphrates unite near Qurna, in Iraq. The combined flow, called Shatt al-Arab, empties into the Gulf.

The Euphrates basin lies 28 per cent in Turkey, 17 per cent in Syria, 40 per cent in Iraq and 15 per cent in Saudi Arabia. The river is 3000 km long, divided between Turkey (1230 km), Syria (710 km), and Iraq (1060 km). Contrary to this linear distribution, the catchment area that produces inputs into the river has 62 per cent in Turkey and 38 per cent in Syria. It is estimated that Turkey contributes 89 per cent of annual flow and Syria contributes 11 per cent (Naff and Matson 1984); one expert estimates the percentage of the Euphrates flow originating in Turkey as high as 98 per cent (Kolars 1992). The remaining riparians contribute very little water.

The Tigris basin is distributed between Turkey (12 per cent), Syria (0.2 per cent), Iraq (54 per cent) and Iran (34 per cent). The

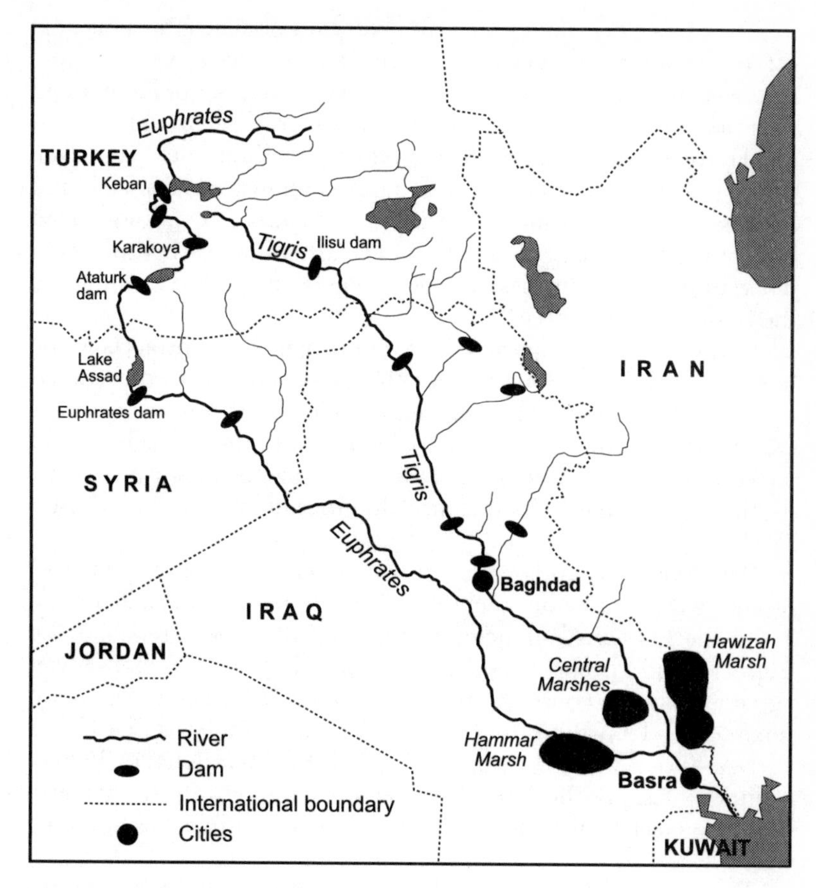

Figure 2.10 The Tigris-Euphrates basin, its riparians and the major civil works

catchment area producing the water of the Tigris is situated in Turkey (21 per cent), Syria (0.3 per cent), Iraq (31 per cent) and Iran (48 per cent). The river is 1850 km long, with 400 km in Turkey, 32 km in Syria and 1418 km in Iraq (Kliot 1994). Turkey provides 51 per cent, Iraq 39 per cent, and Iran 10 per cent of the annual water volume of the Tigris.

Not only the riparian states, but also scientists, have disagreed about the mean annual discharge of both rivers, which were subject to large annual and seasonal fluctuations. Estimates on the total annual flow of Euphrates vary between 28.7–30.5 billion cubic metres. The Tigris' estimated annual total flow is between 43–52.6 billion cubic metres.

These statistics already reveal a certain imbalance, in that only a few states are the source of most of the annual water supply whereas all states participate in its use. However, Saudi Arabia and Iran are not major riparians in this basin. Iran is riparian only to the Tigris; Saudi Arabia is riparian only to the Euphrates. Moreover, the Saudi Arabian sector of the basin is almost always arid, and because of unfavourable geographic and climatic conditions Iran cannot use the waters of the Tigris for agriculture or hydro-power. Therefore, these countries have generally been ignored in studies of the basin (Cohen 1991).

Iraq was the first riparian to develop engineering projects in the basin. The Hindiya barrage on the Euphrates was constructed between 1911 and 1914 to prevent flooding and transfer water to canals for year-round irrigation (Kliot 1994). In the 1950s Iraq built a second Euphrates barrage at ar-Ramadi. Like the Hindiya Barrage, its main objective was flood control and irrigation (Naff and Matson 1984).

Both Syria and Turkey were slow to develop the use of the waters of the two rivers before the 1960s. During the second half of the 1950s the Russians conducted research on the Syrian reach of the Euphrates and proposed a dam at Tabqa. In 1966 a Syrian-Soviet agreement led to construction of the Tabqa High Dam. The dam, renamed al-Thawrah ('the Revolution'), was completed in 1973. A current Syrian water development called the Great Khabur Project is aimed at using the Khabur River for irrigation. Syria has also conducted technical studies for another irrigation project using Tigris water.

Turkey began constructing the Keban Dam on the Euphrates in 1966; production of electricity began in 1974. The Karakaya Dam on the Euphrates was commissioned in 1986, as part of the Southeastern (Grand) Anatolia Development Project (GAP) (Kolars and Mitchell 1991). GAP includes 22 dams and 19 hydroelectric power stations to be built on the Euphrates-Tigris over an area 'as big as Belgium' (Newspot 1992). The Ataturk Dam, the heart of the GAP, was completed in 1992. The GAP also embraces other economic and social improvements, such as transportation, industrial employment opportunities, and improved education and health services. From Turkey's perspective, it is an integrated regional development.

As these examples indicate, each of these riparians to date has tended to develop its water use plans unilaterally, without regard to the needs of the other riparians, the environment, or the actual capacity of the basin. Although there have been some international efforts from foreign investors to coordinate projects to develop the rivers for the interest of all, a common joint development project has never become reality (Chalabi and Majzoub 1995).

On the contrary, the completion of the Keban and Tabqa Dams caused serious tensions among Iraq, Syria and Turkey. Following the start of construction of the Keban Dam, Syria launched a campaign against Turkey. Subsequently Turkey reassured Syria that the dam would not cause any considerable harm (Chalabi and Majzoub 1995). Controversy over the Tabqa Dam was more serious, driving Syria and Iraq to the edge of armed conflict in 1975. Iraq threatened to bomb the dam (Naff and Matson 1984). Both countries moved troops towards their common border. Saudi Arabia, and possibly the Soviet Union, mediated; eventually the threat of war died down, after Syria released more water from the dam to Iraq. Although the terms of the agreement were never made public, Iraqi officials have privately stated that Syria agreed to take only 40 per cent of the river's water, leaving the remainder for Iraq (Naff and Matson 1984).

Tensions again rose in January 1990 when Turkey began filling the Ataturk Dam reservoir. Both Syria and Iraq accused Turkey of not informing them about the cutoff, thereby causing considerable harm. Iraq even threatened to bomb the Euphrates dams (Cohen 1991). Turkey countered that its co-riparians had 'been informed in a timely way that river flow would be interrupted for a period of one month, due to technical necessity' (Turkish Ministry of Foreign Affairs 1995). Turkey also claimed that before the impoundment, it released more water than its commitment under the protocol of 1987 (to a total flow of 500 cubic metres/second [cm/s]) so that the downstream countries could store this additional water. Some independent academics also support the Turkish view that it took all necessary measures to avoid significant harm to the downstream states during the impoundment (Vesilind 1993). Iraq and Syria continue to call the Arab League states to unite against Turkey on the GAP issue; they threatened to take legal action in order to obtain compensation, as well as to boycott British, French, German, and Belgian companies if they build a dam on the Euphrates (Al-Hayat 1996).

The Jordan

Much more information than understanding

What can be said about the Jordan that has not been said already? The Jordan basin has become a very well published topic during the 1990s. Over 30 books have been devoted to the Middle East water resources either totally devoted to the Jordan basin or with a substantial focus on the Jordan basin issues. There were important land-mark studies in the 1940s when ideas needed to be constructed by parties to the up-coming disputes. The assumptions of the Arab communities were set out by a British engineer (Ionides 1953). The foundations of the Jewish case were put in place by the American author Lowdermilk (1944). Retrospective perspectives on the events of the 1940s and the Johnston initiatives were provided by Stevens (1965) and Smith (1965). The surge of monographs and edited books followed the new regional politics which followed the demise of the former Soviet Union. Some of these were also knowledge constructing. For example the study of Hillel (1994) but the majority were attempts to provide overviews and analysis from what the authors and contributors regarded as a balanced perspective (Salameh et al. 1993, Isaac and Shuval 1994, Kliot 1994, Allan 1996a, Elmusa 1997, Soffer 1999). By the end of the 1990s the analyses began to dig more deeply into both the documentary evidence (Medzini 1998) and into the constructed information itself (Amery and Wolf 2000) and to question the fundamental perceptions underlying resource allocation and management in parts of the region (Trottier 1999). International relations and international law have been the focus of a number of important studies (Allan and Mallat 1995, Shapland 1997, Libiszewski 1991, Lonergan and Brooks 1994). Other studies took theoretical positions deriving from international relations theory (Lowi 1993, Sherman 1999, Wolf 1995). Throughout the 1990s attempts were made to provide useful background to the negotiations (Libiszewski 1991, Isaac and Shuval, 1994, Garrfinkle 1994, Elmusa 1996, Lonergan and Brooks 1994, Wolf 2000). A number of studies have been associated with the Peace Process. The multi-lateral track participants have generated some useful documentation on the status and trends in use of the water resources of the Jordan basin (EXACT 1998).

A limited resource

The Jordan is a prominent river but a very small one. Its northern tributaries, the Dan, the Hatzbani and the Banias together supply about 500 million cubic metres of water per year. By the time the river reaches Lake Tiberias/Kinneret it has been augmented by additional flow from the Golan bringing the total average flow to about 800 cubic metres annually. Israel turned Lake Tiberias/ Kinneret into storage by the late 1960s and has since then pumped over 350 million cubic metres of Jordan water upwards and westwards to the coastal plain of Israel in its water carrier. Over 100

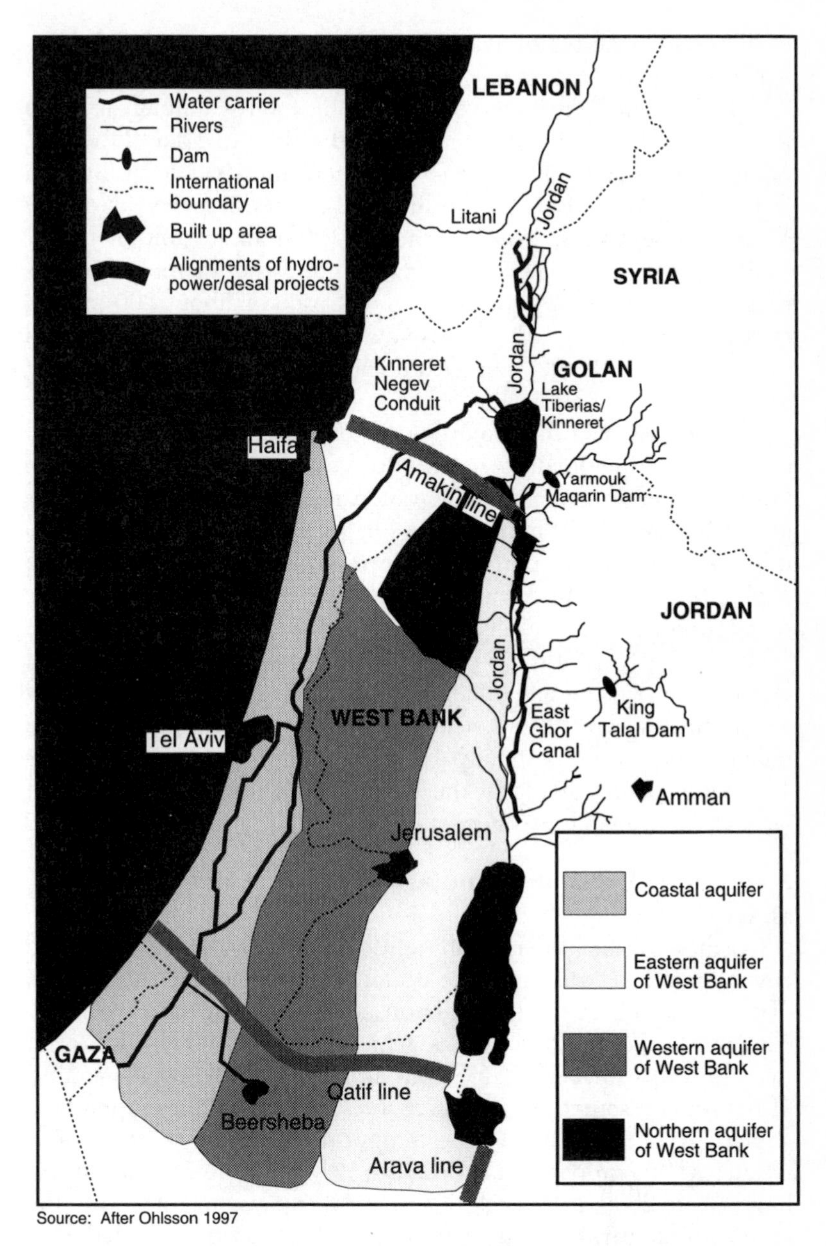

Source: After Ohlsson 1997

Figure 2.11 The Jordan basin and its tributaries, major aquifers, the riparians and water transfer systems

million cubic metres of water are lost to evaporation from Lake Tiberias/Kinneret storage each year.

The Yarmuk contributed over 400 cubic metres annually to basin flow in the 1950s. By the year 2000 the flow to Lake Tiberias/ Kinneret averaged only about 10 per cent of the 1950s level. By then Syria was using about 250 million cubic metres per year of the Yarmuk basin's water and Jordan over 100 million cubic metres.

Below the confluence of the Jordan and the Yarmuk the river used to flow south at an average annual rate of about 800 million cubic metres per year. The flow is now a trickle except in high rainfall years such as 1991–92. The Dead Sea has been reduced in volume and extent.

Groundwater is the major source of water for all the political entities in the Jordan basin—Jordan, Israel and Palestine. Groundwater is the main source of water for Israel and Palestine at 1.2 billion cubic metres per year. Jordan has only about 500 million cubic metres per year of groundwater in or close to the Jordan basin. It has additional groundwater at its southern border with Saudi Arabia.

In the early 1950s the peoples and economies of Jordan and Palestine were using one billion cubic metres of surface and ground water annually. By the end of the century they were using over three billion cubic metres annually. For food self-sufficiency for the over 12 million people living in the Jordan basin three times as much water as is available was needed.

A contended resource and attempts to achieve agreed shares

Advocates of the theories of political ecology contend that the environment, including its water resources, is managed in the interests of the powerful. In the Jordan basin power relations have been very explicit. Throughout Israel has achieved a hegemonic position in military terms since 1948. Without aiming explicitly to take control of the water resources Israel has, as a result of its military supremacy, gained permanent, or at least temporary sovereignty over water resources in the upper Jordan (Medzini 1998) or more generally in the basin (Wolf 2000) as a result of territorial expansion. Integral to the politics of natural resources is the construction of knowledge to reinforce the position of those with power.

The political ecology of water resources and management in the Jordan basin countries in the last half of the twentieth century can be considered by the decade. The 1940s was a period of massive social and political disruption. The 1950s was the decade when the most rapid development of groundwater of any experienced in the history of the area to the west of Jordan took place through Israel's increase in the utilisation of the coastal aquifers and West Bank. Over one billion cubic metres per year of additional water was mobilised by Israel for irrigation. The 1960s was the decade of serious contention over the waters of the upper Jordan (Medzini 1998). Water related armed conflict took place as both Syria and Israel were successful in frustrating their neighbour's intent to divert water. Israel was forced to adopt the very expensive option of building its water carrier from the lower level of Lake Tiberias/Kinneret rather than diverting water from a higher point in the upper Jordan. The 1960s also experienced the June 1967 War which left Israel with control of the whole of the upper Jordan and the West Bank aquifers. Water was neither the trigger for the war not the main goal of any of its adversaries. The outcome of the war did, however, determine the hydropolitics of the next two decades—the 1970s and the 1980s. The 1990s has been a period of dramatic hydropolitical change. Jordan and Israel have made a deal and Palestine and Israel can make one. Water need not be a significant impediment to peace between Syria and Israel. Nor between Lebanon and Israel once peace with Syria is in place. Such circumstances were impossible to imagine from the perspective of 1990.

There is a long tradition of constructing knowledge about the water resources of the Jordan basin countries and political ecology theory does explain the approaches taken by the numerous authors of the 30 or more books about water in the Jordan basin. Lowdermilk's study had the clear agenda of justifying a Jewish claim for the regional water resources. That of Ionides (1953) was inspired by concern for the sustainable use of the limited water resources for economic and social purposes. The analysis also showed concern for the interests of the Arab populations of the Jordan basin.

The armistice which marked the end of military events associated with the establishment of the state of Israel in 1947–1948 left Israel and Jordan with borders different from the UN Partition recommendations and different also from the Mandate borders of the British period of administration (Wolf 2000:75–76). The territorial outcome determined that water resources would be contentious. The long term sustainable management of water would pose great problems, especially if it was assumed that scarce water resources would determine options. In practice it has been the transformation of the economies from a dependence on agriculture to a dependence on industry and services that has determined water policy options.

The 1952–1955 period saw an attempt by the United States to achieve a rational division of water amongst the Jordan riparians. The US Government sent a special ambassador—the Johnston Mission—to attempt a basin-wide arrangement to optimise the allocation of Jordan water for Jordan, Israel and Syria (Lowi 1993). Thinking on water resource management in the North was imbued by two ideas. The first was that of the hydraulic mission inspired by science, engineering capability and the capacity of governments to fund big projects. The second was the need to avoid the consequences of environmental mismanagement which had been experienced in the United States in the 1930s. The Tennessee Valley Authority (TVA) was set up to address environmental, economic and social challenges in a poor region of the United States. The intent was to reverse resource depletion by planning and regulating resource use and by engineering water to minimise environmental damage. The Jordan basin appeared to be a very suitable candidate for the TVA approach with its very beneficial social spin-offs.

The Johnston mission was successful in the technical aspects of resource evaluation. It even came up with numbers that satisfied the water professionals of the three riparians but by 1955 it was clear that an agreement over water contradicted the polarised politics of Arabs and Israelis. The ministers of the Arab countries rejected the Johnston Plan. Despite the failure the numbers arrived at in the Plan have provided a version of how the allocations could be. The proportions established by Johnston have provided a sense of the reasonable during the very troubled Jordan basin hydropolitics of the succeeding half century.

There has been a tendency to assume that water resources would be determining of economic outcomes and would have a significant and predictable impact on the international relations of riparians. Armed conflict was presumed to be an unavoidable element in riparian relations. After Johnston there was some evidence that armed conflict could occur over water. Israel entered the 1960s constructing the idea that it needed to move what it regarded as its share of Jordan water from the Jordan Valley to the coastal plain. New water had to be modified.

Water resources do not determine socio-economic development: socio-economic development determines water management options

The Jordan basin has witnessed a remarkable experiment in the allocation and management of water. Israel, Palestine and Jordan all ran out of water in the 1950s or 1960s. They ran out in the sense that they could not meet their food needs. There was not enough water to irrigate the crops required to feed the modest populations of the 1950s and 1960s.

The performance of the Israeli political economy has shown that it is possible to develop rapidly, at the pace of the Asian tiger economies in the period from the 1960s, in an arid environment without the process being underwritten by revenues from oil resources. The Israeli experience should help to answer the mega-question 'which comes first in the process of achieving successful socio-economic development'. Is economic growth necessary to enable political transformation? Or is political reform a necessary preliminary to enable economic development.

The evidence from the Israeli water sector is that the economic development of the Israeli economy in the three decades between 1955 and 1985 enabled water policy reform to be legislated and achieved by the end of the 1980s. In 1953 Prime Minister Sharrett reinforced the national belief in the central role of agriculture and irrigation (1996:92). By 1985, however, the proportion of the GDP generated by agriculture had fallen to three per cent. 97 per cent of the GDP was coming from five per cent of the nation's water devoted to industry and services. In the three decades wives and children living on farms had obtained jobs in industry and services in the urban sector. There were very few full time jobs on the farms.

There was no general realisation that the allocation of water had been achieved according to principles of economics which advocate the combination of scarce resources in activities which bring a high economic return. That families were prosperous through having found jobs in a diversified economy meant that the Israeli Government had the political space in the mid-1980s to cut the water allocated to agriculture by 30 per cent and to promise to cut by a further 30 per cent. It reversed this policy in the 1993–1999 period. But by 2000 the need to reduce levels of water use in agriculture was back on the agenda.

A flavour of the ongoing very contentious discourse in Israel was captured in the quotations below. Some economists with the support of the media contend that there are even more radical steps to take in the adoption of economic principles.

'Water shortage is a dire threat
Dire forecasts concerning Israel's water resources have appeared in the past, but have apparently failed to awaken the public from its indifference. The forecasts published in today's Ha'aretz show that within a short period of time, Israel is liable to suffer the worse water shortages it has ever known.

This can no longer be written off merely as the assessments of isolated experts voicing undue alarm or making predictions based on insufficient data. Technologically advanced Israel, a society which views itself as one the most developed of the world's nations, may find itself sharing the plight of developing countries, with chronic shortages of water for home and agricultural use.'

Ha'artez, *14 June, 2000*

'The forecast collapse of the Israeli water economy is beginning to penetrate into public consciousness, in spite of the public's obsession with coalition arithmetic ... Israel's water economy is reaching its moment of truth this year, and policy makers can no longer evade responsibility by allowing the agricultural lobby to continue encouraging over-pumping of this scarce resource. This lobby organized only two weeks ago to defeat the government's proposal to raise the water price by 20 per cent. On this issue there was no difference between Meretz, the Labor party, Shas, and the Likud. Where an incentive for overuse of water is concerned, there is national unity among farmers.'

Avi Temkin, *Opinion,* Israel's Business Arena *14 June 2000*

Whether Israel can make the additional response depends on its social adaptive capacity and the hydropolitical opportunities. The 1955–1985 Israeli water sector experience showed that it was economics which determined whether sensible water policy reform was politically feasible. However, the analysis cannot be left there; economic and political forces are known to be interdependent. If water policy reform, a political process, was dependent on the successful diversification of the economy, what brought about the enabling economy wide diversification? The answer to that question is that the Israeli economy could not have developed without political institutions, laws and other regulatory measures put in place in its pre-independence and early independence years. All these social and institutional measures were transaction cost reducing (Coase 1990) and economic development promoting. So the impressive Israeli water policy reforms were dependent on both the political and institutional foundations of the 1950s as well as the general economic diversification thereafter. Israel created the social adaptive capacity (Ohlsson and Turton 1999) to cope with the water resource scarcity.

The water policies of Israel contrast with those of its neighbours and of Middle Eastern economies generally. Both Jordanian and Palestinian professionals are aware of the socio-economic and political processes which enabled the transformation of the Israeli water sector. They realise they do not yet have the political space in which to begin to implement them.

Groundwater: accounted and variably accessible

In an arid region groundwater is a very special element in the water resources of communities and political economies. Both surface waters and soil water are susceptible to natural losses which reduce their availability for human use. Annual potential evapo-transpiration rates, that is the loss of water to the atmosphere, are greater by between two and 20 times the annual rainfall across the region. Only when water reaches sub-surface aquifers is the storage secure from such unwanted losses. Water may be lost on its way to sub-surface storage through evaporation from ground, vegetation and water surfaces. It may also be lost in supporting vegetation that brings no return to local communities.

Groundwater varies in accessibility according to depth, the size of the aquifer and its porosity and transmissibility. The physical nature of the strata in which the groundwater reservoir resides determines the rate at which water can be withdrawn. Aquifer qualities in the MENA region are diverse. They range from the alluvium of the region's river basins such as the Nile and the Tigris-Euphrates, to the widespread limestone formations which are very porous—with many small voids, and pervious—with many chambers which allow the storage and movement of water. Less porous sandstone strata underlie the extensive deserts of the region. Impervious volcanic formations are also widespread. They hold water only in fractures associated with tectonically disturbed zones. These formations contain very little water. Saturated alluvial aquifers can hold 20 per cent of their volume as water and water can be transmitted through them quickly. By its nature, however, alluvium tends to be found in flat areas and there are no gradients to induce flow. Water can be quickly transmitted through alluvial aquifers by pumping. Rock aquifers hold much less water and movement through them is slow, except for the limestones which have features resembling underground rivers. In limestone the volumes stored vary widely across the aquifer according to the arrangement of the dissolved caves. Water content can vary from less than five per cent in parts of the formation to 100 per cent in sub-surface lakes.

Groundwater accessibility mainly depends on depth. The same aquifer may lie at depths of over 1000 metres in one place and flow at the surface at another. The western aquifers of the West Bank, for example, lie at over 500 metres depth to the north and west of Jerusalem. They also support the spring flow at the surface in the Tel Aviv region at heights close to sea level.

Groundwater quality as well as volume determines its accessibility for use. Water can be very high quality, with less than 200 parts per million (ppm) of dissolved solids (tds that is of salts) or it can be as saline as sea water with 35000 ppm of tds. Water which remains in the ground for a long time tends to pick up the chemical qualities of the surrounding rock. Fossil aquifers which have been in situ for between 12000 and 30000 years in the case of many of the central Saharan and Saudi Arabian aquifers are remarkably fresh despite their residence time. In coastal locations, groundwater can be sea water. Freshwater is less dense than sea water. This is reflected in a fresh water/salt water interface in all coastal zones. The interface slopes inland from close to the shore with the freshwater lying over a wedge of intruding sea water. The wedge intrudes further if the overlying volume of freshwater is reduced in volume by pumping.

The history of the development of groundwater is a recent one. There was relatively little exploitation of groundwater except where it flowed naturally at the surface as springs until the early decades of the twentieth century except for the 'horizontal wells' of the *qanat* systems of Iran. These springs could be substantial and could turn into rivers. The Khabur river, a tributary of the Euphrates, rises in Turkey as a spring and flows into Syria. The tributaries of the upper Jordan—the Dan, the Banias and the Hasbani—all originate as springs fed by the sub-surface storage on the flanks of Jebel Sheikh/ Mount Hermon. All these springs generate flows of over 100 million cubic metres per year. These levels of flow have declined in the past two decades.

Most springs are small and frequently the flows are unrecorded by government departments. Their capacity and seasonal flows are well understood, however, by local communities. Local customary regulation, completely independent of central government, is effective in arranging access and water shares for communities in for example Morocco (Wolf 2000), the West Bank (Trottier 2000) and Yemen (Handley 1999).

The volumes of groundwater of adequate quality available to the economies of the Middle East are shown in Table 2.6. Figures

Table 2.6 Groundwater availability in the MENA sub-regions—cubic kilometres (billion cubic metres)

	Total freshwater	Groundwater	
	km3	Km3	%
Asian MENA			
Arid Asian MENA			
Levant and Gulf	22	13	59.0%
Iraq	35	1	0.3%
Iran and Turkey	325	69	21.2%
Total Asian MENA	**382**	**84**	12.5%
African MENA			
Total African MENA	**85**	**22**	25.9%
Total MENA	**467**	**106**	22.7%

Source: FAO 1997a and 1997b

extracted from Table 2.3 show how the groundwater resources figure in the water budgets of the MENA sub-regions.

Table 2.6 provides an approximate notion of the relative significance of groundwater in the different parts of the MENA region. The countries of the arid Arabian Peninsula together with those of the Levant, except Lebanon, are poorly endowed with water resources. Of this poor endowment about 60 per cent is groundwater. Predictably these are the economies which have tended to over-use their renewable groundwater and draw down their non-renewable fossil water resources. Jordan, Israel, Gaza, Saudi Arabia, Yemen and the Gulf states have all endured water deficits and find it difficult to meet the needs of the water using sectors, especially the irrigation sector.

Table 2.3 shows that unlike MENA surface waters the majority of which crosses international boundaries, MENA groundwater is almost all used by the communities which live in the region in which the waters are recharged. The only major exceptions are in the West Bank, a very important exception, and in the border area between Jordan and Saudi Arabia. The authors of Table 2.3 estimate that the western and northern aquifers beneath the uplands of the West Bank provide underground flow to Israel of 700 million cubic metres per year. Palestine contests this allocation. The Qa Disi aquifer in a region of very flat terrain in southern Jordan slopes underground beneath neighbouring Saudi Arabia. Both Saudi Arabia and Jordan

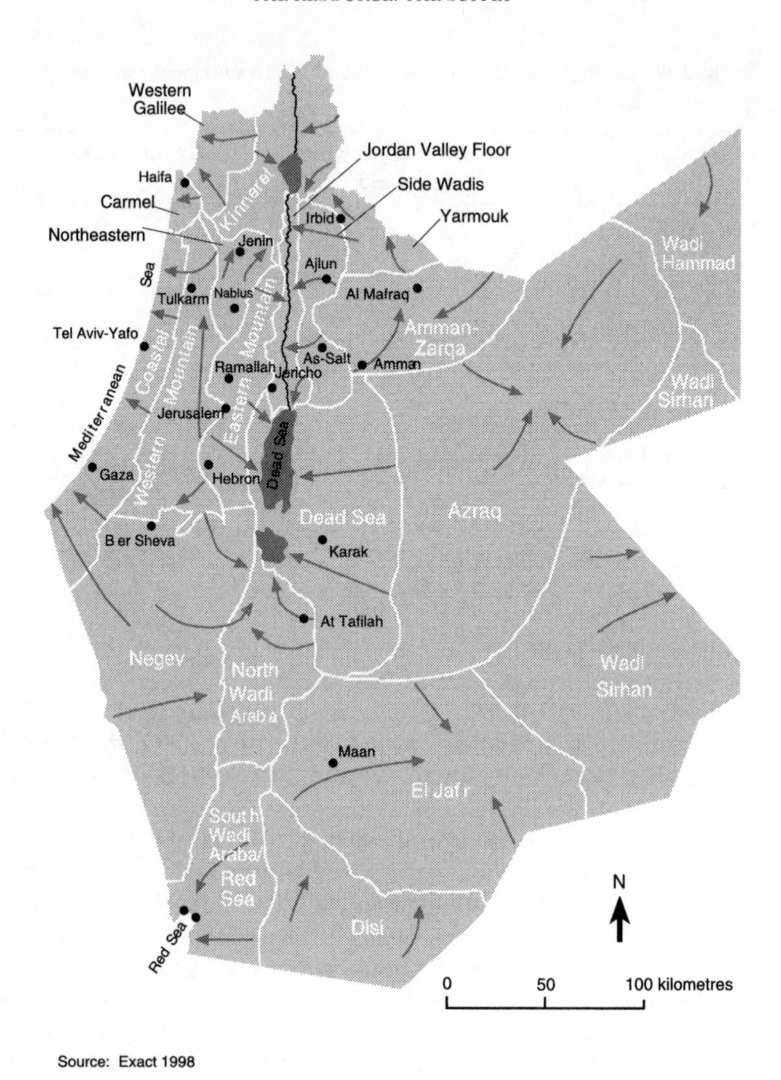

Figure 2.12 Groundwater flows in Israel, Jordan and Palestine

have set up agricultural schemes to utilise the water. Use of Qa Disi water, at rates very much above the rate of recharge, explains the increase in the trend in water use in Jordan since the late 1980s.

The temptation to use groundwater is not resisted anywhere. Coastal areas are particularly prone to the impact of over-pumping because they are usually flat and often have soils suitable for irrigation and cultivation. The cultivation of these coastal tracts in

Libya, Egypt, Gaza, Israel, Lebanon and Syria has led to the serious depletion of the coastal aquifers and sea water intrusion. The same is also true of the states on the shore of the Gulf. Leaders of all political persuasions have not been able to get their farming communities to slow down the rate of pumping. Only the complete destruction of the aquifer brings irrigation to an end as it has in parts of Libya, Gaza and in some of the Gulf states such as Oman. Nowhere are the coastal groundwater aquifers able to sustain the demands placed upon them.

Aquifers located beneath regions far from the coast exist in the deserts of the region. None of them are recharged significantly by rainfall. They were little utilised until the 1970s. Utilisation was only possible then because pumps could raise water from hundreds of metres below the surface and oil revenues were available to service the costs of such pumping. Desert aquifers fed centre pivot irrigation schemes in Saudi Arabia and Libya to raise wheat and fodder. Saudi Arabia's cereal production increased so rapidly in the second half of the 1980s that it became a major wheat exporter. Libya's experiments in the Sahara involved cereal schemes and then livestock rearing. None was successful. So the emphasis shifted from in situ desert agriculture in the 1970s to moving the water to the coast for municipal and industrial use via the Great Man-Made River (GMMR) in the 1980s and 1990s (Allan 1989:236). The promise of devoting GMMR water to agriculture has always been prominent. The volumes of water actually allocated to agriculture in the two decades after the inception of the GMMR project in 1980 have been negligible. The promise to use GMMR water in agriculture was consistent with the ideas constructed on how people and politicians would like things to be. The unwillingness to allocate water to agriculture in Libya despite the rhetoric anticipating such allocation shows a remarkable ability to play simultaneously the difficult games of the politics as well as the economics of water management.

In a few locations settlement is heavy in mountainous areas outside the well watered regions of northern Iran and of Turkey. The two economies that face the worst water resource challenges are Jordan and Yemen. Neither economy has enjoyed oil wealth and most of their populations live in high areas and depend substantially on groundwater. The major cities of Amman in Jordan and Sana'a and Ta'iz in Yemen are mainly dependent on groundwater. Amman will become progressively more dependent on water pumped up from the Jordan Valley. It will become dependent on water from the southern Qa Disi aquifer. Sana'a and Ta'iz in Yemen are so high at about 2000 metres above sea level that there is no foreseeable solution to their worsening water resource predicament (Handley 1999).

At the same time provided water resources are devoted to domestic and industrial uses and not to agriculture there is no community in the MENA region that is so short of water that cannot address their social and economic challenges. Only the isolated highland settlements in Jordan and Yemen dependent on groundwater face serious problems. All the other major settled regions are either coastal or along rivers.

Soil water, re-used water, manufactured water and virtual water: sometimes unaccounted but accessible water

Soil water

The MENA region is the least well endowed region in the world in terms of soil water. In temperate latitudes enjoyed by Eurasia and much of North America the bulk water needed by farmers comes from rainfall. All the farmers have to do to have sufficient water for a farming livelihood is be prepared to live where the rainfall occurs.

The soil water available each year to MENA farmers is incalculable. In the geographical core of the MENA region the volumes of soil water range from negligible in Egypt and countries of the Arabian Peninsula, to less than 20 per cent of the national water budget in countries like Israel and Jordan. In the northern part of the region Turkey, Syria and Iraq enjoy levels of rainfall which provide soil water resources for winter season dryland farming and grazing. This soil water supports significant proportions of the agricultural outputs of the respective economies.

Soil water is not taken into account anywhere in the national water budgets of the MENA economies. If soil water can be complemented with water from a more secure source such as groundwater storage then the value, albeit unaccounted, of the soil water to an economy is very greatly increased. In the steppe regions of southern Turkey, northern Syria, northern Iraq and northern Jordan supplementary irrigation can secure the harvest in low rainfall years and increase yields threefold or more. In these circumstances where the farming has become apparently economically viable those who started out supplementing soil water tend to grow dependent on groundwater (Lancaster et al. 1998). The farming practices shift so that the soil water in some locations becomes the minority water supporting crops.

It is the soil water in other parts of the global system that is strategically essential for the MENA economies. This water is accessed via trade in water intensive commodities such as grain raised with the soil water in the soil profiles of the mid-west of the United States, Canada and the economies of northern Europe.

Re-used water: in agriculture

The re-use of water has played a major role in the use and management of water for millennia in areas where topographic circumstances allowed water to drain from irrigated fields to those located at lower levels further down the system. Water applied for irrigation is used in the process of crop development. Water is essential for the development of plant tissue and comprises a high proportion of that tissue throughout the various stages of the growth of the plant. Water is also 'pumped' through the plant and is transpired into the atmosphere through the leaves of the plant. The water used for these biological purposes is called consumptive use. Water not captured by the roots of the plant and not held in the root zone by the forces of surface tension around the soil particles will drain through the soil into neighbouring soil profiles or into groundwater storage. The water used in productive crop development and the transpired water are lost to the local irrigation system. The water draining through the soil into neighbouring soil profiles and into groundwater storage is available for second or even further use in irrigation.

This re-use was not discussed in the era of modern irrigation, that is since the mid-nineteenth century, because irrigation engineers were professionally committed to design efficient systems that minimised the volumes of excess water draining from an irrigation system. To draw attention to the existence of drainage water was to draw attention to what was considered by the professional community to be an inefficient system. By the mid-1990s, Keller et al. (1995). pointed out that the environmentally favoured, that is the flat well drained soils of the Nile lowlands, made the Egyptian irrigation system an efficient agricultural water re-using system calculated at the national level. It was argued that inefficiency at the field level was not necessarily significant. Excess water draining from field level activities was useful water which enabled additional irrigation further down the system. The work of the Kellers and subsequent studies in Egypt suggest that the water allocated to irrigation in Egypt is used 2.3 times (Abu Zeid 1997). The average depth of application of irrigation water across the country as a whole has been calculated as 1.3 metres reflecting the requirements for two or more cropping seasons (Abu Zeid 1997). At the national level Egyptian water utilisation in agriculture, through re-use, is over 70 per cent. Such levels of efficiency are rarely achieved anywhere, even with advanced water application technologies at the field level.

Water re-use in irrigation is very significant in some river basins in the Middle East and North Africa. In the Nile, water re-use in Egypt's irrigated sector increases the water available at the field level by about 40 per cent for users of water for irrigation downstream of those who have made first or second use of water for that purpose. Re-used irrigation water is as invisible and until recently as unaccounted as soil water and virtual water.

Outsider professionals, with experience of hydrological circumstances less favourable than those enjoyed by Egyptian farmers, are misled by their assumptions about the importance of field level efficiency and by the invisibility of the re-used water which is not taken into account. Their concentration on field level activity is misleading. Their approach is valid, however, where water in excess of the consumptive use for crop development cannot be recovered for re-use. Re-use may not be possible because there is no shallow aquifer to receive and make available for re-use the water drained from the field. This is the case in many parts of the region where the aquifers are far beneath the surface, and the terrain is sloping, elevated and underlain by pervious, normally limestone, strata. The West Bank and northern Israel are examples of this situation. Israelis are particularly prone to assume wrongly, that Egyptian water use is inefficient.

The re-use of water drained from irrigated fields is also impossible when the underlying aquifers and soil profiles are saline. the Tigris-Euphrates lowlands of Syria and Iraq are not blessed with robust and fertile alluvium such as enjoyed by Egypt and the Sudan. The soils of the Syrian and Iraqi Tigris and Euphrates lowlands are difficult to irrigate and the excess water used beyond crop growth needs drains into generally low quality near surface aquifers. In Libya some tracts are underlain by saline aquifers caused by the chemistry of the strata in which they lie. The Surt coastal region of Libya is affected by serious problems of this sort. Irrigation development there cannot be based on local groundwater because of its salinity (Pallas 1976, Allan 1981a). The poor economics of applying expensive Great Man-Made River water to crops in this region are aggravated by the loss of high quality drainage water to a saline aquifer. In these circumstances field efficiency is important.

Coastal tracts throughout the region are affected by the same problems. Soils on flat coastal plains which characterise the region often overlie shallow aquifers which themselves overlie an intruding wedge of saline ocean or Mediterranean water. Use of the coastal freshwater for irrigation with modern pumps has everywhere led to interference with the sub-surface balance between fresh and salt water. The sea water intrudes further and takes up positions closer to the surface. Pumps begin to pump salt water. Such saline water intrusions have been shown to extend twenty kilometres in coastal north-west Libya and to greater distances beneath the Nile delta. Israel and Gaza are seriously affected by such intrusions. All the coastal agriculture of the Gulf states is also seriously affected. Irrigated agriculture on the coastal tracts of the arid MENA region does have to be carried out with strict observance of the principles of water use efficiency at the field level. Re-use is usually prevented by the contamination of the coastal aquifers.

Paradoxically there is a need for the freshwater wedge which keeps out the intruding saline sea water to be augmented by additional freshwater. The trade-off between the economic use of water avoiding losses to the ground and the simultaneous need to maintain a secure freshwater resource by permitting the infiltration of freshwater to keep back the freshwater/saltwater interface places water managers in an impossible position. Farmers will only cease to pump when their wells become saline. With the end of pumping the rate of water withdrawal falls and perforce the environmental considerations and the long term sustainability of the water resource will be addressed. The social impact of the end of irrigated farming is serious unless the farming is a hobby for a family with its main income off farm. This last is not an uncommon situation in the Gulf and in the countries which have Mediterranean shores.

The current water re-use practice is consistent with the perceptions of farmers and water policy makers. The farmers are not conscious that they are using re-used water except where they are pumping from a drain which they know is sourced from nearby irrigated fields. For water professionals the re-use of drainage water is becoming a more formal and accounted element of water policy in those regions, such as Egypt, where it is considered a viable practice. Such water is not, however, going to increase in volume from a very approximately estimated 20 billion cubic metres used each year across the region. About 70 per cent of this re-use of irrigation water is achieved in Egypt. Re-use irrigation water cannot be regarded as a future source of additional new water for the MENA region.

In contrast re-used urban waste water is very evident water. The plants which treat the water to primary, secondary and tertiary levels of quality are expensive and require resources and political will to put them in place. If the economic returns from subsequent use and/or the calculated environmental benefits of such treatment are not politically acceptable then such treatment and use will not take place. The political context is always changing. The perceived value of water in the environment is constantly increasing. The social value of agricultural activity is also changing. Tools of social analysis are not yet available to predict how perceptions and approaches will change. Neither are the 'tools' with which to address the challenges of improving social adaptive capacity. This social adaptive capacity has been identified by Ohlsson and Turton (1999) as the main bottleneck to matching water availability to socially, economically and environmentally sound water management.

Re-using municipal and industrial water

Water use in the municipal sector, including the domestic use by urban populations, and the industrial sectors rarely exceeds ten per cent of a national water budget. Its importance as an input to urban society and to the livelihoods of people living in urban areas is very high indeed and much greater than the proportional use suggests. At the end of the twentieth century urban populations were rising at a much higher rate than the regional average for the whole MENA population of about three per cent per year. The population of some cities was increasing at over eight per cent per year (Handley 1999).

The water delivered to and used in cities will double by 2030. Most of the water needed will come from water re-allocated from agriculture but for many coastal cities and urban settlements the water will come from desalination plants. By the year 2000 about 20 billion cubic metres per year were being used for municipal including domestic use in the MENA region. Desalinated water is expensive, although its costs are falling. The MENA region is in global terms relatively prosperous and its urban population can afford the one US dollar per cubic metre which it cost by the year 2000 to desalinate water. Desalination is discussed in more detail later in the next section.

Domestic and industrial uses are different from agricultural uses. The overall losses to transpiration are very small indeed in urban areas compared with those in the vast irrigated tracts of the region. The maintenance of green amenity in cities leads to inevitable losses to transpiration but these losses are comparatively modest in the national water budgets of the economies of the region.

Most of the high quality water delivered to homes and industry is not used for consumption. It is used to clean and cool. Very little of this high quality water is lost during use and most of the water is, or could be, returned to the system, albeit at a lower quality. Over 70 per cent of the water delivered to municipal users is returned to the system and if treated could be used again. Because of the cost of treatment the extent of subsequent use is limited. About ten per cent of the national water budget of Israel was derived from treated urban waste water by the year 2000 comprising about 15 per cent of the water used in irrigation. If all of the urban waste water of Israel could be treated the volume of new water available for agriculture would be about 500 million cubic metres per year. If the expected levels of municipal and industrial use in Israel and Palestine come to pass through the increase of population then the potential volumes of treated urban waste water could reach 700 million cubic metres per year. As the population of the region to the west of Jordan will rise beyond ten million then the volumes of treated urban waste water available to agriculture could approach the current levels of water use in irrigated farming—1400 million cubic metres per year. The economic viability of achieving this level of treatment has yet to be demonstrated. The availability of treated urban waste water does mean, however, that there will always be irrigated farming in the MENA region in the environs of urban areas. Israel is already securing crop production in the dry tracts to the south of Tel Aviv. In the past it was thought that water brought from Lake Kinneret/Tiberias would service these southern farms. By the year 2000 it was clear that these tracts would be watered by reliable treated water from the Tel Aviv metropolitan region.

Manufactured water: the increasingly important option desalination

As for most water management approaches it is the political economy that determines whether desalination is an option to be implemented. The technology and the economics of desalination changed significantly over the second half of the twentieth century. At the same time the MENA region had a particular need for the technology because of its poor water resource position and rising water demand. The region was also well endowed in some economies with the necessary method of energy resource input needed for the thermal desalination.

Most of the installed desalination capacity in the world is located in the Gulf countries (Dabbagh et al. 1994). By the 1990s the cost of desalinating water had fallen to a level that almost all domestic and industrial water users could afford the price of between one and two dollars per cubic metre. This price reflected the low cost of thermal energy which characterised the 1990s.

The costs of desalination are expected to fall further despite the increase in the cost of energy, already experienced by the last year of the 1990s and continuing into the first decade of the twenty-first century. Figure 2.13 provides a perspective on past trends and the expected future trajectory of desalination costs. The figure shows that it is by driving down operating costs, especially management costs, that the cost savings will be achieved (Ministry of Energy and Infrastructure 1995).

It is very difficult to be definitive about the costs of delivering desalinated water. The costs of desalination vary according to whether the water being desalinated is brackish, with just a few thousand parts per million of dissolved salt, or sea water, with over 35,000 parts per million. It also varies according to the availability of cheap thermal energy. In economies where the cost of extracting oil and gas is low, as it is in the Gulf, and where energy would otherwise be flared and wasted, then the desalinated water would in these circumstances be produced very advantageously. Similar cost advantages are achieved where desalination is associated with local power generation or other industrial activities which produce energy that would otherwise be wasted.

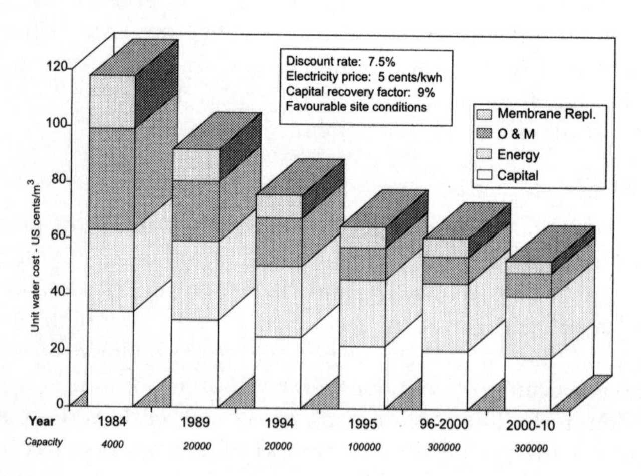

Figure 2.13 Evolution of desalination costs

Because the costs are so difficult to e
of the desalination activity is concentrated
in a privileged world of oil enrichment i
reliable opportunity costs or definitive predic
will be in the future. The views of desalinati
converged, however. They estimate that desal
manufactured for about one dollar per cubic
thermal and the membrane systems. They also pre
will fall further and levels of about 0.5 dollars per
anticipated.

Thermal desalination is a process which depends o
of salt by raising the temperature to separate the salt fro
The membrane system 'filters' the salt from the w
membrane system is cheaper than the thermal system bu
very careful supervision of the membranes which can eas
not managed properly.

The experience of Malta in the central Mediterranean is re
to the MENA region. Its population of 300,000 and its millio
so tourists a year depend on desalinated water for about 50 per c
of their domestic and industrial water. Malta is not a rich econom
but it has managed to install and successfully operate its desalination capacity. By the second half of the 1990s it was realised by its professionals and consultants that the installed capacity was greater than its real needs. Water was being lost in the distribution system. By reducing leaks, introducing tariffs which drew attention to the value of water together with a public awareness programme higher water use efficiencies were achieved. The extent of the water savings have been such that one of the desalination plants was put on the market in 2000 and another was not working full-time.

Desalination is a process which is eminently appropriate for the supply of domestic, industrial and service sector water. The volumes of water needed in these sectors rarely requires more than ten per cent of the available national water. At the level of the individual the cost per head of providing water for an individual's domestic and jobs needs is at a maximum no more than 100 US dollars per year, assuming water delivered at about one US dollar per cubic metre. For those living in a strong economy enjoying full employment, or in an oil economy where public services such as water are a government responsibility such costs are negligible. For those living in weak economies on low incomes such desalination costs are a problem. The contrast between Israel and Gaza exemplifies the contrast. Israel has long had the option of desalinating water at costs

n could be easily accommodated. Its politicians and essionals have hesitated to build desalination plants because the e would have an impact on the peace talks which preoccupied 1990s. If Israel demonstrated its capacity to meet domestic and dustrial water needs with desalinated water at costs which ompared with that of pumping of water from Lake Kinneret/ iberias it would call into question the economics of diverting water out of the Jordan basin. Water from the lake, unburdened by the costs of pumping to the coastal plain of Israel, could be delivered in the Jordan valley at costs affordable by irrigators.

The other reason that Israel has been reluctant to demonstrate its capacity to desalinate is that the future is recognised as a period when desalination will be a viable means of meeting rising water demands for industry and domestic use. Israel wants assistance to take on the investment burden of the desalination programme. In the peace talks of the late 1990s Israel confirmed its position by insisting that only future new water could be discussed in the negotiations. If the Palestinians were to accept this argument Palestine would become a very unequal partner in the desalination programme. Meanwhile the international community discerning the intent of Israel were very unwilling to being drawn into investments which would underwrite the Israeli desalination project.

In late 1999 the frustration of part of the Israeli water policy community was revealed by the Minister of Infrastructure, Ariel Sharon. He indicated that one possible scenario was to desalinate 800 cubic metres of water per year over the first two decades of the twenty-first century (Sharon 1999). Sharon proposed that a two phase project would 'protect Israelis, Palestinians and Jordanians from water shortages in the future'. Sharon said that the water crisis would peak by 2010, when Israelis and Palestinians together would total 10 million. The initial phase of his proposal called for a 50 million cubic metre per year desalination plant in Gaza for drinking water and domestic consumption; desalination of 50 million cubic metres per year of brackish water to supply Jordan in the Jordan Rift Valley; and desalination of 50–100 million cubic metres per year along the Mediterranean for use by Israel. In a second phase, a large-scale desalination plant with a capacity of 800 million cubic metres of water would be constructed for use by all three partners (Sharon 1999). The total capacity of one billion metres per year is a number to conjure with. It is half the total use of Israel and Palestine together in the year 2000, and one third of the total use of Israel, Palestine and Jordan.

The proposal should be seen for what it was. Sharon was constructing the idea that Israel had the technological capacity to manufacture water for the region at volumes equal to the rising demand for water. At the time Germany held the Presidency of the EU and by proposing the project he hoped to advance the idea that it was a suitable candidate for EU investment to advance the water security of all the economies of the Jordan basin and the prosperity of the weak economies.

By April 2000 the Israeli ministerial committee for the economy decided to build a facility capable of desalinating 50 million cubic metres of salt water annually, with an option of doubling the figure. The decision was reached after a discussion on how to deal with the ongoing water crisis. The cost of the facility was estimated at $150 million (Cohen 2000). These published comments reflect the discourse going on in all government circles in Jordan and in Palestine as well as in Israel. The desalination programme will move more quickly than suggested in the statements of Cohen above. But the projects will evolve more slowly than those who advocate desalination as the main solution to the need for new water.

The reason for the uncertainty about the pace of the expansion of desalination capacity is that there are other possible solutions. For example the committee meeting in April 2000 also decided 'to open negotiations with the Turkish government to receive quotations for a five year contract to purchase an annual quantity of some 45 million cubic metres of freshwater. The committee also decided to accelerate projects for using partially treated sewage for reuse in agriculture, for building the sea water desalination plant, for purifying polluted wells, and steps to improve home water conservation' (Cohen 2000). This outcome was a disappointment to farmers and officials from the agricultural sector. They had wanted a much higher figure for desalination but the Ministry of Finance argued that about 300 million cubic metres of water could be recovered from treated urban waste water, brackish water partially treated sea water and the move to major desalination should be delayed until after 2010 (Cohen 2000).

Water desalination in Israel-Jordan: challenges, symbolic projects and real solutions

A study in the mid-1990s of hydro-power and desalination options provided valuable insights into the economics of coastal desalination compared with other water and power manufacturing projects for the Jordan basin countries. The study is reviewed here in some detail not because it was a definitive study but because the comparative approach of its analysis provides a relatively comprehensive set of insights into a very complex subject. It is also a subject of very great strategic importance to the Jordan basin economies.

Only the coastal option is likely to go ahead but the review of some of the other Jordan Valley scenarios provided important insights into the economics of sea water desalination. The study reassessed the old notion of major schemes for moving water from the Mediterranean Sea and the Red Sea to the Jordan Rift Valley to generate power and produce fresh water (Ministry of Energy and Infrastructure, 1995). The study was completed at a time when the intensity of the negotiations associated with the Peace Talks initiated in 1992 provided special international relations, political and economic contexts. Some, albeit mainly engineers with power generation interests, believed circumstances would lead to the construction of one of the schemes (Ministry of Energy and Infrastructure 1995). Others were and remain extremely sceptical (Arlosoroff 1995). The reason for giving space here to the estimates that resulted from the study of the proposed Dead Sea water transfer schemes is not to promote them. The studies are discussed because they highlight the relative costs and impacts of some of the options being considered.

The proposed Dead Sea Hydro Projects would generate power by moving water from the Mediterranean or the Red Sea to the deep Jordan Rift Valley. 800 million cubic metres of fresh water could be produced annually if the project, were to be operational after the twenty years planning and construction phase. Such a volume is very large compared with current annual water use of two billion cubic metres by the economies in the southern Jordan basin. For those who believe, as many have in the past, that the provision of new water is the logical way to address a water deficit, then the proposals appear to be timely.

However, there is no longer a consensus in the region that providing new water is the solution for such economies. Israel has already demonstrated conclusively that the implementation of an alternative policy based on the assumption that gaining sound economic returns to scarce water is a more sustainable approach. This strategy has enabled Israel to show it to be possible to cut its water consumption in the high water using sector, agriculture, by 30 per cent since 1986, and further reductions in 1991 (Voice of Israel 1991) a position which was well on the way to the targeted 60 per cent reduction which was being talked about in Israel in the 1991–1992 period (Ben Meier 1992).

The trajectory of water use in Jordan was very different from that in Israel in the same period. Whereas water use in Israel fell by 30 per cent between 1986 and 1992 and then rose again to its mid-1980s level Jordan's water use increased steadily by almost 30 per cent. For Jordan which has not been able to achieve a secure supply of water to its urban centres and not yet having adopted principles of demand management in agriculture the availability of new water appeared attractive. The delivered cost of desalinated water of about US$ 0.5 per cubic metre was not consistent, however, with agricultural use even if it were to be applied in situ deep in the southern Jordan valley.

It has to be emphasised at the outset that the freshwater produced by any of the Dead Sea Hydro Projects would be available at an estimated US$ 0.5–0.6 per cubic metre for potable water, and ten cents less for lower quality (500 mg/litre) water suitable for industry and agriculture. The freshwater would be produced at elevations deep in the Jordan Valley from where it would have to be pumped to users at higher levels. Israel has some three decades of experience of moving water from low to high levels where it can be put to economic use and is aware of the energy costs attracted by such operations. Its Water Carrier, which since the mid-1960s has lifted about 350 million cubic metres per year from the Lake Kinneret/Tiberias region of the Jordan Valley has claimed a significant proportion of the Israeli economy's electrical power in the process. The costs of such water are well understood in Israel. By the time such water reaches the parts of Israel which neighbour Gaza the costs of delivery are about US$ 0.6 per cubic metre. This cost was almost three times the level which the Israeli Government had been able to negotiate with Israeli farmers by the mid-1990s for the price for water delivered to farms for irrigation. US$ 0.6 per cubic metre is affordable by industrial and domestic users.

Another important economic trend to which those keen on augmenting the water supplies of Israel, Jordan and Palestine with new desalinated water have drawn attention is the falling cost of desalinated water. In the early 1980s the unit cost of water was close to US$ 1.2 per cubic metre. By 1994 it had fallen to between US$ 0.6 and 0.7. It is calculated that by sometime in the first decade of the twenty-first century the costs per cubic metre will fall to close to US$ 0.5. Such data are difficult to compare for different plants in different production contexts. It is necessary to know what has been included in the costs and especially if the energy being used has an opportunity cost (See Figure 2.13).

The discount rate on the capital invested is one of a number of crucial variables which have a dramatic impact on the calculated costs of delivered desalinated water (Dabbagh et al. 1994:226–227) Dabbagh et al. (1994:235–236). Some very important comparative data based on 1991 prices suggest that the cost of providing desalination production capacity could be US$ 1687 per cubic metre per day, or US$ 615,755 per cubic metre per year. This would suggest a cost of US$ 3.7 billion for a capacity of 800 million cubic metres per year at 1991 prices. This figure was much higher than the US$ 1.86 billion estimated by the Israeli Ministry of Energy and Infrastructure (1995). This is because of the cost of an 800 million cubic metres per year capacity coastal desalination plant (based on an estimated investment cost of only US$720 per cubic metre per day), although this figure would not include the cost of providing the necessary power generation capacity.

Dabbagh et al. (1994:236) provided the useful global comparative figure that operating costs can be estimated to be about ten per cent of the plant construction costs. The low estimated costs of the Israeli estimates reflect the generally held optimism of desalination engineers and the industry in general. They judge that the efficiency of technology and management of the coming decade will witness significant reductions in the levels of operating costs. It is also anticipated that the industry generally will continue to grow. The International Atomic Energy Agency (1992) anticipated that the demand for desalinated water will double each decade over the next 25 years implying an annual growth rate of 7.18 per cent (International Atomic Energy Agency 1992).

The interests of the participating economies

> 'The ... project will have far-reaching benefits for all the people in the region since it will facilitate some crucial and practical responses to the most pressing problems we face, namely the production of desalinated water supplies—sufficient to satisfy the needs of the Jordanians, the Palestinians and ourselves: the revitalisation of the Dead Sea; and all the incumbent ancillary projects which will provide the necessary foundation to create a vibrant society in the current deserts of the Rift Valley area including mariculture, tourism, etc' (Moshe Shahal, Minister of Energy and Infrastructure 1995:1).

It will be shown in Chapter 6 (page 243) that Israel's water allocation policy has shifted dramatically in response to both environmental and international relations events. There are elements of its bureaucracy and technocracy which can see advantages in the proposed Dead Sea projects. The quotation above reflects the views of Israeli officials concerned with power generation. There are equivalent interests in Jordan, especially with respect to the Red-Dead proposal which would bring benefits to Jordan particularly.

The conflicting inspirations, which drive the water allocation and management policies of Israel and Jordan, will make the discourse on whether to pursue the Dead Sea projects difficult. The discourse will be further complicated by the contradictory positions adopted by economic sectoral interests within the respective economies. The interests of irrigators using 90 per cent of water but only producing between three and seven per cent of Israeli and the Jordanian GDPs respectively conflict with national interests. The national economic interest relates to the water use efficiency of the industrial and service sector users. These non-agricultural sectors are responsible for the other 97 and 93 per cent respectively of the Israeli and Jordanian GDPs.

The proposed alignments

The three alignments have comparative advantages and disadvantages. In this section the advantages of the different schemes will be described and assessed. In the following section the economic and strategic advantages of the various Dead Sea Hydro Project(s) will be examined in terms of alternative investments on water desalination plants located at sea level on the Mediterranean shore.

The three Dead Sea Hydro Project schemes are (See Figure 2.12 for the alignments):

- The Qatif alignment—the Mediterranean-Dead Sea route via Qatif
- The Arava alignment—the Red Sea—Dead Sea route starting at the Gulf of Aqaba
- The Amakim alignment—the Mediterranean-Dead Sea route starting near Haifa or Hadera

Designs were evaluated on the basis of a final capacity, after construction phases spread over 20 years, of 800 million cubic metres per year of potable water. This level of potable water production would require the transmission of about two billion cubic metres of sea water annually. The design capacity has been set at 2.7 billion cubic metres per year in order to address the problem of the level of the Dead Sea. The level of the Dead Sea has been falling for decades as a consequence of the utilisation of Jordan and Yarmuk waters by Israel, Syria and Jordan.

The environmental impacts of the proposed schemes require more detailed study. The impacts on the Mediterranean and the Red Seas were they to be used for the 2.7 billion cubic metres annually would be negligible in both volume and quality terms. The impacts on the Jordan Valley, it is argued by those proposing the schemes, would be positive in that there would be the opportunity to restore many of the environmental characteristics impacted by the water management schemes in the upper Yarmuk and Jordan.

The costs of implementation

The economic comparison of the three alignments has been carried out according to the criterion of net present value which produces an index derived from the following costs and benefits:

Costs

- investments in the water conveyance and hydro-power systems.
- the operating and maintenance costs of the conveyance and hydro-power systems.

Benefits

- the value of the electricity produced.
- the cost savings achieved by adopting the Dead Sea projects compared with the alternative of Mediterranean coast sited scheme. The financial benefits of the Dead Sea projects would come both from the cost savings in the production of desalinated water and also in savings in conveyance costs to some users in Israel, Jordan and Palestine would also benefit from the water desalinated in the Jordan Rift Valley.

The assumptions on which the costs and benefits are based are critical to the validity of such analyses. The three most cost-effecting factors in an economic analysis of this type are first, projected energy prices, secondly, the geographical location of anticipated water demand and the levels of such demands, and thirdly, and most important, the discount rate assumed on the capital investment.

On energy prices it has been assumed that current costs of electricity 4.9 cents per kilowatt hour should reflect the current international cost of energy. Such an assumption appears to be pessimistic in that it is unlikely that energy prices will fall in real terms over the six or more decades of any of the proposed projects and electricity produced by hydro-power plants will be at a comparative advantage.

The pattern of water allocation and management following the development of the production of new desalinating plants is difficult to predict especially in the light of the non-rigorous economic approach of those managing water by some of the proposing partners and beneficiaries.

It is evident that the Dead Sea projects could only go ahead on the basis of soft-financing. The calculations of the Israeli Ministry of Energy have been completed assuming discount rates of three and five per cent, a very low valuation of the cost of the capital inputs. It is assumed that such funds would be available from multi-lateral and bi-lateral agencies.

Comparison of the benefits—energy and water

The comparative investment costs of the three Dead Sea Hydro Project alignments are shown in Figure 2.11. The estimated investment costs of the coast sited plant show it to have initial advantages over the Dead Sea projects although the investment costs for the desalination plants themselves would be cheaper in the Jordan Valley. It is the costs of sea water conveyance which makes the Dead Sea projects more expensive overall. However, all the economic indicators with respect to operating inputs—for example energy inputs—and for the costs per cubic metre of water produced show the Dead Sea projects to have advantages.

Likelihood of implementation

It is on the issue of implementation that the analysis becomes difficult because the returns to the expensive water depend on the capacity to deploy the desalinised water in economic or social enterprises and activities which bring sound returns. Agriculture must be excluded as a destination for the high cost water. The likelihood of large urban centres developing in the uncomfortably hot environs of the Dead Sea is small. The establishment of industries requiring relatively high volumes of high cost water is also unlikely. The options in the Jordan Valley are limited. The need for industrial water should also be seen against the levels of water use in industry in Israel where only about five per cent of total national water use is industrial reflecting an annual use of much less than 100

DOUGLAS COLLEGE LIBRARY

Table 2.7 Some comparative economic indicators of the Dead Sea Hydro Projects

Alignment	**Total investment** (Excludes desalinated water transmission costs) bn US $	**Cost potable desalinated water*** @ 3% discount rate (Excludes delivery) US$/m3	**Cost non-potable desalinated water**** @ 3% discount rate (Excludes delivery) US$/m3
Qatif	3.5	0.38	c 0.28
Amakim	3.7	0.42	c 0.32
Arava	4.2	0.45	c 0.35
Alternative unit at coast site	2.8	0.53	c 0.43

* High quality water at 250 mg/litre - potable.
** Water for industry and agriculture at 500 mg/litre.
Cost of electricity - 4.9 US cents/kilowatt hour.
Desalinated water transmission costs of between 4 and 8 US cents per cubic metre.
Source: Ministry of Energy and Infrastructure 1995

million cubic metres per year. Israel's share of any Dead Sea desalination plant water would be far in excess of its foreseeable industrial water needs.

To pump a proportion of the proposed 800 million cubic metres per year of desalinated production out of the Jordan Valley to Amman at a cost of between 0.5 and 1.0 US dollars per cubic metre, which Water Carrier experience suggests as a probable cost, makes such water an unlikely source for the Amman metropolis. Israeli cities would lie close to the proposed alternative desalination plant sited at the coast. Few costs would be incurred in introducing desalinated water from a coastal site to the Israeli metropolitan water systems in the Tel Aviv region.

The preliminary studies of the Dead Sea Hydro Projects, such as the 1995 evaluation by the Israeli Ministry of Energy, concede that much more detailed studies are required if a full range of economic and environmental outcomes are to be thoroughly understood. It is clear that the projects need to be framed to meet a range of economic and environmental goals. These goals would have associated criteria with respect to the volumes and qualities of water needed for feasible and environmentally sound economic activity in the productive and service sectors. Instead of the high volumes of desalinated water being proposed for production to meet unforeseeable economic purposes by the current Dead Sea schemes, modest volumes of water would be programmed to meet the water needs of economically and environmentally viable projects.

The Med-Dead and the Red-Dead Hydro Projects have been a focus for the lobbying interest of elements of the Israeli and Jordanian bureaucracies and technocracies. The projects will also be of great interest to many international companies with the capacity to design and build the exciting and expensive structures required to transfer water, build desalination plants and high cost water distribution systems of unprecedented scales. The costs of the proposed projects make it difficult to find economic advantages once the power generation benefits have been accounted for. A number of factors make the Dead Sea schemes unattractive. The terrain of the region in which the water transfer pipelines would have to be constructed would attract costs. The elevations of sea water sources, the heights at which the majority of users would live and work and a number of environmental uncertainties make them less attractive than the coastal option.

The Dead Sea schemes will attract continuing attention because they represent potentially powerful symbols associated with the production of water and energy which are central to the economic and social development of the poor partners in the proposed enterprises, Jordan and Palestine. They are equally potentially valuable symbols for the international donor community with goals of reinforcing the Middle East Peace Process and establishing infrastructures which cement interdependent economic relations. In political economies in which such symbols play a role it is impossible to determine to what extent the intangible symbolic benefits perceived by the international community could be important in mobilising capital on soft terms. This is the case despite the at best rather ambiguous predicted economic outcomes pressed by strong supporters of the Dead Sea projects. The projects will not be given further consideration if the very persuasive opposing analyses are widely canvassed (Arlosoroff 1995).

Research on desalination

The commitment of the MENA governments, their officials and scientists to desalination is reflected in the decision to set up in 1997 the Desalination Research Center in Muscat in Oman. (MEDRC

2000) Located in the Gulf the Centre was close to the majority of operational desalination plants in the world. Its success as an engine of respected research depends on funding and leadership.

The international community and the Peace Process

Since it would be international finance, both grants and loans, which could alone mobilise the necessary capital to construct any of the projects, the approach of international bodies such as the World Bank to developing water resources are important. The Bank has evolved a clear position on the provision and use of water since 1990 and especially on water allocation and management in semi-arid regions such as the Middle East. It strongly advocates an economic approach to water allocation and use based on principles of allocative and productive efficiency (Serageldin 1995:v). Only Israel has firmly adopted such principles although both Jordan (Haddadin, 1996:64–66; Salameh 1992) and officials on the West Bank (Haddad 1996:11; Nasser 1996:50) have acknowledged their unavoidable role of these principles in future approaches to water allocation and management.

The position of major patrons such as the United States has been important but not determining. US policy has, since the days of the review of water in the Jordan by Johnston in the 1950s (Lowi 1994), been inspired by a belief that USAID could play a significant role in the economic development of the region. The US tried to influence the international relations of the Jordan riparians in particular by encouraging cooperation over water in the belief that such cooperation would lead to cooperation in other areas.

The four decades since the hostilities of 1956 have in practice been characterised by the capacity of Israel to assert a hegemonic position over water, with the tacit, although sometimes unwilling, support of the United States. There has been no formal cooperation over water amongst the Jordan riparians and only a limited amount of informal cooperation between Jordan and Israel during the years when they had not officially ended their war. Water has not proved, however, to be a difficult matter over which to reach agreements in the Peace Talks initiated in 1992. These agreements included articles on water between Jordan and Israel (September 1994) and Jordan and the Palestinian Authority (September 1995). The relative ease with which water was addressed surprised many but not those who were aware of the water use options available to the respective economies and the economic value of water in relation to the total economies. These analysts had emphasised the changing perspectives on the value of water being adopted on the allocation and use of water in arid and semi-arid regions (Allan 1993; Fisher 1994, 1995a and 1995b).

It is unlikely, however, that the US development assistance agency and other development assistance agencies might be persuaded by the symbolic significance of the Dead Sea Hydro Project to be a way to meet some of the economic goals associated with power and water provision. It is possible, but again unlikely, that the economic activity associated with, and multiplied by, embarking on one of the projects would be considered to be generally beneficial. That US companies would benefit from the consulting and contracting might also dispose decision makers to support the project. On the other hand there is evidence that USAID can when it chooses be the most determined advocate of sound economic principles. It has argued for two decades in Egypt against the reclamation of new land and has opposed the use of scarce Nile water on sandy soils where the productivity of water and water losses associated with agricultural use are economically unsustainable.

The significance of symbolic engineering projects should not be underestimated, however (Allan 1983). Such projects are politically feasible. Not only are they politically feasible but they do ameliorate water resourcing problems. Economic reform and the re-allocation of water out of agriculture would also ameliorate the water supply problem but they are not politically feasible. Spectacular hydraulic projects can gain wide support where there is high awareness of the social value of water and no awareness of its economic value or any will to manage it as a commodity.

The interests of powerful political constituencies, imbued with the 'hydraulic mission', ally naturally with communities with traditional views on water and water rights. At the same time the precision of estimates on the economic outcomes from investment are always uncertain. In the case of water management involving pumping the estimates are extremely unpredictable because energy prices are so potentially volatile. In these uncertain circumstances it is impossible to provide definitive economic estimates. The 1990s experienced both very low oil prices and for the last two years of the decade high prices. The global oil market is volatile. Advocates of the Dead Sea Hydro Projects, which are substantially independent of world energy prices, argue that the schemes provide the Jordan Valley riparians with an infrastructure which would to some extent free them from their hostage position vis-à-vis energy.

'Virtual water'

Quantifying the volume of invisible virtual water impacting the water balance of the region is impossible for a number of reasons. First, estimating the volumes of water needed to produce water intensive commodities such as wheat is difficult. For example for wheat the estimates range from about 1000 cubic metres per tonne of the author to about 1300 cubic metres of the Food and Agricultural Organization of the United Nations. Secondly, wheat, though a very important commodity in the regional import basket, is only one of hundreds of items all of which require varying volumes of water in their production. Thirdly, the region also exports virtual water in exported commodities.

The pattern of trade indicates that the MENA virtual water imports vastly exceed exports. By the mid-1980s MENA grain and flour imports had risen rapidly from about seven million tonnes per year in the early 1970s to over 40 million tonnes by the mid-1980s. This volume was equivalent to about 20 per cent of the region's total freshwater use by the late twentieth century. At the 1000 tonnes (cubic metres) of water per tonne of grain estimate of water content the regional imports of virtual water by the mid-1980s were equivalent to the annual flow of the Nile into the Egyptian agricultural sector. The Nile is the region's single biggest water resource.

The demographic trends of the region point to a doubling of population by about 2030. Taking into account possible improvements in water productivity in MENA agriculture the need to import water intensive staple food commodities will remain. It is likely that the imports of virtual water will increase four or five-fold by 2030. The remedy to this situation lies in the wider economy through the use of water in sectors which provide livelihoods in a diversified economy. The role of virtual water will be discussed at length in chapters 5 and 6.

Who needs to know what about Middle East and North African water? Insiders and outsiders

In the second half of the twentieth century very extensive regions of the world, including scores of major and minor economies in Africa and Asia, entered unprecedented circumstances of water deficit. The issue is interpreted in very diverse ways because the indicators of water shortage are not always clear. If all Middle Eastern countries were to be totally self-sufficient in water including the massive volumes of water needed for food production they would require twice as much water as the region possesses. This apparently impossible deficit has been remedied without stress by accessing water in the global system via trade in food. 'Virtual water', embedded in water intensive commodities such as grain, has been readily available at subsidised prices for decades.

The MENA region should provide the basis for research on how economies and the water users in those economies respond to the painful transition to water deficit. In practice the information on the water deficit has been suppressed. The main water users have not wanted to know about the predicament. Government leaders have not wanted to generate lethal political problems for themselves by disturbing the misconceptions of the big farming constituencies using the major proportion of the region's renewable and stored waters. It has not been politically feasible to disseminate and share the bad news about the regional water deficit.

In brief insiders are coping with the bad hydrological news of the late twentieth century. It has been possible simultaneously to keep in place traditional approaches to water use and management as well as addressing the water deficit. Global virtual water has provided the politically non-contentious and politically feasible ameliorative alternative. The outcome is definitely a second best solution. According to political criteria the second best approach has worked.

Outsiders meanwhile are increasing the profile of their commitment to policies based on environmentally and economically sound principles. The message for outsiders to hear is that they should understand the political ecology of the MENA region. They should grasp the politically determining perceptions of MENA insiders rather than the hydrology of the MENA region. They should understand the politics that explains these perceptions and especially the capacity of key groups to achieve their political goals with respect to water resources.

The analysis which follows will show that information on water resources and 'new knowledge' on how to achieve water policy reform can take many decades to move from the existence of the new idea to its incorporation into national environmental and infrastructure policies in MENA countries. Hydrological theory is not enough. In practice it can be misleading. Such theory tends to generate notions that the status of water resources, their strength or weakness, determine economic, social and political outcomes.

It will be shown that the impact of unprecedented water deficits are very different in different political economies. A diverse and strong political economy with strong social adaptive capacity responds differently from a political economy weak in its economic and social adaptive capacities. At the same time it is essential that those devising remedies to water deficits take into account political and economic futures different from those that exist now. The future will frequently be a socially and economically developed future with new and different water management options. Secondly it will be shown that global economic systems are very important indeed with respect to local periodic water shortages. Global economic systems are especially important for those political economies which now, or will in future, face longer term permanent water deficits. The global system also has the capacity to meet any local periodic water shortages provided that economic and political systems to move supplies to the stressed locality are enabling rather than impeding.

Note

1 A university in a central city site in an industrialised economy can be concentrated on a hectare. In the mid-1990s it used 10000 m3/yr of water costing $15000 for water (and sewage services), and employed 1000 full-time and part-time staff, that is 1000 jobs, educated 3500 students per year, provided international library, consulting and other specialist services and turned over $45 million per year in 1996. If it had been a wheat field the activity would also have used 10000 m3 per year of rainfall falling on the site and would only have turned over about $1200 per year and have provided a twentieth of one livelihood. (Data for 1995 for SOAS, University of London, located in Central London.)

PART TWO: ECONOMIC AND ENVIRONMENTAL IMPERATIVES IGNORED: AND WHY THEY ARE IGNORED

CHAPTER 3

Water is an economic resource: the economist's view

'Not all waters are equal: some are more valued than others.'

Valuing water, as yet in only parts of the hydrosphere, has proved to be as confusing as it has been helpful.

Water is an economic resource even if the believers are few

'Economic analysis shows that water ownership is equivalent to a right to the monetary value that water represents. By valuing water in dispute (including social value), parties can evaluate the possibility of trading off water for non-water benefits... The value of the water in dispute is surprisingly low ... .'

Fisher 1995:1 MIT economist

'Opportunity costs of water are generally ignored.'

Morris, J. 1998:229 UK economist

'As old rules lose their effect it becomes hard to gain the necessary acceptance for new ones.'

Kinnersley 1995 p 272 UK water sector consultant

At first sight it is not conceptually challenging to consider water as an economic resource. Some water is a very expensive consumable, for example chilled bottled water. Water in farming is less obviously valuable even if it is an essential input to farming. For an economist water is an element in the 'land' factor of production. Farmers also certainly value a naturally well watered, and well drained, tract. The effectiveness of the drainage, though usually taken for granted, is as important as the infiltration of water into the soil profile following rainfall. The presence of water usually very significantly enhances the 'land' factor of production in an economic farming enterprise.

The ownership of a tract of land brings with it the ownership of the soil water status and the drainage status of that tract. The soil moisture qualities of a tract of land result first, from adequacy or not of rainfall, and secondly, from the mechanical qualities of the soil profile that hold and transmit water at rates favourable or not to plant and crop growth. The comparative advantage is high for those who own well watered tracts in regions with favourable climate regimes for crop production. Their water is 'free'; their competitors' water may cost 60 US cents per cubic metre to mobilise. Where the climate enables cropping in two seasons, as in some tropical and equatorial regions, the 'advantage' of farmers with land with year round soil moisture is greatest. In tropical and Mediterranean regions where rainfed cropping is possible in only one season, crops can be raised in a second season with irrigation. Farmers in temperate regions can also only raise one crop per year. Temperatures are, however, too low in the winter season for crop production. It may even be necessary to supplement rainfall with irrigation in the one growing season per year in these temperate regions. In rainless regions, a category into which most of the Middle East and North Africa—MENA—region falls, the comparative disadvantage is greatest. Cropping is only possible with expensive engineered water rather than with natural rainfall. Engineered water is always more expensive to deliver than free rainwater.

This preliminary discussion has focused on water for crop production since it is by far the most water demanding activity. Water, albeit low quality water, used in supporting crops and vegetation is the dominant human and environmental use in the terrestrial environment even in the water short Middle East. Crop and vegetation use is the dominant consumptive use of water (see glossary). This dominance is true whether measured in terms of the water needed per head of human population or in terms of the proportion of total water use world-wide in supporting vegetation and crop canopies on which humanity depends. About 90 per cent of human water use is devoted to producing food needs.

Where water is unreliable or permanently inadequate for crop production in one or more seasons, as it is the case in the Middle East and North Africa—MENA—region, farmers and engineers have for millennia diverted water from channels and lifted it from the ground on to fields. The link, expressed as entitlements and legal rights, between a tract of land and access to an irrigation source for all or supplementary water for crop production has tended to be very strong. It is usual for 'water rights' to be legally associated with a tract of land although the nature of the link and the formality of the association varies from community to community and from legal regime to legal regime (See chapter 7).

In the Middle East and North Africa—MENA—region the Islamic tradition is pervasive and the fundamental and ancient legal significance of water is enshrined in the word for law itself, *shari'a*. 'Water lay at the heart of the legal system. Before it simply meant law, the Arabic word for Islamic law, *shari'a*, denoted the law of water.' (Mallat 1995:128) The provisions on water in the *shari'a* were, however, mainly concerned with water for drinking. On the selling of high volume low quality water in agriculture the message is clear and very negative with respect to placing monetary values on such water. Water can only be sold in a container. Flowing water and water in channels cannot be sold according to the *shari'a*. On sustainability the traditions of Islam are equally clear. It exhorts that natural resources, including presumably water, be used carefully as if the water using cultivator 'would live for ever' (Mubarak 1891).

Arising from the Qu'ranic guidance on the legitimacy of selling water is one of the most interesting water discourses in the region namely over whether charges can be made for water in channels and pipes. The outcome of the discourse will be that charges for water in channels will become commonplace. When such charges will become commonplace is, however, by no means clear. Meanwhile the charges based on the areas of cultivated land will continue. The availability of technology, the economic capacity to install that technology and the transformation in perceptions of those involved in the farming sector to 'permit' water charging involves reform which might be achieved in ten years in Tunisia, Morocco and Jordan. It may take three decades in other economies, such as Egypt, where the technical, economic and social milieu is different (See Perry 1996).

The sense that water rights should be associated with a physical tract is a natural transition in perception 'inherited' from the association of rainfall and a tract in the natural order. If nature does not deliver water free to users then users tend to assume that other agencies, such as governments, should emulate nature. However, this inclination to consider all water to be similar to rainfall, that is free, has very powerful consequences with respect to the perception of the value of the water compared with the real costs of engineering the water to points where it can be used. Those who use engineered water to irrigate crops are particularly prone to valuing water below its delivery cost.

Engineered water is not the same as rainfall. Engineers and farmers can choose where to put engineered water. They can deliver water to different points both within a tract owned by a farmer and between tracts if tradition and/or the legal regime allow it. In practice, despite it being logical to economists, once a water right has been acquired the transfer of water from one tract to another by sale is unusual. High volume water, such as used in agriculture, rarely figures in markets. (Young 1986) Where water is transferred from agriculture to municipal uses the arrangements are not the result of willing sellers and buyers coming together in a 'market'. Buying and selling of large quantities of water occurs most commonly when state institutions create an enabling commercial environment. (Dellapenna 1995) Although it is certainly the case that spontaneous water markets have sprung up in the MENA region in response to periodic water crises in the domestic water sector. Ta'iz in Yemen in the 1990s was such a case (Handley 1999).

It is consistent with this train of thought to conceptualise water as 'environmental capital' (Allan and Karshenas 1996:117). Environmental capital comes free to the first generation of users. They and subsequent generations can degrade or enhance environmental capital over time. It has become part of the fabric of the ideas held by many water users that water rights are a form of environmental capital. There is no consensus, however, that it would be fair for one generation to gain a monetary advantage from converting their right into financial capital. The owner of the 'rights' acquired them without cost. The purchaser of the right and subsequent generations would have to bear the burden of paying for them and there are usually stressful associated politics (Chatterton 1996:358).

In the MENA region water has been abundant for agriculture and high quality water sufficient for other human needs for millennia, provided people were willing to live close to the abundant resources. The migration and adjustment processes have not equipped communities well for the challenge of managing water as a scarce resource according to economic principles. Morris (1996:230) has helpfully pointed out that water is:

an unreliable renewable natural resource:
- a 'fugitive', re-usable good
- a common property/public good
- a climatically dependent stochastically supplied resource

an essential factor in ecological systems and an essential factor of production ensuring the viability of human and other biological systems and is:

- an essential, life supporting commodity with no substitute
- subject to economies of scale in provision and disposal
- endowed with many non-market, environmental qualities

In other words water is essential, unreliable and extremely hard to value comprehensively. Valuing water, as yet in only parts of the hydrosphere, has proved to be as confusing as it has been helpful. The rest of this chapter is about why economic approaches to water management would be useful and why their application has been very partial and often unsatisfactory.

The challenge of valuing water

> ' Soil water is "economically invisible" and considered to be costless. Yet a cubic metre of soil water is just as useful as an equivalent volume of engineered water. The consequences of misconceptions about the "economic value" of differently sourced waters profoundly influences MENA discourses on water allocation and management, water policy and water policy reform.'

If we review the diverse forms in which water occurs it becomes clear that some waters are much easier to conceptualise as economic resources than others. Falkenmark (1997) pointed out that water can be categorised as 'blue water' and 'green water'. Even more important was her question 'who owns the rain?' (Falkenmark 1986, 1997:39, McCaffrey 1997:57). 'Blue water' is the water people seem to value the most. It is the freshwater that occurs at the surface, either in motion or in natural or engineered surface storages, as well as in groundwater reservoirs. 'Green water' is the water in biological systems which supports natural vegetation and crops. Blue water is relatively easy to measure. Green water is impossible to monitor and is usually ignored by resource scientists, including Falkenmark in her calculation of water stress, and by economists and especially by engineers and politicians. 'Brown' water is the water in the soil.

If we look at the whole hydrological cycle it is clear that humanity depends directly on a very small proportion of the global hydrosphere for survival (Table 3.1). The directly useful hydrosphere includes the freshwater—that is high quality water—in surface water flows and storages, the water in the ground which includes soil water in soil profiles and water in the ground at various depths of storage.

Table 3.1 The 'directly useful' hydrosphere and the total hydrosphere: measurable, marketable, accounted and ignored waters

	Accounted	Non -accounted	Globally available per cent
Measurable, Accounted, Marketable			
Directly useful—'Blue water'			
Surface flows in rivers & streams	Accounted		0.00012
Lakes	Accounted		0.0072
Groundwater (where high quality and accessible)	Accounted		0.592
Cannot be monitored, nor accounted, nor marketed: are ignored			
Indirectly useful water—'Green water and brown water'			
Land			
Soil water—brown		Not ac'd	0.0051
Biological water—green		Not ac'd	0.00008
Atmosphere			
Atmospheric water—potential blue water		Not ac'd	0.00094
Cannot be monitored, nor accounted, nor marketed: are ignored			
Not directly useful water—but essential in the cycling of the 'hydrological system'			
Ocean		Not ac'd	97.50
Ice—ice caps/glaciers		Not ac'd	1.98

Sources: Falkenmark 1997:24, UNESCO 1998, Shiklamanov 2000, WMO 1998, Ward and Robinson 1990
Note: Freshwater comprises only 2.5 per cent of global water. 97.5 per cent of water is in the world's oceans. Especially note how much greater (500 times) is the soil water than the freshwater in rivers. The remaining groundwater resources are substantial but expensive and difficult to engineer.

The depth at which the groundwater lies determines the potential availability of such groundwater.

Only a tiny proportion of the world's water is directly relevant to human survival. The Middle East and North Africa—MENA—region has a minor proportion of that water. A very small part of the useful MENA freshwater is susceptible to commodification and accounting procedures necessary for the deployment of economic instruments such as valuation, cost accounting and the pricing of products.

The susceptibility of the water sector of the Middle East to economic conceptualisation, as well as to the introduction of economic instruments, depends on the extent to which the hydrosphere is measurable, quantifiable and manageable. Can water inputs be quantified and valued?

Regions which are not water short have little reason to discover the tangible and transactional costs of mobilising additional water to raise food. Millions of people, and the farmers who provide their food, in the humid and tropical temperate regions, together comprising at least two thirds of the world's population have given little thought to water scarcity with respect to the major water consuming sector, crop production. Uncertainty about the annual level and distribution of rainfall is part of their expectations. Meteorological statistics confirm that a tract that receives 1000 millimetres of rainfall annually will experience variations of no more than ten per cent on 95 per cent of occasions. In the MENA region where 500 millimetres is a high level of annual rainfall the variation will generally be over 30 per cent on 95 per cent of occasions.

The Middle Eastern and North African—MENA—region is very different from rainfall favoured regions. The MENA economies were the first to run out of water. Their farmers and water managers were the first to discover the problems of not being able to substitute one source of water for another. Or if they have attempted it they have been amongst the first to discover how costly such substitution can be. Farmers throughout the region have learned how expensive and how limited is the option of substituting for shortages of soil water by accessing, for example, groundwater. In Libya coastal groundwaters were driven down to disaster levels during the 1960s and the 1970s (Allan 1971) and the experiments in the south of the country at Kufrah, Sarir and Fezzan were socially and economically unsustainable based on the fossil groundwater beneath the Sahara. In upland Yemen groundwater development has been pushed to unsustainable levels with serious economic and social consequences (Handley 1996 and 1999, Ward 1998). In Saudi Arabia the experiment with grain self-sufficiency of the 1980s was shown to be technically feasible but environmentally and economically unsustainable. In Oman the obsession with non-economic crop production damaged coastal aquifers as seriously as anywhere in the region. In Syria the manageability of groundwater in comparison with unreliable rainwater has led to unsustainable levels of groundwater extraction (Allan 1987). In Jordan it has proved impossible to develop a groundwater management regime that observes the limits of the groundwater potential of the Badia region in the marginal north-east of the country.

Where it has been possible to substitute surface (blue water) flows for the deficit, and economically invisible, rainfed (brown) soil water—in for example Egypt and in the Jordan catchment economies—the supply has been insufficient to meet needs. By the 1950s the Jordan catchment economies had run out of such water. Egypt did so in the 1970s. The Tigris-Euphrates economies, the Sudan and Lebanon were, however, still in river water surplus by the end of the twentieth century.

The position is similar for sustainable (blue) groundwater. The only recharging groundwater resource managed, albeit very controversially, with a close awareness of the relationship between rates of recharge and rates of withdrawal, are those of the West Bank of Palestine. Elsewhere fossil groundwater has not been subject to any evaluative procedures and is withdrawn according to short term 'need' rather than its overall contribution to the economy in the long term.

The issue of the valuation of the MENA water resources is of fundamental importance to the adoption of an economic approach to the allocation and management of water in the region. Table 3.2 shows where the current assumptions of farmers, populations generally and of politicians responsible for water policy, differ from those of economists. The assumptions of farmers, food purchasers and governments are driven by beliefs based on long experience of water surplus circumstances, albeit in the constrained water endowment of semi-arid and arid environments. The conclusion of farmers with generations of experience of using nearly free irrigation water is not that water should be subject to economic analysis. They conclude livelihoods depend on supplies being increased and oppose the introduction of water pricing which would exacerbate already difficult farm level circumstances. The economist on the other hand calculates the costs of water delivery, the value of outputs in the activities where water is an input and all 'economic' factors, including 'exogenous' factors, relevant to understanding their definition of the value of water in an economy.

The differences evident in Table 3.2 between the current assumptions by MENA users/water policy makers listed on the left and those of water economists are important. Those on the left inform the political economy of MENA water discussed in the

Table 3.2 Assumptions in currency in the MENA region relevant to the allocation and management of water and contrasting the assumptions necessary for the development of water policy based on economic principles and the deployment of economic instruments

Assumptions current in the region of farmers, other water users, officials, legislators	**Assumptions necessary for *economically inspired* water policies, water policy makers, policy reform and water management**
Soil water	**Soil water**
• soil water is economically invisible and it follows that:	• soil water be regarded as a resource **to be supplemented by 'blue water' accounted at the economic cost of delivery**
• soil water is costless and should not be valued or considered considered to be an input to to crop production	• ***environmental economics*** can value 'soil water' but such methods are not in currency • **some tracts and some farmers have the comparative advantage of access to 'free' soil water**
Green water [water used by vegetation and crops]	**Green water**
• green water is not counted nor is it considered an element in crop producing sectors	• economics also does not account for green water
Blue water	**Blue water**
• it is a state's duty to pursue 'supply management' policies to provide new 'blue water' for use in productive, especially irrigated agriculture, and service sectors	• newly mobilised 'blue water' **must be part of an economic, unsubsidised enterprise**
• 'new water' will be free or highly subsidised	• 'new water' to be costed as an input
• water use efficiency will be achieved via investment: • in improved water engineering	• water use efficiency will be achieved via investment: • in improved water engineering • **in engineering and in institutions that re-allocate water to higher return activities**
Virtual water	**Virtual water**
• 'vw' does not exist	• **'vw' is not a part of the conventional economy but it is recognised as a key element in the political economy of regional MENA water**

following chapter entitled 'Water is not an economic resource'. Those on the right inform water economists. They claim that 'water is an economic resource'. The instruments inspired by the principles of economics which economists advocate to steer water policy reform, water policy making and the behaviour of water users are the subject of much of the rest of this chapter.

Free water: in the interests of whom?

> Outsider Northerners had scarcely digested their 1980s ideological re-orientation of Damscus road proportions, that water was an economic resource, before they were impatiently forcing it on the South.

For non-economist farmers, for governments responsible for water and food security policy and even for individual food consumers there is an intuitive appeal in the notion that water for farming should be free. Rainfall is after all free, no one pays for water in the major rivers that flow into many countries in the region. In addition it is intuitive for the mass of food consumers in the urban Middle East and North Africa that food should be as cheap as possible. That idea has proved to be more powerful in the Middle East and North Africa region than the other notion that farmers should derive a good living for their families from their farms.

The idea of free water is compelling. Food production is the majority use of all types of water, including soil water, the supplementary water in surface flows and groundwater. It is assumed that everyone benefits from the incorporation of this free water into agricultural products. The price of the products reflects the free, or very low priced, water input. Aware of these widely held assumptions governments judge that the political costs (World Bank 1997a) of introducing agricultural water charges, reflected in higher food prices, would not be worth paying.

The economists' reaction to this is to draw attention to the factors which determine the price of food. They point out that exogenous economic instruments such as agriculture sector subsidies in the United States and Europe determine world grain prices (ABARE 1989 and 1995, LeHeron 1993 and LeHeron and Roche 1995) including the prices at which staple grains enter the MENA region. These trading activities in water intensive commodities are not evident to food consumers. As a result the underlying economic impacts are only part of the political economy in a negative sense because they are invisible. Circumstances are therefore misleading.

The impacts of the exogenous 'virtual water' are only detectable by those with an economist's insight. To everyone else it is politically invisible. Being politically invisible virtual water is irrelevant to politically influential MENA food producers who are heavy water users. Virtual water is even less significant to the even more politically influential food consumers, and to the politically attuned legislators who relate to them as constituents.

For non-economists, who are not farmers, such as public servants, teachers and the increasing army of the employed involved in service sector activities, residing in urban centres, the perception of the value of water is very different from that of farmers. Farmers need high volumes of water to ensure secure livelihoods. Those involved in urban sector jobs are often not even aware that water contributes anything to their livelihoods. When cost recovering prices are charged for water used in urban jobs the economic transactions are unstressful and un-noticed because the volumes of water used are so small.

Those who hold the notion that water should be free cannot subscribe to the argument of this chapter that water is an economic resource. Understanding the nature of the discourse on water in the MENA region requires awareness that the notion that water 'is' an economic resource, and the idea that 'it is not', are simultaneously and firmly woven into the discourse on water. This is true of the region as well as of the world, including many economies in the North. In the MENA region at the end of the twentieth century the advocates of the notion that water is not an economic resource were firmly in the ascendancy.

The idea that water is an economic resource is traceable to the founding Ricardian concepts of classical economics in the early nineteenth century. Humankind has conducted its affairs for six thousand years without operationalising his insight that environmental resources, such as water, should be valued as inputs. Ricardo's analysis was ignored for over 150 years by everyone, including neo-Marxist and neo-classical professional economists in the North. By the 1970s the green movement had constructed in the North the idea of natural resource value based on the principles of environmental sustainability. Professional agricultural and project economists will be remembered in the first six decades of the twentieth century for their abuse of economic principle. They deployed a number of pseudo-economic *ex-ante* and *ex-post* approaches to evaluation, and to the planning and development of water resources in the United States, which were specifically designed to deliver water to farms below the economic cost of its delivery. The period between the 1940s and the 1970s was a time when 'if the Corps of Engineers builds a dam, and calls it a flood control dam, the water is free.' (Reisner 1986/1993:174) Until the 1980s US water use in arid region agriculture had been under-pinned by the assumption of cross-subsidy from hydro-power generation (Reisner 1986/1993:134–135).

Economists in the water rich temperate regions in the post-industrialised North only adopted the principle of water being a resource which should have its value closely scrutinised, including its environmental value, in the 1980s. This analysis is being written in the year 2000. Two decades is a very short period in the measureless march of socio-economic change. Before 1980 the non-economic approach to agricultural water in the United States was summed up by the term 'the cash register dam'. According to the unfathomable economic alchemy of the Corps of Engineers and the US Bureau of Agriculture low priced water was provided for both the needy small farm as well as for the canny private corporation able to manipulate the legislators over the siting of diversion works. (Reisner 1993:134 and 171) These outsider Northerners to the MENA region had scarcely digested their 1980s ideological conversion of Damascus road proportions, that water was an economic resource, before they were impatiently forcing it on the MENA region and elsewhere in the South.

With the idea that water is an economic resource being so new it is not surprising that full cost recovery of water in agriculture is everywhere elusive even in the diverse and strong economies of the North (Cheret 1998). In the MENA region the idea that water is an economic resource has made some, but little headway. The idea arrived in the early 1990s. At the time of writing, in the year 2000, the idea had been assimilated at all levels of the water managing and using communities in Israel. In Jordan, Morocco, Tunisia and Cyprus water ministers and water officials had adopted the idea and economic principles were beginning to play a part in water policy making (World Bank 1998b). These developments show that the position was very dynamic and there were emerging signs of response to the economists' message that water is an economic resource. The diffusion of the idea of the economic value of water is gathering pace.

A diversity of water sectors: responses to the demand for water according to volumes and quality needed

All waters have delivery costs, have or could have a price for delivery, and they have an influence on the financial performance of the enterprise in which water is used. Water also has an economic value in an economy, even if this last is difficult to define with precision.

In most economies six types of water play a role in their environmental, productive, municipal and domestic sectors. These waters operate in, or could potentially operate in, separate economies in each of which water is priced/not priced and valued/not valued very differently. Within these economies water can be characterised by supply and demand circumstances which can be quantified according to the water use in various sectors as well as by the costs of delivery of the different water types. (See Figures 3.1–3.3)

Six types of water are identifiable. There is a seventh, environmental water, which is not considered in this discussion because it would require more space than is available in one chapter on the economics of water to deal with the environmental economics of water. California has made an attempt to include a value of water in its accounting for water allocation but the imprecision of estimates has made the exercise controversial. There is no example in the MENA region where water has been given an environmental value expressed in volume or economic terms.

The economic features which distinguish the six types of conventionally valued waters are first, the characteristics of supply and/or demand, secondly, the quality needed by the user, and thirdly, the consequent costs of delivery and price. The types of water are illustrated in Figures 3.1–3.3, namely bottled water, drinking water,

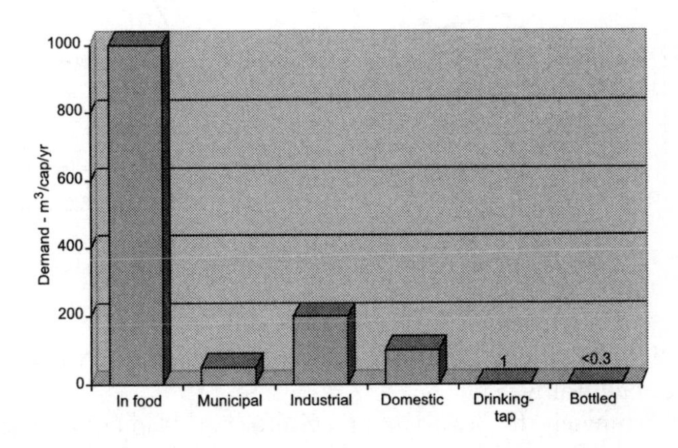

Figure 3.1 Demand for water (indicative for semi-arid circumstances) — cubic metres per head per year

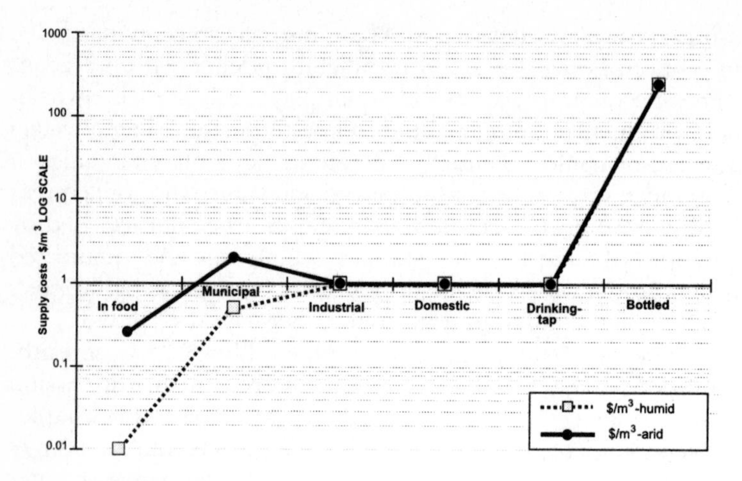

Figure 3.2 Supply costs of water (indicative) for humid and arid regions ($/cu.m) —semi-log diagram

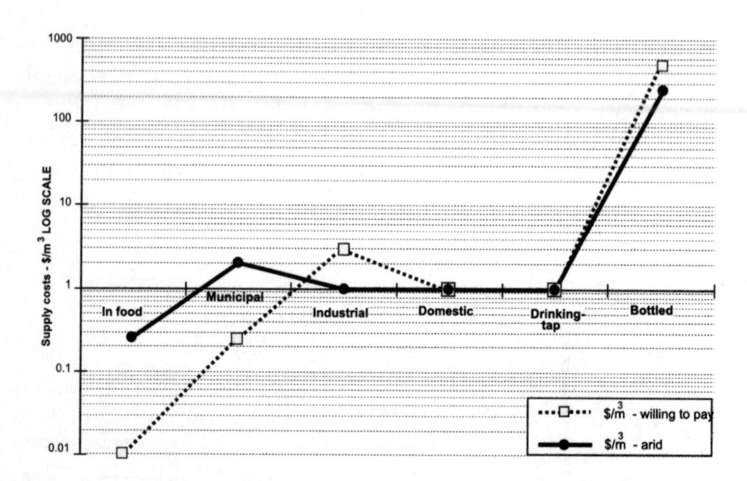

Figure 3.3 Supply costs of water (indicative) and levels of willingness to pay

domestic water, municipal water, industrial (including services) water and agricultural water.

Unfortunately, for a variety of cultural and traditional reasons, when water allocation is discussed it is assumed by almost all users, except possibly the willing buyers and the willing sellers involved in the market for bottled water, that water should be free or nearly free.

Figures 3.2 and 3.3 reveal that the levels of demand in the different using sectors differ greatly in terms of the delivered costs of water and of the capacity of the different sectors to bear the cost of different waters. The users of the highest volumes, the farmers with irrigation systems, can only afford prices of a few US cents per cubic metre. Such prices are much lower than the delivered cost of most water. Meanwhile the users of small volumes of water are prepared to pay one US dollar per cubic metre for water delivered to the home. For bottled water prices lie between $US 100 and $US 3000 per cubic metre.

The volumes of water use per head of population and of the relative use of the different types of water are very important numbers to have in wide currency. An individual requires only one cubic metre per year of drinking water and about 100 cubic metres per year of domestic water, although in some economies the usage can run at many times this level. But 100 cubic metres per year is an appropriate number in the light of current water technologies. They also need over 1000 cubic metres per year of water for the production of their food needs. The implication of these numbers is that one million people would need one billion cubic metres of water per year to be food and water secure. For the approximately 300 million people of the MENA region about 300 cubic kilometres (billion cubic metres) of freshwater are needed annually. The region can only access, but does not utilise all of, about 200 cubic kilometres annually plus the unmeasurable rainfed soil water which could amount to a very variable volume of about 100 cubic kilometres per year. Of the 200 cubic kilometres of surface and groundwater available each year only about 120 are mobilised. Some countries have mobilised all or a very high proportion of their freshwaters, Egypt, Libya, Tunisia, Israel, Jordan, Palestine and the economies of the Arabian Peninsula. Other economies had at the turn of the twenty-first century as yet unutilised water, Turkey, Syria, Iraq and Iran.

One of the reasons it is difficult to attribute a value to water and reflect this value in a price is that it is found in many different forms within a region or a national economy. It may be invisible in a *groundwater* reservoir or not considered as part of the water budget at all if it is in *soil profiles.* Freshwater can also be at the surface in *streams, rivers and lakes.* The surface water is evident and measurable without too much difficulty. Groundwater use and levels of storage can only be measured with methods of less than adequate precision. Volumes of usable *soil water* on the other hand are extremely difficult to measure and monitor. Yet in many environments it is the soil water which is the major national source of water. In temperate regions it is usual for 90 per cent of a national water budget to come from the soil profiles. In the case of North America and parts of Northern Europe there is sufficient 'surplus' water in the soil profiles to grow crops beyond local national needs. Surplus 'soil water' can be exported. Exported commodities, such as grain, with huge volumes of embedded 'virtual water'—1000 tonnes/cubic metres of water per tonne of grain—are transferred via the global trading system. The volume of water involved in the transactions amounted to about one quarter of the water needs of the Middle East and North Africa.

It is an unfortunate political fact that perceptions of the economic value of water are impaired by non-awareness. There is no awareness of the role of soil water in an agricultural economy; no awareness of the economic value of soil water. The political impact of non-awareness is compounded by the extreme difficulty of devising a methodology for valuing soil water. In addition there is no awareness of the power of global commodity trading that in effect moves massive quantities of low quality and low value virtual water to water short regions.

How can virtual water be valued? Should it be regarded as a free factor of production as no direct effort has to be expended to make use of it in rainfed crop production? Economic logic encourages default to this position. The only costs of making use of soil water are some notional transactional costs. These elusive costs could include unlikely elements such as the higher than average suicide rate amongst farmers and the shortage of spouses in some remote farming regions in the Northern economies. Both are the result of some farmers and their families having to live in often very isolated and stressful locations where the rainfall occurs. The rainfall determined isolation plus the unpredictable challenges of weather, markets and the disregard of government can be beyond an isolated individual's bearing.

Alternatively, should the value of 'virtual water' be counted as the opportunity cost of the water, that is the cost of mobilising the water, in the economy to which the virtual water is imported? Estimates could differ from almost nothing for water falling on the fields of Ohio or the Paris basin where the water is incorporated into the crop, to 60 US cents or more per cubic metre at which water could be valued in a Middle Eastern importing country. Such opportunity costs could be estimated for Israel or upland Jordan, where new water costs this sum to mobilise.

The analysis in the preceding paragraph reveals that water like food may command a modest price at source, for example at the farm gate in terms of agricultural products and at the spring, pump or canal outlet, in the case of water. In food markets, however, the value added to the food product in the process of preparation for, and distribution in, the market can be massively enhanced in value and in the price commanded. Examples in the case of food commodities are legion. Most of the cost of many packaged food products is added in the processing and especially in the manufacturing and marketing of the commodity. Even the water sector has become rich in examples of water which may cost only a few cents a cubic metre at source being packaged, distributed and sold at prices which are 1000 times the cost of delivery at source. And 'designer' water can cost even more.

It is not being suggested that all water should be subject to the painful allocative pressures exerted by pricing disciplines which may be appropriate in strong economies. It is a generally accepted development goal that the basic needs of about 100 cubic metres per person per year should be made available to individuals and communities at prices consistent with the economic capacity of the respective communities. A uniformly high economic cost for domestic water is not politically acceptable. The case of Israel and its neighbours demonstrate very starkly the inappropriateness of uniform water prices across the economies of Israel, the West Bank and Gaza based on economically inspired analytical principles. Economic circumstances and purchasing power are different at a point in time in immediately neighbouring economies. The Israeli economy generates approaching US$ 20000 per capita per year for Israelis; the Palestinian subordinated economy generates only US$ 1000 per capita per year.

Economic circumstances also change over time in a single economy. Over a period of forty years up to the mid-1980s the Israeli economy moved from being an agricultural economy to being a post-industrial economy. By the mid-1980s 97 per cent of the Israeli GDP was being generated in sectors which used very little water, in industry and especially in services. Less than five per cent of national water usage produced this 97 per cent of GDP. 65 per cent of the water was used in agriculture which only produced three per cent of the GDP.

A strong and diverse economy such as that of contemporary Israel can subsidise water in agriculture, but still charge 25 US cents per cubic metre on the farms for water. This water could be valued in the national economy at 60 cents per cubic metre. Israel had the economic strength by the 1980s both to observe the stressful economic rules which point to the transfer of water out of agriculture while at the same time delivering subsidised water to most farms. By contrast the economic contexts in the West Bank and Gaza are very different. Not only are much smaller quantities of water used per head but much more important the using communities cannot afford the economically determined price per cubic metre for domestic water never mind that for irrigation water in agriculture.

Relevance of economic instruments in the water sector?

> Linking water and jobs to achieve economic and strategic security is not intuitive to peoples who have for generations been used to linking water and food to achieve that goal.

> Getting more crop per drop (IWMI mission) in a water short region is less important than getting more jobs per drop.

Economic instruments are part of a suite of water allocation and management tools which can help water users and water managers to achieve improvements in *water use efficiency*. These improvements are often urgent and essential. The idea that water has an economic value and that it should be managed universally via economic systems is, however, a very new idea. This 'new knowledge', that hydro-economics (Merrett 1997) should inspire water policy and water management practice only gained currency in Northern economies in the 1980s. The United States President, Jimmy Carter, tried to shift the planning assumptions on water to accord with sound economic and environmental principles and encountered damaging opposition in the late 1970s (Reisner 1993:320ff). This evidently politically costly approach was projected on to the political economies of the South in the 1990s by international agencies such as the World Bank and the European Investment Bank (World Bank 1993, Serageldin 1995). The new approach, established via a very stressful discourse in the North, does not fit with long held beliefs by traditional users of water who in many cases have accessed free water for millennia. More importantly many of these users are poor and are not economically or politically equipped to adopt the 'new outsider knowledge'.

Table 1.7 in chapter 1 is a framework which situates economic instruments amongst the institutional tools which address political, economic, equity, safety and environmental priorities of water policy (Winpenny 1991) and water management. The framework is provided to demonstrate how many other influences are active in the water sector. In practice economic instruments are as yet not very influential compared with measures inspired by the need to stimulate agricultural production for goals such as food self-sufficiency and by equity, that is to supply water at 'affordable' prices or free. 'Affordable' means the price acceptable to current water users which if left unchanged brings a political dividend. If the acceptable price were to be exceeded opposition would be generated and dangerous political stresses.

Table 1.7 identifies the five goals which have to be addressed by water planners and policy makers. The group of *economic* goals lead via processes of planning and policy making—inspired by economic principles—to the development of institutional instruments and the deployment of engineering instruments. Economic instruments are associated with 'economic' goals. They are deployed to influence the allocation and management of water, as opposed to other instruments which reflect other *goals.* These goals are to achieve equitable water availability, improved and safe water quality and patterns of sustainable water use. One goal is inspired by the dangerous notion of food self-sufficiency, dangerous at least for a water deficit economy. The economic instruments reflect just one of the numerous policy priorities that have to be addressed.

Those advocating economically inspired policies have to struggle for attention against those whose intent to influence policy are different. Importantly the non-economically inspired policy priorities accord more readily with long held community concerns. For example water quality, affordable water for the food producers, affordable water for the poor and the importance of amenity water, especially unencumbered by obligations to achieve cost recovery, are politically less stressful than the introduction of some of the economic pricing instruments advocated by the economically inspired.

The economic goal of achieving high efficiency in water use is achieved by both engineering and institutional instruments. The goal is inspired by *principles* of *water use efficiency* (WUE). *WUE* can be achieved via two basic approaches. The first is technical or productive efficiency. The second is economic or allocative efficiency. Productive efficiency is achieved by investing in *technically effective* measures which bring *productive efficiency.* Improvements in productive efficiency are achieved by structures, such as dams, reservoirs, channels and tubes. These were the means as well as the symbols of the hydraulic mission. They were the techniques deployed in the first supply management phase of the region's water managing history up to 1970. By the 1950s the hydraulic mission augmented its approach to water distribution. Sprinkler and trickle irrigation and fertigation technologies featured amongst the means by which engineers reinforced the technical efficiency of water use in agriculture. Trickle irrigation could improve returns to water by 100 per cent compared with water spreading methods. Some of the old technologies were first developed in the ancient Middle East. Both old and new productively efficient technologies were important in the MENA region which by the end of the twentieth century was still deeply involved in its hydraulic mission.

These *technically (productively)* effective 'instruments' are essential but they achieve very modest improvements in returns to water compared with the second approach, that of economically *or allocatively* inspired approaches. Productively or technically efficient approaches operate at the level of the irrigation service and the farm. Economic/allocative instruments on the other hand operate in the institutional and organisational domains. Their aim is to re-allocate water to high economic return activities. They encourage the allocation of water to higher value uses either within a farm or the agricultural sector or they tend to shift water from one sector to another. Flowers and horticultural products can bring returns to water of ten or more times those of cereals. Shifting water from one sector to another can bring increases in returns to water of 10000 times. For example a higher educational and other service institutions occupying sites as small as a hectare in urban locations can provide thousands of full-time and part-time jobs, educate thousands of people each year, provide information, library and consulting services at an international level and turn over millions of US dollars per year while using only 10000 cubic metres per year of water. If the sites were to be used for the production of wheat, the productive activity would also use 10000 cubic metres of water per year but the output would only be valued at about US dollars 3000 annually and would provide the livelihood for one family. There are flaws in this simplistic example because the 'returns to water' are achieved by very different levels of other capital, labour and technology inputs. However, the parable does get across the point that economic entities, firms, regions and national economies, which are extremely short of water can nevertheless be associated with the generation of economically sustainable livelihoods. Put even more simply productive efficiency is about getting 'more crop per drop' (IIMI 1997). Allocative or economic efficiency is about getting 'more jobs per drop'. Getting more crop per drop (IWMI mission) in a water short region is less important than getting more jobs per drop.

The MENA economies have already moved significantly, and without any stimulus from water resource planners, towards the 'more jobs per drop' model. Jordan for example derived 93 per cent of its gross domestic product from the non-agricultural sectors by the mid-1990s, which only use about five per cent of the national water budget. Agriculture used over 80 per cent of national water to produce only seven per cent of the gross domestic product. The reality of the increasing dependence on high efficiency water practices in industry and service sectors is not matched in the region by awareness of the power and success of the re-allocation which has brought this about. Neither communities nor legislators are aware. The successes have been achieved as a concommitant of socio-economic development. The successes were not planned. Linking water and jobs in diverse livelihoods to achieve economic and strategic security is not intuitive to peoples who have for generations been used to linking water and food via agricultural livelihoods to achieve these goals.

Allocative practices based on *economic principles* can be very powerful indeed in gaining higher returns to a scarce resource such as water. Re-allocation is, however, generally politically stressful, resisted by existing users, and handled with great caution by political leaderships. These leaders normally hold the same very strong beliefs about water systems as those of the large numbers of traditional users of high volumes of low quality water in the rural sector. The families of legislators are often still engaged in rural livelihoods.

The instrument used to achieve allocative efficiency is water pricing in water markets or in other forms of water transaction. Tariff policies and subsidies of many kinds also exist but these have only indirect influence on water use efficiency. They are normally negative in effect with respect to efficiency. Most tariffs are introduced to meet political and equity priorities rather than to accord with economic principles.

Water is rarely delivered to agricultural sector users anywhere, and certainly not in the MENA region, at a price which takes into account the cost of water delivery. Nor is it priced at a level which takes into account comprehensively water's other very real values to the economy as a whole nor in recognition of its intangible value as an environmental resource (Winpenny 1991). This is especially the case in the MENA region. The economic value of water to the economy as a whole is unfathomable to the individual water user and especially to users of water on farms. The environmental value of water is also a challenging concept. This description of environmental values is more likely to have been part of the experience and intuition of a community than that of economic valuation. However, the environmental value of water is particularly difficult to incorporate into the national accounts even of those nations with legions of professional economists on hand. (Lutz et al. 1991) The difference between the value of water calculated according to comprehensive principles—economic, environmental and equity—and what is actually paid for the water is a subsidy to the user. It is also a subsidy to the consumers of the production enabled by the water.

Ignoring for the purposes of this discussion the volumetrically small, but economically significant market for bottled water, where there is a large population of willing buyers and very responsive willing sellers, the big issue in a review of economic instruments in the water sector is to what extent have economic instruments found their way into water allocation and management practices in agriculture? Another important issue is the question, to what extent has the pricing of water enabled the transfer of water out of agriculture into industries producing higher value products?

Water resources in another continent, in the semi-arid West of the United States, were the first to be subjected to a revolution in perception of their economic value as well as of the intrinsic value of water environments. In the 1960s water marketing was a theoretical policy prescription devised by economists to provide a remedy to the palpable misallocation and inefficient use of water (Tarlock 1997:191, Hartman et al. 1970, National Water Commission 1973:260). The belief system of the farming communities, of the politicians who represented them and especially of the agencies responsible for mobilising heavily subsidised new water, the US Bureau of Reclamation and the Corps of Engineers were proof until the 1980s against the new way of approaching the value of water. Water users, their state legislators and the water providing federal agencies were locked into a political nexus of what proved to be very unsafe federal and state politics with respect to the stewardship of the environment.

The concept of the water market made no significant headway in the otherwise market friendly United States until the 1990s. Until then the urban water supply interests in California, for example, did not want to erode political support for new publicly subsidised water projects. It was the stormy politics created by the environmental movement combined with the protracted Californian drought in the four years up to 1992 that provided a 'window of opportunity' (Kingdon 1984) during which the new ideas were given temporary substance, in the form of water banks and water transfers (Gray 1994:132).

In other semi-arid regions, the Mediterranean (Mendeluce et al. 1998), the Middle East and North Africa (Fishelson 1992), Mexico (Garduño 1998) and Australia (Langford 1998) there have been decades of discourse between enthusiasts of economic instruments and resistant water using communities, officials and politicians. Only political economies that are strong and diverse can contemplate a tough approach to water pricing in agriculture. Even highly industrialised economies which have been relatively successful in introducing water charges for irrigation water, Israel and France, have only managed to achieve a 40 per cent recovery of the economic value of the water (Cheret 1998).

In a politics free world it would be easy to identify the impact of economic instruments. In the real world where politics dominate their role is difficult to identify and to analyse.

Despite the persuasive economic logic for the deployment of economic instruments it is unusual for the full capital costs and delivery costs of water to be reflected in the price of water charged to users. This is particularly the case in irrigated farming, the major water using sector world-wide. Water pricing is resisted by most users in the sector. There are deeply held beliefs on the part of farmers that water should continue to be freely available, at least with respect to price, even if significant levels of transactional costs have been traditionally part of making water available. For example farmers using water for irrigation have often had to devote significant family labour and animal power to lift water on to their fields in Egypt. Small pumps now do this job. Not to speak of the time and sometimes resources devoted to ensuring that officials responsible for water allocation perform for them as they need (Radwan 1998).

Water prices in the agricultural sector reflecting a significant proportion of the costs of delivery to the farmer can only be introduced when they are politically feasible. This political reality applies also to the valuation of water in an economy. Politically feasible circumstances occur when an economy has achieved a high level of economic diversity and strength. The presence of water pricing instruments reflect socio-economic circumstances—economic strength and diversity. They can be viewed as a 'response' in a political-economy rather than an innovation.

Subsidies, on the other hand, are readily accepted by farmers and other beneficiaries. The introduction of subsidies to irrigated farming will only be resisted by Ministries of Finance and their political allies. The achievement of strategic water and economic security through livelihood creation in diverse sectors has only been adopted consciously and comprehensively and explicitly by Israel in the MENA region (Katz, A. 1994 quoted in Katz, D. 1994). In Turkey the goal is also explicit. However, Turkey is exceptional in having a relatively rich water endowment which increases its water policy options.

Policies of economic security via diversity are gaining ground in Jordan, Syria, Egypt and the non-oil Maghreb economies, Morocco and Tunisia. In the oil economies of the Gulf, and in Libya and Algeria the goal of economic strength via diversity is increasingly explicit. In the immediate future the Yemen lacks the enhancement of oil revenues and it faces the most acute water policy challenges (Ward 1998, Handley 1996). The Sudan has not yet made any significant strides in transforming its economic base from agriculture to a dependence on a more secure economic mix.

Why are economic instruments potentially important?

> 'When enough livelihoods have been created to make both opinion formers and the majority of the employed population comfortable with the notion of de-emphasising agriculture then agriculture can be de-emphasised'
>
> (Feitelson 1996)

Water is part of the economic fabric of a political economy. It is a factor of production in the productive sectors; it is an essential source of environmental capital for an economy. Water also has political significance because of the socio-cultural nature of a political economy. Traditional perceptions determine the value of water as well as its cultural significance. The latter is often powerfully reinforced and therefore legitimised by commentary in religious texts (Abdel Haleem 1989:35, Mallat 1995b:128). The wonderfully expressed notion that we 'should cultivate our world as if we would live for ever' is a Hadith—a commentary on the sacred Qur'anic text. The words constitute an ancient aphorism endowed with religious significance in the Islamic tradition.

Contradicting this wisdom, there are in contemporary currency, widely and deeply held perceptions of entitlements to access water sources, which are believed to be limitless. A limitless supply of freshwater for drinking and for basic domestic uses in a pre-industrial society is a perception confirmed to be sound by empirical observation. Sufficient empirical observation has not, however, been carried out to confirm whether there is enough water globally for human needs of future global populations. Empirical observation does show that there is not enough low quality, high volume, water in the MENA region to grow the food needs of its population. Nevertheless expectations are deeply embedded that there is enough water in the region for all human needs.

Politicians know that belief transcends empirically observed 'fact'. Scientific truth after all is merely the point at which those seeking such verity mutually agree to seek no further. In the event scientific truths, especially outsider scientific truths are just an element in the streams of information which vie to be included in the beliefs which interact in the political domain. The extent to which scientific observation makes a contribution is politically determined by the process of the 'network of consensus' described by Foucault (1966 and 1971) and usefully clarified by Eco (1987). Put simply scientific truths, economic and environmental, find a place in determining the way users and policy makers perceive and allocate water according to the political price that would have to be paid to give operational substance to them. Policies which bring a high political price tag have to wait until the economy has diversified before they can be afforded and accommodated. When enough livelihoods have been created to make both opinion formers and the majority of the employed population comfortable with the notion of de-emphasising agriculture then agriculture can be de-emphasised (Feitelson and Allan 1996).

When freshwater of adequate quality is available in the soil, the groundwater, the rivers and lakes the assertion of free entitlement is not difficult to accommodate. When demands exceed supplies adjustments have to be made. The adjustments chosen are often assessed by economists according to economic principles alone. Such ill-founded and culture-free assumptions lead to much righteous frustration in the profession of economics.

The expectations and beliefs of water users are based on experience, history and culture and are not much encumbered by economic principle and scientific analysis. Politicians are alert to signals of disapproval from these numerous water users who believe that water supplies will be as adequate in the future as they have been in the past. Water short users do not include in their evaluation of the situation a role for the notion of water use efficiency. They fix on issues such as water entitlements and impaired livelihoods. They respond vigorously to threats to their traditional secure supplies of cheap or free water even if their water demands have increased significantly. They are particularly exercised by the removal of subsidies, such as when subsidies come off diesel prices in regions where diesel pumps are used to lift water. This was the case in Yemen in 1998 (Ward 1998). Similar opposition occurs when it is proposed that water should be priced for the first time or increased in price, as in Israel in 1986 and subsequently in Jordan. All unpriced water, and all water which does not attract a cost-recovery price, is subsidised. Economists argue that such politically determined prices have very serious immediate and longer term consequences for the economy and the environment. Water users are blind to these consequences and they suppress awareness of them in current discourses.

Water pricing

Despite the difficulties of introducing economic instruments such as pricing such measures are important because where introduced they have sent signals to users that water is not a free good. Israel is the only economy in the MENA region that has introduced water prices in agriculture reflecting not only the cost of delivering the water to the farm (Aslosoroff 1996) but also reflecting a proportion of the value of water in the economy. There have also been studies of the impact of pricing water on the levels of productivity achieved by farmers following the water price increase (Arlosoroff 1996). The water crisis of 1986 which led to a re-allocation of water and a hike in the price of water to over US cents 20 per cubic metre had a significant impact on water productivity (See Figure 3.4).

Pricing water at its full cost of delivery would be a very powerful signal but such prices are for the moment not feasible anywhere in the MENA region at least for agricultural water. Putting a value on environmental water would also, in theory, be useful because it would draw to the attention of users competing for access to water that the water supplies are limited. The water assumed by farmers to be best allocated to them for irrigation has important, possibly essential, alternative use in maintaining minimal surface flows, safe groundwater levels or other intangible but nevertheless important environmental amenity. Like economic values for water, environmental values are non-intuitive. It is difficult to get legislators, officials or users to adopt the economic concept that water has

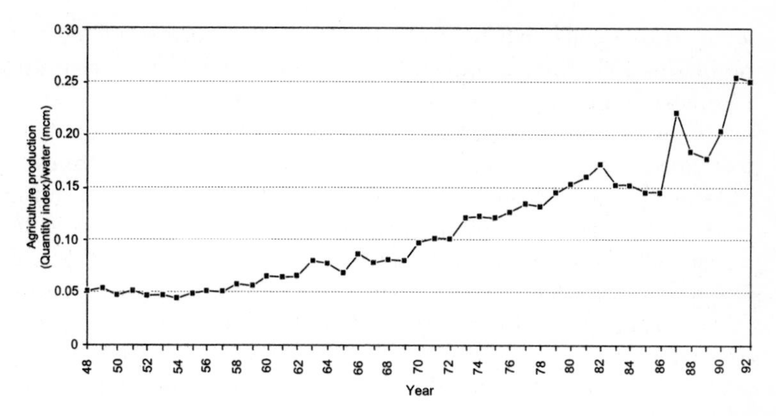

Figure 3.4 The agricultural sector in Israel—water use efficiency and returns to water showing the impact of increases in the price of agricultural water in the 1980s.

environmental value. Fortunately there are regional religious traditions that favour a sound approach to environmental resources.

Water pricing is considered to be a sound economic instrument by economists. Subsidies, on the other hand, are abhorred by the profession and considered to be distorting and even dangerous when they become deeply embedded to the point of permanence in a political economy. In agricultural economies water pricing is virtually absent world-wide. Subsidies, on the other hand, are the rule although they vary in form and application from economy to economy. Prices send constructive signals if they reflect the value of the commodity in question. Agricultural subsidies send the wrong signals. That they are particularly prominent in the globally influential economies of Europe and the United States is proof of their political importance. Removing them is proving to be a two or three decade political challenge. In addition to having little success on the discourse on agriculture the outsider professional economists who urge water pricing on the MENA economies have had little success at home. They have not been able to persuade politicians in their own countries to listen to them rather than to the farmers. They have not been able to lodge economic principles securely in the network of consensus on how water should be valued in the North.

Confusingly it is from the economists nurtured in these political economies that come the most strident criticism of subsidies. The World Bank epitomises the principled ideological position of the North's economists on other people's subsidies (World Bank 1993, Serageldin 1995). The economically inspired counsel would come better from a collective North, which in biblical terms had 'cast out the beams [of massive and distorting USDA and EC agricultural subsidies] from the eyes of legislators in these colossus economies, before the counsel was focused on the motes in the eyes of legislators [in relatively economically frail] developing country economies'.

Water pricing for the non-economist MENA insider is a serious issue. To be influenced by the price of water the user of water has to register a significant change in the price charged. Perry (1998) has shown that in Jordan the cost of the water service to farmers is about 2–3 US cents per cubic metre. 2–3 cents is much higher than the price currently being charged. If 2–3 cents per cubic metre were to be charged Perry calculated that price would only amount to five per cent of the gross value of production. He also calculated such a charge, while it would certainly stimulate some very tough politics, would have no impact on demand. The reason is that the average gross value added for water in Jordan is about one US$ per cubic metre of water. Put another way the value of water to the farmers in Jordan and elsewhere in the MENA region is a lot more than the operational cost of delivery. Perry suggests that in order to have an impact on demand the price would have to rise to a level which the farmer notices but which is at the same time politically unacceptable. He judges that the price would have to be between 20 per cent and 40 per cent of its value to be noticed. At these levels the price of water would have a significant impact on farm incomes as Israeli experience has shown (Figure 3.6). The impact on the poor would be particularly hard.

Another reason pricing water in agriculture is unfeasible is that the volumes in use are so huge. The number of users and the number of individual management units, that is agricultural fields, is so large in for example Egypt, that operationalising pricing is economically and politically prohibitive. Pricing requires the monitoring of water use with a precision acceptable to provider and user. There is no sign that affordable monitoring can be installed.

Subsidies and water

The sector which uses most water, agriculture, takes between 65 and 90 per cent of water in dry regions, is everywhere subsidised, especially in the affluent Northern economies. The subsidies to water are particularly prominent in the arid tracts of Northern economies (Reisner 1993:320). In addition it is hard to argue that water should not be subsidised in dry regions when in well favoured humid temperate regions agriculture is also hugely subsidised, by a raft of subsidies on inputs, farm incomes and on export terms of trade. Subsidies play a major role in economies in which agriculture generates less than five per cent of jobs. How can it be argued by staff of agencies funded by these economies that a tough economic rationale should apply to agriculture in poor economies where the sector provides 50 per cent of livelihoods? And a lot more jobs in downstream industries.

Tariffs and water: global and regional

Tariffs are also widely used economic instruments. The prices of commodities on the world market are substantially determined by these instruments. The very misleading subsidies and non-economic prices in the global trading system are confusing enough. For example the price of wheat on the world market has been falling for a century partly because of subsidies in North America and Europe. The price has fallen steadily in the past half century (Dyson 1994) with aberrant spikes in the mid-1970s and in the year after the establishment of the World Trade Organisation in 1995. Decision makers at the national and the farm level are faced with even more complex circumstances by the compounding impact of tariffs. Governments attempt to protect their own producers from the competition of commodities which are being sold to them on the world market at prices reflecting subsidies. Governments such as those of Egypt enjoy the low world price for staple grains but have to protect their own wheat producers lest the economy becomes over dependent on the unrealistically priced grain in the global trading system. The long term secure availability of strategic grain supplies is beyond the control of Egypt and other MENA economies in current circumstances where strategic water is accessed in the global grain trade via 'virtual water' (Allan 1996a).

One thing is certain and it will be beneficial for the MENA economies. The subsidies on grain and therefore on 'virtual water' will take a very long time to unpick in the muscular agricultural politics of the North.

Willingness to pay and capacity to afford. What is a reasonable proportion of an annual family GDP to devote to water purchasing?

Those advocating financial instruments tend to assert that a market mechanism signalling the value of water to users will lead to the moderation of demand to fit the available supplies of water (Pigram 1997). An underlying assumption is that water consumers have the capacity to pay a price reflecting the 'value' of water. But they will be willing buyers only if the prices are affordable and they regard the potential service, or the already experienced service, to be worth affording. (Littlefair 1998) In an economy such as that of Israel affording 100 cubic metres of water per year at one dollar per cubic metre is easy for domestic consumers. Israelis enjoyed average levels of GDP per head of US$ 20000 per year at the end of the twentieth century. The cost of water was a fraction of one per cent of the total annual individual income. In households in the neighbouring West Bank and Gaza the lower estimated 50 cubic metres per head per year of water use could not be afforded at one dollar per cubic metre. At that price water would cost about five per cent of an individual's average annual income of about 1000 US dollars per year.

Advocates of water pricing come from the economics community and its disciples who inhabit the international and bi-lateral agencies in the North and the economists in the South who have adopted the tenets of economic science. (Prakash 1994) A significant effort has been made by the economics community in the North to understand the contexts in which private sector markets can be found in poor communities in the South. These studies focus on the concept of willingness to pay for water recording the evidence of prices actually paid for water in water stressed localities, such as Yemen (Handley 1996 and 1999). However, the research in the MENA region is very limited indeed compared with the scope and depth of the studies conducted elsewhere (Diner et al. 1997, Goldblatt 1996, Murugan 1996, Rogerson 1996, Schur 1994, USAID 1996, Whittington et al. 1989, 1990, 1991).

The willingness to pay literature has been stimulated by the international agencies' wish to demonstrate that there is a potential flow of revenues that would fund, partially or totally, the initial investments necessary to provide piped water supplies as well as the resources necessary to support their sustainable operation. (Briscoe 1997) The economist inspired studies of the World Bank and other agencies have resulted in findings which tend to overestimate the amount individuals would actually be willing to pay for water whilst government agencies tend to underestimate (Rogerson 1996, Winpenny 1994 and 1997). For those who might be tempted to apply the concept of willingness to pay to the MENA region Rogerson's (1996) observation that in order to gain a safe understanding of feasible water prices rather than crudely determined willingness to pay much research is required at the household level. He judges that the failure of many projects results from inappropriate pricing of water. Project failure has been observed to result from a mix of unmet needs including reliability and timeliness of supply (USAID 1996). The nature of consumer demand tends to be misjudged by funding agencies reflecting the distance between planners and beneficiaries (Altaf et al. 1991).

The willingness to pay studies and the debate surrounding them is just one of many which encounter the very strongly argued idea that minimum levels of water supply is an 'entitlement' of all communities including those that are very economically challenged (Schur 1994). This equity inspired notion is just one of a number of beliefs that puts the politics into the political economy of domestic water supplies. Another is the importance of the environmental value of water (Kessler 1997, Winpenny 1994, 1996 and 1997).

The World Bank has suggested that an individual should normally pay no more than two per cent of disposable annual income for water. Five per cent is an unacceptable level (Serageldin 1995). There are many poor communities in urban areas in Africa, the Middle East and South Asia where families have to find 20 per cent of their annual disposable income to pay for water often of poor quality (Whittington et al. 1989, Ward 1998).

The explanation for the paucity of MENA willingness to pay studies is that over half the economies of the region are oil enriched. A very early priority of the governments of these economies was the delivery of safe domestic water and sewage services. The populations of the oil enriched economies are mainly small amounting to only about 30 per cent of the region's approximately 300 million total population. Egypt has no problem finding the five or so billion cubic metres annually for its domestic water supply systems. Meeting the substantial challenge of the rapidly expanding needs of the rising urban population has been major ongoing achievement. Addressing the freshwater and sewage needs of the Egypt's rural population, about 50 per cent of the total, is a challenge which will take a number of decades to address comprehensively. The ideas of willingness to pay could be of particular relevance to Egypt in circumstances where it needs to mobilise capital to finance the improvement and expansion of the reticulation system. Substantial international finance has been made available to address the urban water supply and disposal systems of the capital, Cairo (Ambric reports 1980–1985, Roberts et al. 1985 and 1995). The ambitious rural provision with which so much progress was made in the 1960s during the Nasser leadership needs rehabilitation and improvement.

A case which exemplifies the problems of meeting domestic freshwater needs in the arid and semi-arid MENA environment where populations are rising rapidly and available water is limited, is that of Yemen. In Yemen the community which is most severely challenged is the 400,000 people of Ta'iz in the mountainous south of the country. The urban population, provided for the most part with a piped water domestic supply system, endured in the 1990s a deteriorating water supply available at between 20 and 40 day intervals. Collecting fees for this unreliable supply was hard to justify and water bills tended to be unpaid. Domestic water was available from private vendors plying trucks. Drinking water was available from corner shop mini-distillation plants. The solution was extremely job creative. The charges for water of about three US dollars per cubic metre for tankered water and of about 30 US dollars per cubic metre for the five and ten litre containers served the purpose of signally the cost of delivery of water. Handley (1999) recorded the delivery costs of the different types of water in the Ta'iz area (Figure 3.6). The data show how a community responds with economic and market institutions to fill the gap left by institutional failure stemming from the community's and governmental incapacity to regulate access to the region's limited water resources.

The Ta'iz case is replete with corroboration of the power of a strong society vis-à-vis a weak state and the negative consequences of attempts by a central agency, backed by international agency funding and expertise, to address the technical, socio-political and

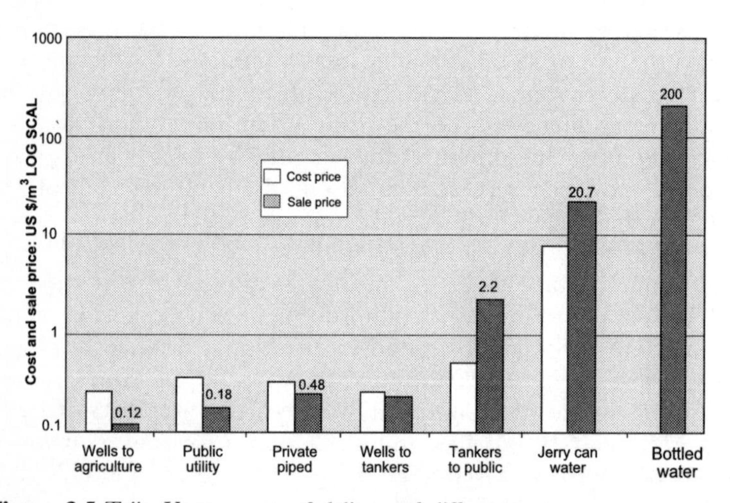

Figure 3.5 Ta'iz, Yemen: costs of delivery of different waters

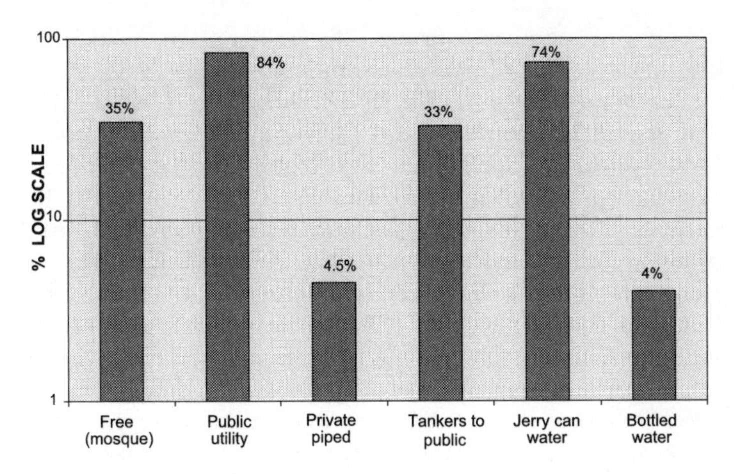

Figure 3.6 Ta'iz Yemen: prices of different waters

institutional problems (Ward 1998, Handley 1999). On the subject of realistic pricing which users are prepared to pay, albeit without other options, the Ta'iz case is informative. The experience in Ta'iz is also very informing concerning the capacity of local private institutions to meet the basic needs of urban populations as well as of needs substantially beyond the basic.

Valuing environmental water: precaution and sustainability

> '… the principles of precaution and sustainability might be better served by using economics to determine efficient ways of delivering politically and legally defined standards of service and quality than for defining the principles in the first place'.
>
> (Morris 1996:201)

The 1980s was a decade when the environmental value of water gained a prominent place on the agendas of those responsible for water management and policy making especially in industrialised economies (Meir 1994, Arlosoroff 1996, Tarlock 1997, World Bank 1993). Water pricing was amongst the instruments advocated to steer demand in the agricultural sector to reduce that demand partly in order to give priority to environmental priorities such as maintaining minimum flows in rivers and halting the reduction in wet lands. The justification of these moves was the need to protect ecological capital.

The world-wide campaign to raise the profile of the value of the environment was a political phenomenon of the 1970s and the 1980s. It culminated in the Rio UNCED Summit (United Nations Conference on Environment and Development) in 1992 and the preparatory meeting in Dublin. The Dublin meeting enunciated what have become known as the Dublin principles which have been prominent in the Northern discourse stream in the parallel international water discourses of the late 1990s. Biswas argues that the principles have been severely attenuated in the Southern discourse in the years since 1992 (Biswas 2000). The Dublin principles were inspired by notions of economic and environmental sustainability and equity. The principles were first that water resources should be managed holistically because water occurred in natural river basin systems, secondly that water was an economic resource, thirdly that women played an essential role in the effective management of water and fourthly that water was an environmental resource.

The tension implicit in the simultaneous consideration by a policy maker of developmental and environmental priorities is often uncomfortable. In a period when the analytical emphasis has shifted from achieving indicators of socio-economic improvement to protecting the environment and species biodiversity the level of political tension has escalated.

Some comfort comes from economists who have observed that political economies deal with the conflicting development and environmental priorities in different ways according to the length of time an environmental resource has been subject to 'development' for socio-economic purposes. Those involved in the first stages of combining an environmental capital resource with other factors of production in a 'developmental' enterprise are likely to degrade the resource. When the economy has achieved diversity and strength its policy makers can more easily listen to the advocates of the 'green' message. In due course they may even be able to devote resources from the present and future economy, which has been enriched by the extraction of wealth from national environmental capital in the past, to repair the damage caused over decades by the process of economic development.

The term, the environmental Kuznets curve—EKcurve—helpfully conceptualises the cycle. Kuznets (1973) had highlighted other changes that took place in an economy with respect to the distribution of income over decades of socio-economic development. He observed that measures of income distribution worsened in the first part of the cycle. After ceasing to worsen they then tended to return to the early levels as the politics of the political economy shifted and the interests of the poor could be articulated and measures taken to address them. The idea that equity with respect to income distribution could influence policy in a later rather than an early phase of the developmental cycle appealed to a number of those dealing with the political economy of environment and development (Grossman et al. 1994). The notion that the environment could be repaired by the fruits of socio-economic development gained currency in the early 1990s especially in relation to trade (Bhagwati 1993, Cairncross 1992).

The utility of the concept of the EKcurve to politicians responsible for water policy worried greatly those in the green community eager to have a green beginning, middle as well as end to the development narrative of the environmental resource of water (Ferguson et al. 1996). Hamilton and O'Connor (1994) have shown that the loss of natural resources has already over-ridden the financial gains in countries which have to rely on the export of non-fuel primary commodities. Critics of the EKcurve argument argue that the costs of environmental degradation should not be paid at any stage of the socio-economic development process (Brechin et al. 1997, Martinez-Alier 1995, Shafik 1995). The evolution of the precautionary discourse in the South is in places elemental. In South Asia the anti-dam campaigns have introduced unprecedented evidence of the capability of campaigners to construct alternative ideas in contention with those that inspire governments, engineers and other interests. The campaigners have gained such influence in civil society that international agencies, such as the World Bank, have withdrawn from major projects and notional and provincial governments have slowed down construction (Roy 1999). The factors which synergised to bring about the movements led by Medha Patkar and embraced by Arundhati Roy (Roy 1999) have no parallel in the MENA region. The hydraulic mission is still alive and flourishing in the region. The big players, Egypt, Turkey and Iran, are all engaged in major hydraulic projects. Only the Turkish government is facing opprobrium. But the criticism is from outsiders and orchestrated from London and other Northern centres. Pressure is exerted by the Northern NGOS on Northern funding bodies such as the UK Department of Trade and Industry, not directly on the Turkish Government (Crawford et al. 2000).

Another important conceptual contribution came from Karshenas (1994). He emphasised the political logic of water policy makers being able to reach for environmentally inspired policy options during a phase of economic diversity and strength built on the foundations of a period when no such options were available. The power of the concept to identify and explain the contrasting circumstances of the Israel, West Bank and Gaza political economies and their environmental policy options is illustrated in Figure 3.10.

In the last half of the twentieth century the MENA region experienced a case which seems to confirm the Karshenas theory that politically and economically tough water regulation and pricing policies can be introduced when an economy is strong and diverse – see Figures 3.8 and 3.9 (Allan and Karshenas 1996b). The case of Israel, which moved from being an agriculturally oriented economy to that of a highly industrialised economy in only three

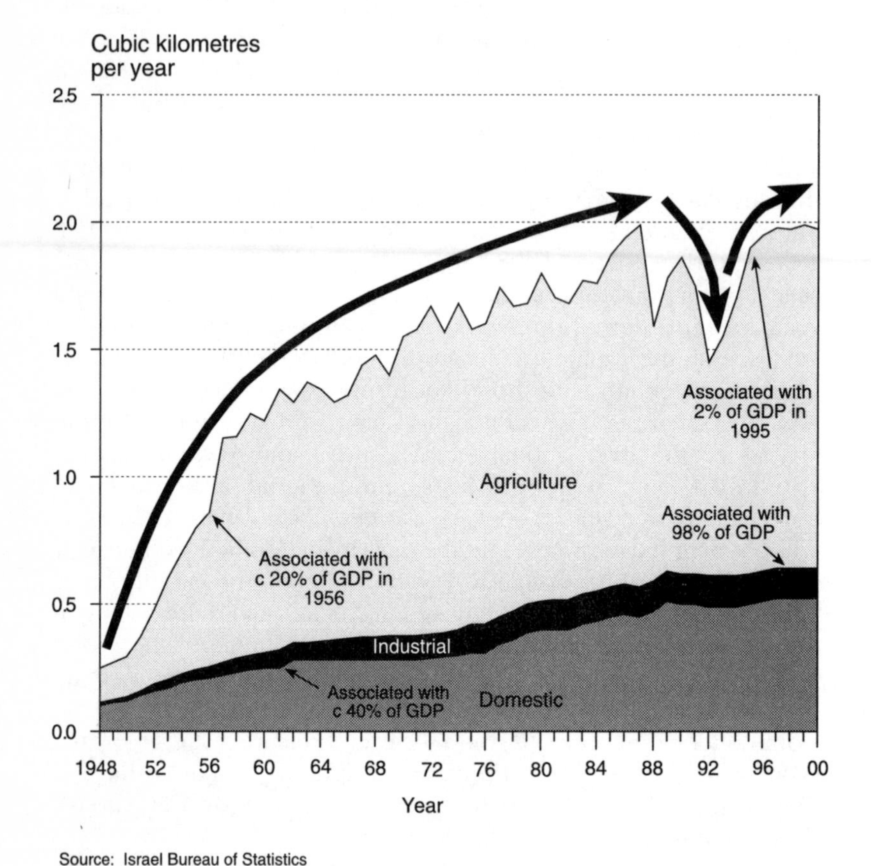

Source: Israel Bureau of Statistics

Figure 3.7 The water sector and its changing role in the political economy of Israel

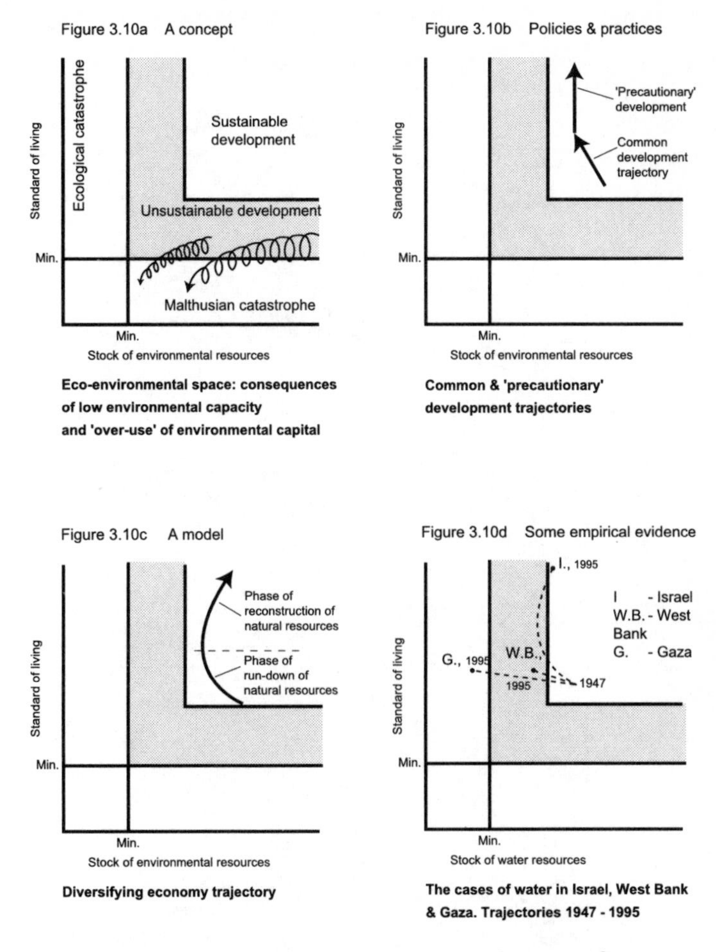

Figure 3.8 Karshenas curve diagrams—Israel, West Bank and Gaza

decades between the mid-1950s and the mid-1980s, is useful in illustrating the phenomenon. Figure 3.7 shows the difference in the relevance of heavy water using agricultural sector at two points in Israeli economic history. In the mid-1950s the agricultural sector was a major force in the national economy. The economic and social significance of agricultural water use underpinned national identity, economic policy and international relations. By the mid-1980s agriculture was a minor element—less than three per cent—of the national economy. It was even less by the year 2000. Despite the statistics the actual role of agricultural water in the political economy remained politically influential. The notion of water being important

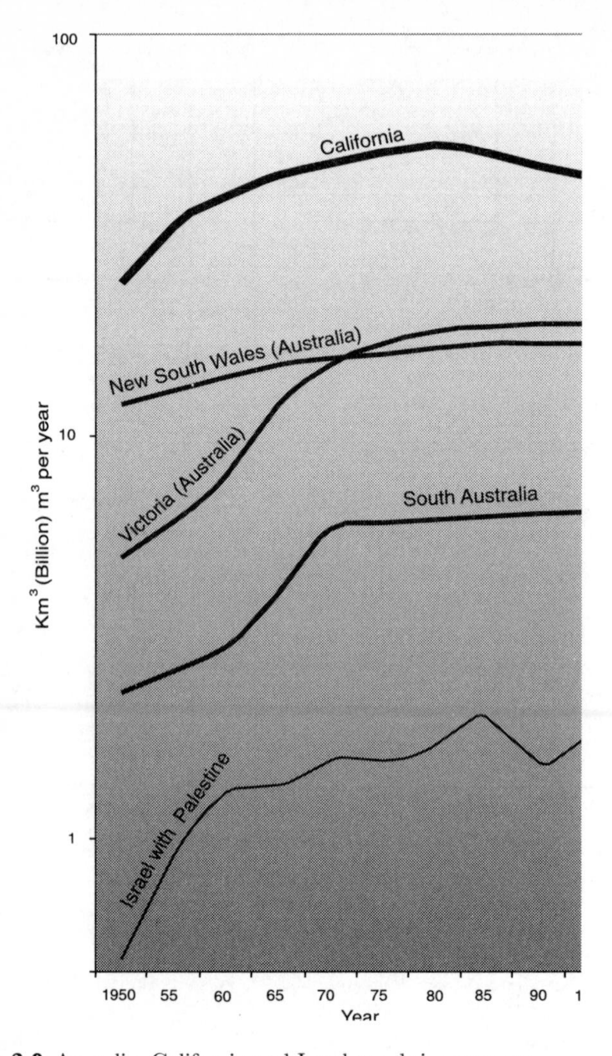

Figure 3.9 Australia, California and Israel trends in water use

persists as one of the unshiftable foundation symbols of nation building so essential and legitimising in the decades before and the decade after the creation of the state in 1947.

In Israel it became politically feasible to regulate demand by quotas by the 1980s and to re-allocate by increases in the price of irrigation water in the late 1980s and 1990s. Such measures were possible because the agricultural sector was embedded in an economy, which had the capacity to subsidise agriculture, including through a 60 per cent subsidy on water. Domestic users of water could easily afford the real price of water delivery. In poor economies such as the West Bank and Gaza neither public sector nor private sector users could afford the real costs of water delivery. Neither agricultural nor domestic users were economically competent to afford the delivery of economically priced water. Nor could they mobilise resources to respond to environmental priorities determined by hydrological principles, namely avoiding overdrawing the West Bank groundwater resources beyond sustainable levels of use. Nor could they manage water so that the quality of the resource was not impaired through for example the over-use of coastal aquifers. Strong and diverse economies, such as California, Australia and Israel, can deploy economic instruments; weak political economies do not have the option (Allan and Karshenas 1996, Migdal 1988).

The Israeli case seems to point to economic transformation being a necessary preliminary to otherwise impossibly stressful regulation of the economy. Economic diversification made possible the political initiative to reform the allocation and management of water in agriculture in the 1980s. It will be emphasised in later chapters that it was the development of national political institutions, however, throughout the period since the 1940s that enabled the multiplying diversification of the Israeli economy as a whole. This multiplied diversification enabled water policy reform.

If the environmental value of water is to be amongst the factors taken into account by water policy makers it would be helpful if its economic value could be established. Environmental values could then be accounted when trading off benefits deriving from the productive use of water for power generation and crop production, versus environmental uses such as reducing sedimentation, maintaining amenity and ecological diversity (Winpenny 1996:201). But putting values on untraded, especially non-user benefits is conceptually difficult (Morris 1996:231); operationalising a policy making system recognising such benefits is impossible except via campaigns such as successfully mounted by the green movement. The methods devised by economists to place values on the social and environmental impacts of water resource development are universally imprecise. This realisation has led one economist to conclude that the principles of precaution and sustainability 'might be better served by using economics to determine efficient ways of delivering politically and legally defined standards of service and quality than for defining these in the first place' (Morris 1996:201). In other words economic concepts, and the instruments from which they derive, should be confined to accelerating appropriate economic and environmental outcomes. They are unlikely to make a significant input to the policy discourse.

Some mega-projects: economic and environmental trade-offs

In the MENA region there are some mega-environmental management controversies as a result of past, current and proposed engineering interventions. These projects highlight the trade-offs which can occur when development priorities take precedence over environmental ones.

The High Aswan Dam

> 'Politically [the dam] had the advantage of being gigantic and daring, thrusting Egypt into the vanguard of modern hydraulic engineering. Moreover, during its construction and after its completion, it would be highly visible and fittingly monumental.' (Waterbury 1979:101)

Aswan, close to the southern border of Egypt with the Sudan has been the preferred site for minor as well as major structures since the turn of the twentieth century. The last, and most ambitious scheme, the High Aswan Dam—HAD—was finally given substance in the 1950s by the revolutionary government led by President Nasser. The dam was commissioned in 1970. The structure imposed a total control by creating a reservoir, Lake Nasser-Nubia, which stored three times the annual Egyptian share of the average flow of the Nile. The Egyptian share of 75 per cent of an assumed flow of 84 cubic kilometres per year, that is 55.5 cubic kilometres per year, was agreed with the Sudan in 1959 (Nile Waters Agreement 1960). The other riparians did not endorse this agreement.

The positive economic impact on the economy of Egypt has been profound, in the form of increased production and productivity from the irrigation sector. The irregular flow regime was regulated to provide a predictable annual flow. New tracts were claimed for agriculture, albeit not without trouble (Waterbury 1979, Hunting Technical Services 1979) and more important the 15 per cent of non-perennial irrigated area was converted to perennial irrigation. The comprehensive control of the Egyptian Nile has enabled an extension in the irrigated area and secure water throughout the system even if this has been achieved by unnecessarily high levels of delivery at the farm level (Radwan 1998).

The benefits of the HAD have been little mentioned by outsider analysts of the project. Rising water tables, water-logging, soil salination, increased channel scour, coastal erosion and the diminishment of the Mediterranean fishing have all attracted generally inadequately or partially researched criticism where they refer to the Nile (McCully 1996:275). The critical evaluations are all made by non-Egyptian scientists generally with ideologically green axes to grind. The review by Abu Zeid (1990) provides a concise version of the Egyptian view of the project. But even this useful study does not provide a measure of the economic benefits against which to compare the environmental disbenefits. To Egyptians the economic benefits of the HAD are as unarguable as the entitlement of Egypt to the Nile waters themselves. This author believes that the economic benefits to the Egyptian economy in water securitisation, production and productivity gains and in job creation far outweigh the environmental disbenefits.

The benefits include the escape from two phases of low flow. The impacts of the 1970s low Nile flows were avoided; the even more serious low flow impacts of the 1980–1987 period were also almost completely avoided. Minor cuts were imposed in 1986 and major ones would have been imposed at the end of 1987 if the Nile had not delivered its second highest flood on record. The economic consequences of the Egyptian agricultural sector absorbing the 1970s and 1980s low flows would have been devastating. A very cursory examination of the economic benefits would total in late 1990s prices at least one billion US$ per year and a drought protection of at least five billion US$ in the 1970s and the 1980s. Flood protection afforded in 1987–1988 would amount to between one and two billion US$. In addition Aswan hydro-power was crucial to the Egyptian economy throughout the 1970s and the 1980s supporting livelihoods in all sectors of the economy.

The message of the success of the High Dam (Allan 1981) has been difficult to get into currency. Non-Egyptians find it difficult to cope with the might of the symbolic significance of the Nile, and the even greater symbolic significance of securing its flow. Regulating the flow and ensuring that it is not diminished by upstream riparians have been of symbolic importance for millennia. Egyptian leaders can mobilise national unity and enhance their public standing with predictable ease by recognising the importance of the Nile; even recognising its potential is sufficient to bring a political reward.

President Nasser was rewarded with adulation of crowds chanting, 'Nasser, Nasser, we come to salute you: after the Dam our land will be paradise' (Waterbury 1979:116). Waterbury also notes that 'Politically [the dam] had the advantage of being gigantic and daring, thrusting Egypt into the vanguard of modern hydraulic engineering. Moreover, during its construction and after its completion, it would be highly visible and fittingly monumental' (Waterbury 1979:101).

Three decades later President Mubarak mobilised the same predictable Nile waters to gain national support for the development of the New Valley in the south of the country. The New Valley is an old channel of the Nile to the north-east of Aswan. In the spring months of 1997 President Mubarak launched the project to use water from an overfull Lake Nasser-Nubia to provide supplies for irrigate farming and an ambitious suite of urban industrial projects in the New Valley. The power of water to attract investment is evident throughout the region. Shaikh Zaid of Abu Dhabi is the leading advocate of using water to multiply economic activity and green the desert. His own country has been transformed. He has given generously to projects in Arabia, for the Marib Dam in the Yemen, for example. His latest major involvement in water resource management is in the New Valley Project. The canal linking the pumping station near the reservoir to the settlements in the New Valley is named after him to mark his generosity in making the early phase of the project possible. Egyptian economists have found in favour of the project at the same time as unease is expressed privately by officials central to water planning in Egypt. The international community is united in scepticism about the project on grounds that the excess water in the Aswan reservoir is not secure and especially because earlier attempts to develop the oases in the New Valley have failed in the 1960s phase of Egyptian land reclamation. Any hesitation amongst the informed is swept along by enthusiasm of the Egyptian population more generally for a Nile related project.

In a chapter about the economics of water and the arguments of the professional economist, the New Valley Project is a fine example of the way that economic principles are used selectively in the political domain. In Egypt's political economy the main player in the New Valley Project is the politically influential presidential inner circle of Egypt. Its members are deeply aware of the political ecological stakes, both national and international, although they might not be familiar with the concept of political ecology. Their agenda is to strengthen the negotiating position of Egypt vis-à-vis its upstream riparians. By allocating Nile water, or rather over-committing it, to the New Valley Project any future assertion by an upstream riparian that Egypt has spare water will be confounded. The advocates of the New Valley Project reach meanwhile for economic theory to justify the re-allocation on water use efficiency grounds. The five cubic kilometres of water needed annually to water the proposed new agriculture and industry in the New Valley will be provided from the reduction of the production of sugar and rice in the old lands. The reduction in the raising of sugar in southern Egypt is certainly economically rational if politically stressful. The reduction of the rice area will be more difficult to achieve because rice is mainly grown in the northern part of the Nile delta with re-used and low quality water. The extravagant application of low quality water in the coastal delta is sensible because the infiltration of the excess irrigation water helps to keep the toxic sea water intrusion wedge at bay.

The New Valley Project is a political project. But this is nothing new. The hydraulic works which were a feature of the New Deal in 1930s USA, which were being actively pursued via political processes in the United States until the early 1980s, were also political projects. Egypt is still very much imbued with a hydraulic mission and it has the necessary economic muscle to realise its political goals in the face of the disapproval of outsider Northern professionals and international agencies. The New Valley Project may take its place alongside many other land reclamation projects which appeared to fail at first but then became integral parts of the Egyptian economy (Allan 1983a). Or it may fail like many others that proved in the 1960s that land reclamation is one of the most challenging of social and economic challenges (Wilcox and Craig 1913, Allan 1983a). It is less likely to fail for the reason that the lesson was learned in the 1960s that land cannot be reclaimed without substantial investment in both the engineering, agronomic and the social infrastructures. The achievements in Egyptian land reclamation in the 1980s and the 1990s bear witness to the learning. Very important is the diversity of approach to investment which is designed to protect the Egyptian economy from the economic and social costs of failure. Most important is the increasing strength of the Egyptian economy which will constantly expand the options available to government.

The Sudd wetland

Another project on the Nile has had a very different history. The Jonglei Canal project, first proposed in the early twentieth century to increase the flow of the Bahr al Gazal stretch of the White Nile system (Collins 1990), was eventually embarked upon as a jointly financed project by the Sudan and Egypt in 1978. It was abandoned in 1983 when almost complete as a result of the armed intervention of southern Sudanese groups. They opposed the Government of the Sudan in Khartoum and its partner in the project, the Egyptian Government, for what they regarded as an invasion of their regional sovereignty. They were mainly exercised by the anticipated environmental impact of the drainage of the Sudd wetland (Southern Development Investigation Team 1955, Howell, Lock and Cobb 1988, Howell and Lock 1995:260) which would have affected traditional livestock rearing livelihoods (Garang 1981, Howell 1988).

A past or future Jonglei project would increase the flow of the Nile through the Sudd swamp in a first phase by four billion cubic metres per year. This phase of the project was aborted in 1983. A second phase could increase the flow by a similar volume. The Jonglei project is still regarded by northern interests, that is by governments in Egypt and northern Sudan, as a major potential water prize to be gained. The prize requires joint action envisaged by the 1959 Nile Waters Agreement via shared investments to achieve increased volumes of Nile water for the irrigation schemes in northern Sudan and in Egypt. To the leaders of the sixty million people of Egypt and the approaching 20 million of northern Sudan the Sudd symbolises a wasted water resource to which an engineering solution could bring a rich water return. The Sudd wetland figures as a different resource and a very different symbol for the peoples of southern Sudan. It proved to be a symbol so potent that it precipitated armed action which terminated the project (Collins 1990).

The symbolic significance of the Sudd wetland has also changed for the international community. For a century dominated by the ambitions of the hydraulic mission the Sudd had been viewed as a potential source of irrigation water by colonial interests—the British administrations in Egypt—until 1922—and in Sudan until its independence in the late 1950s. By the mid-1980s the international community had become attuned to the green principles of precaution, sustainability and the conservation of biodiversity. The second biggest wetland in Africa was no longer a candidate for severe reduction in area and other hydrological modification. The shift in nomenclature, from 'Sudd swamp' to 'Sudd wetland' in the international commentary on the region signifies the shift in perception of the international and bi-lateral agencies.

There are no methods with sufficient precision to make clear the superior benefits of economic returns, derivable from 40 per cent increase in the flow of the Bahr al Gazal in two ten per cent phases compared with maintaining the environmental *status quo*. Those who prefer to continue to have the potential of that *status quo* prevailed in 1983. Despite the protracted military struggle with Khartoum forces local interests seemed likely to be able to achieve the goal that the people of the south of the Sudan will be able to pursue traditional livelihoods, and opt for economic developments of their choosing.

Economic analytical tools and economic instruments in changing political economies: social and political factors

> 'water transfers are more like diplomatic negotiations than market transactions.'
>
> Sax 1994:13

> Put simply scientific truths, economic and environmental, find a place in determining the way users and policy makers perceive and allocate water according to the political price that would have to be paid to give operational substance to them. More [political] stress less [operational] thrift.

The next chapter will examine more closely the role of politics in the political economy of water in the MENA region. The discussion in this chapter of economic principles at the macro-level, that is at the level of the national economy, and at the field level of the user at the level of the farm has shown that 'economic explanation' plays an ambiguous role. It has also shown that 'economic instruments' though in theory available are not recognised to be politically acceptable. The major explanatory contribution of economics that of the role of virtual water in maintaining the social, economic and political stability of the MENA region cannot be included in the discourses that lead to policy formation. The notion exposes the perceived insecurity of a national economy. Economic analysis that shows that spectacular economic gains would be achieved by the sectoral re-allocation of water is almost equally unacceptable in MENA water policy making processes.

The achievement of allocative efficiency will certainly involve difficult politics and there are few signs yet that recognition of the necessity of such measures outweighs the awareness of the political stress of introducing them. Economically efficient water use will involve a shift in the assumptions underpinning policy and these will involve some or all of the following activities. Those which could be brought into policy and their prioritisation will depend on the pace at which political acceptance of them can be achieved:

The adoption of principles which will improve water security and the economic use of water

- the adoption of principles of allocative efficiency—gaining sound returns to scarce water
- the adoption of assumptions of equitable rights to a minimum volume of water per person for domestic use
- the development of awareness of the economic value of water and of water markets which will track that economic value in prices for water at an international level

The adoption of practices based on economic principles, which will achieve improved returns to water

- the relinquishment of supply management policies and practices for water
- the adoption of demand management policies and practices for water
- the extension of the adoption of water treatment of urban waste water
- the extension of waste water re-use in agriculture
- the extension of water charging for water for domestic use
- the initiation of water charging for water use in agriculture and the increase of such charges to reflect the cost of delivery

Whether a political economy adopts economic instruments, which strengthen and diversify the economy, is dependent on the social adaptive capacity of a political economy economy (Ohlsson and Turton 1999). Communities believe that water should continue to be made available on terms which are advantageous to users. Economic instruments are inspired by very different notions. Economic instruments are concerned not with continuity; they bring dislocation. They are re-allocative and stressful. Compared with engineering instruments which tend to increase the water available and extend and improve availability, institutional instruments are politically unwelcome both to users and to politicians who are being urged to deploy them. In order to introduce changes which bring political stress it is necessary to afford the institutional capacity to enable reform (North 1990).

The political leaderships of poor water deficit economies are very sensitive to the belief systems of the majority water users in agriculture. There exists in these political economies what political scientists call a 'sanctioned discourse' (Tripp 1997). The congruence of the water belief systems of the majority users of water and those of their legislators—elected or not—has been achieved by the constant discourse between water users and legislators. There is no space in their discourse for consideration of economic instruments to re-allocate water to benefit the national economy by using water in higher value productive activities. Legislators recognise that their continuation in power depends one the congruence of their water policy with that of the belief systems of the majority water users.

Conclusion

> 'One does not learn from experiences which one refuses to examine.'
>
> Dwight Macdonald

Economic instruments, in the form envisaged by economists from the science community and international agencies, are of potential rather than immediate significance. Meanwhile markets for water devised by the local entrepreneur to meet the local domestic and drinking needs of the water short, whether rich or poor, have sprung up in a number of locations. The economist sees the spontaneous resort to market mechanisms as confirmation of the power of economic principles. For outsider economists it is the entrenched public institutions which prevent the development of market systems that would enhance water services and even the distribution of irrigation water.

The preferred comprehensive piped water schemes, on the other hand, can only be introduced sustainably into strong and diverse economies with effective social adaptive capacity. The investment resources needed to construct them, the management institutions needed to regulate them and deliver a flow of funds to operate and then sustain them can only be mobilised in economies that have the economic and institutional competence associated with comprehensive social and economic development.

Even where these favourable circumstances exist the instruments have not been introduced in a full-blooded fashion to achieve total cost recovery in all sectors. Domestic and industrial/service sector users can everywhere accommodate easily in the long run to water charges which cover operating and investments in the future service. By contrast, in no agricultural sector has a full cost recovery system for water been installed. Not even in strong, diverse, industrial economies has it been possible to develop systems which recover more than 40 per cent of the economic value of the water used in irrigated agriculture (Arlosoroff 1996, Cheret 1998).

Politics enable innovation. Politics also slow down innovation. Politics especially impact the creation of innovative water markets. Water markets can be created in extremely poor communities where the prices paid reflect the priority which consumers have to give to their drinking and basic water needs. That consumers are prepared to pay high prices for drinking and basic needs water has been noted by the international agency economists (Whittington et al. 1989 and 1990). This professional awareness has converged with the agencies' preference for private sector solutions. There has been an opposing convergence. International and local NGOs converge with national and local interests in the MENA region to oppose the pricing of water, water markets and privatisation. During the consultative process preparatory to the World Water Forum of March 2000 in The Hague the MENA regional meeting exposed the two contending perspectives—the outsider Northern view and the insider MENA view (GWP—MENA 1999). The resulting report conveyed the insider anti-market and anti-privatisation perspectives of the MENA regional professionals and not the outsider preferences of the Northern professionals from the international agencies who finally edited the report.

Meanwhile the course of a water market never runs smooth. The water markets in poor economies in which the urban poor participate are evidently unsustainable when judged by any standard of equity. In strong economies those which have temporarily sprung into existence to address a temporary water shortage, for example, at the height of the 1992 drought in prosperous California, were only possible because there was an enabling state authority at hand. The state authority reassured the very uncertain urban water authorities and the very anxious farmer with a water right that they had mutual interests that would be gained by buying and selling water. It is argued that there would have been no deal without the participation of the California state authority (Dellapenna 1995).

CHAPTER 4

Water is not an economic resource: the political economy of some MENA waters

'Economics are an illusion; politics are real.'
Mark Reisner 1993

Contexts and perceptions

Not all waters are equal: some are more political than others

'Water flows up-hill to money and power.'
Mark Reisner 1993

Water becomes political if competing users or potential users perceive that access has become, or will become, restricted or even denied. The user may be the individual subsistence farmer competing for seasonal surface or groundwater; or a community of urban users having to compete with irrigators for their domestic water; or a major, newly developing economic sector, such as tourism, competing for water with another sector also expanding its water demand for example for crop production. At the level of the nation, riparian states are particularly prone to forms of conflictual hydropolitics as are riparian entities within nations.

The MENA region has examples of all these circumstances. Individual irrigators in all the irrigation schemes of the region, if taking water from the head of the irrigation canal, have secure access to water. Individual tail-enders endure unreliable water supply as shown by Radwan (1996) in delta Egypt. Adjustments have to be made by individuals, often with serious economic and social consequences for the disadvantaged unless different technology enables access to alternative water, either from the ground or from drainage water (Radwan 1996). At this lowest scale of interaction between individuals conflict over water is the most likely to lead to violence. At the level of the nation state armed conflict has involved artillery, a few aircraft sorties and and very small armed land based forces. Even this low level armed conflict was absent from the last three decades of the twentieth century

Tail-ender riparians at the national scale are particularly sensitive to the security of their water supplies from a shared river. Egypt, Syria and Iraq are major tail-ender water economies which periodically signal their concerns over their respective rivers.

As the scale of hydropolitical interaction increases, from the individual to the community and to the sectoral levels competition for water is just as real but the chances of violent conflict are reduced. The reasons would appear to be socio-political as well as economic. In the Middle East and North Africa state institutions have responsibility for almost all domestic water supplies as well as for a high proportion of water for the productive sectors of the economy, agriculture, industry and services. Water ministries and municipal authorities, manage and deliver water to domestic users more or less effectively in oil rich economies—Saudi Arabia and the Gulf economies, Iraq, Iran, Libya and Algeria. Public authorities also deliver water to users in non-oil enriched economies such as Yemen, Jordan, Israel and Palestine. For the last three, measures were initiated in the late 1990s to transfer some functions to the private sector. In Israel the move was initiated by the Israeli Government. In Jordan and Palestine [Gaza] the transition was a conditional element in a package of assistance provided by the World Bank.

When public systems fail recent history has shown that people do not confront the non-providing institutions violently. They behave in the way predicted by economists. They seek substitutes, in this case alternative, unregulated, private sector supplies. Such responses are well documented in Yemen, in Ta'iz (Handley 1996 and 1999) and in Jordan (Schiffler 1995). In 1995 in Ta'iz, domestic users received mains water only once every 20 days. For the 15 to 20 days per month when there was no water families had to purchase water from tankers which plied the suburbs. Low quality water for washing and domestic use generally was for sale at US$3 per cubic metre (Handley 1996 and 1999). Drinking water was purchased from corner shop desalination plants in five and ten litre containers at ten or more times the price of bulk deliveries. In Amman the price of low quality tankered water was the same as in Ta'iz (Haddedin 1995).

Conflict has been recorded in the Ta'iz area. Conflict did not arise in the under-provided city. It was farmers, the users of the same aquifer as the 400,000 domestic water users in Ta'iz who were forced to compete for water. In 1996 farmers confronted the drillers proposing to drill new wells into the aquifer for water for Ta'iz. They succeeded in facing down the military sent to support the drillers.

But such confrontation has been rare and was a consequence of the extreme circumstances in Ta'iz. It was not the urban consumers of water who confronted the farmers. Their strategy was to seek a solution with local entrepreneur water vendors. It was the state authority which mobilised the drilling rigs to drill for water for the city. The state also sent military support when the drillers were challenged in the upper part of Wadi al Haima. In the Ta'iz area the universal MENA problem of rising population was the main forcing factor. But in Ta'iz the water resources were particularly restricted and the geological circumstances were unfavourable. Above all the population lives above 2000 metres high. For water users in the Ta'iz region the remedial options available were and remain very restricted indeed. Ta'iz cannot desalinate bulk water. Ta'iz consumers have adapted very quickly to the logical option of distributed desalination. They were already desalinating the small quantities of water needed for drinking by the mid-1990s.

Yemen and Jordan are both elevation challenged. Most of the population, over 80 per cent, lives in high tracts, at over 1000 metres. It is expensive to lift water to these locations. When major urban centres exhaust their local aquifers as Amman and Ta'iz have and as Sana'a, also in the Yemen, is well on the way to doing, the supply options are always uncomfortably expensive.

These extreme Yemeni circumstances apart inter-sectoral competition for water has generated a deal of trenchant debate but no recent armed confrontation in the MENA region. The reasons are that state institutions have generally been able to provide enough domestic water, or at least minimal supplies, augmentable by private small-scale water markets. Provision has largely kept pace with population growth. Even the installation of sewage disposal systems which everywhere lag behind the provision of domestic water have been installed in challenging mega cities such as Cairo and Istanbul. Waste water treatment, which lags even more behind provision and disposal, is on the agenda. Amongst economists and engineers there is talk of sole first use water by domestic users in extremely water short economies such as those of Israel, Palestine and Jordan (Arlosoroff, 1996, Shuval 1996, Elmusa 1997). Israel already derives over 20 per cent of its irrigation water from urban waste water treatment plants. It has even been argued that urban waste water re-use will enable farmers in the Palestinian Authority to increase the area irrigated after the year 2000 as water treatment plants are installed beside major settlements in the West Bank.

At the scale of the whole national economy facing a major and growing water deficit the remedy to the shortage has been unspectacular in terms of political conflict but has rivalled the New Testament miracle of the feeding of the 5000 in terms of impact and effectiveness. Where major water gaps have occurred for national economies as a whole it has been possible to fill the national water gap with 'virtual water', water embedded in for example imported wheat. This remedy has been as important to minor economies such as Yemen and Jordan as to major regional economies such as Egypt, Iran and Iraq (See FAO 1997 *Production and trade statistics* also on http://www.fao.org).

Armed conflict over water has been absent from regional international relations, despite the trenchancy of many politicians in asserting their national rights to water. Not since the early 1960s have shots been exchanged as a result of water development initiatives. Both Israel and Syria intervened to frustrate the development of water in the upper Jordan tributaries in the 1962–1965 period (Wolf 1995a and 1996b, Medzini 2000b). But since 1967 there has been no sign that disputed waters were sufficiently important issues over which to take military action. The major June 1967 conflict itself was not precipitated by a water conflict (Haddedin 1995, Wolf 1995a and 1996b, Medzini 2000).

The most political waters are those perceived to be in current or imminent short supply. But even the highly political waters of the Jordan catchment serving some of the most water stressed communities of the region have not precipitated nor escalated to an armed conflict. The waters of the Tigris-Euphrates have generated much discussion, but the water resources of Turkey and Iraq are not at a crisis point (See Figures 2.4–2.7). Of the Tigris-Euphrates riparians Syria is vulnerable to the consequences of outstripping of water supply through population growth. But Syria is water secure in the lowlands of the Euphrates and needs to deploy Euphrates water in other parts of the country, a process which it is achieving. The peoples of the Nile Valley, especially the 60 millions in Egypt, have found easily accessible, economically advantageous and politically benign water to meet their growing water deficit in global systems.

The capacity of global systems to provide water to assuage the economic and political stress of even the most political of the MENA region's waters, those of the Jordan and the highlands of Yemen, has been spectacularly successful. The volumes of 'virtual water' involved in meeting the needs of the Jordan basin countries were minor in terms of the total regional 'virtual water' imports. For the major 'virtual water' importing economies such as Egypt, Saudi Arabia and Algeria, global water has also been sufficient to meet their needs.

Anxiety, awareness and blind spots

> 'Simple truths: vital lies.' '[Governments and peoples] … can protect themselves from painful realisations by diminishing awareness. … This trade-off between anxiety and awareness creates blind-spots of self-deception.'
>
> (after Goleman 1997)

> Politically acceptable traditional approaches to the allocation and management of water and to family fertility are difficult to change, or at least to change fast enough for the water gap to be addressed with indigenous resources.

For water to play a political role it must be *evident*. Also users must be *aware* of aspects of its availability which signal uncertainty, insecurity and the possible need to compete for access. Individual users are more likely to become exercised about water insecurity because their systems of resource use have neither the technologies nor the economic competence to locate and develop alternatives. Hence the violence between individual farmers that can occur on the canals and distributaries of South Asia and Africa as well as in the MENA region.

The purpose of the following discussion is to demonstrate that the MENA peoples and governments have adopted a 'blind-spot' approach to the nexus of the water/demography/economic development challenge. They protect themselves from what could be the painful consequences of their water predicament and its unpleasant policy imperatives by consigning the issue to a blind-spot. Politically acceptable traditional approaches to the allocation and management of water and to family fertility are difficult to change, or at least to change fast enough for the water gap to be addressed with indigenous resources. The psycho-political option is possible for MENA communities because of the exogenous factor of virtual water. A workable second best option has been adopted.

Traditional systems proved themselves adequate in the past throughout the region. The continued existence of traditional systems of irrigation water regulation such as exist widely in Iran, based on the underground channels, the *qanat*, testify to the possibility of developing sustainable water management systems of limited water resources which were, and remain, integral to societies and local circumstances (McLachlan 1988). The *qanat* system was overtaken first, by changes in technology such as the introduction of tube wells and pumping equipment. Secondly, it was impacted by demographic changes through the increase in population and therefore of water demand. Thirdly, the qanat system was overtaken by new economic circumstances such as the transformations associated with Iran becoming an oil economy inducing rural to urban migration and the general dislocation of rural systems. The regulatory systems which evolved around the *qanat* society have not been proof against all the disruptive forces. Some of the disruptions have been familiar throughout history, for example changes in climate which led to fluctuations in the water flow in the systems, or damaging floods and earthquakes. The recent socio-economic events of the 1970s oil boom have, however, been the most disruptive. But there were still 50000 operating *qanats* in the 1990s supporting agricultural outputs and livelihoods in much of Iran's irrigated area (McLachlan 1988).

The most extreme example of adjusting to 'anxiety' in the water politics of the region is in the 'vital lies' diligently circulated denying the real reason for the successful adjustment of the MENA region to its water deficit. Success has been achieved mainly by importing global water via trade. This simple truth is politically unacceptable. The vital lie is that the economies have enjoyed economic stability and have even expanded because of the efforts of the region's farmers and engineers to manage an adequate supply. In practice the main source of current and future economic stability is not in the agricultural sector.

Social theory can contribute a great deal to understanding the way communities and governments perceive water and their water predicament. Communities that have not experienced progressive water deficits through five or more millennia of residence in a region appear to adopt a cerebral strategy very much like that of the individual facing new personal challenges in relations with another person. Trade-offs between awareness and anxiety have been identified as part of personal relationships (Goleman 1997).

Such trade-off behaviour is also strikingly present in national water politics. Water users, professionals in water providing institutions, political leaders and politicians concerned with the water sector appear to conspire to suppress evidence and diminish awareness. They particularly suppress awareness of bad news about current and future water availability.

As in personal relationships these water sector players lie and all parties adjust to the lies. Then via a complex discourse they achieve a politically acceptable, if not a totally comfortable, accommodation. So comprehensive is the acceptance of such lies, as for example that there is enough water in the Nile for the Egyptian economy, that outsiders who attempt to contradict them are unwelcome and despatched with trenchant versions of the accepted truth. The much more elegant and impressive truth that the Egyptian economy will have the capacity to access the water it needs by a variety of means because of the increasing strength and diversity of the economy is an idea that has not found its political moment.

Important elements of the adjustment to note for those wanting to play an innovative role in water policy development is that it is neither politically feasible to acknowledge the problem nor the actual solution currently working wonderfully. For the politicians of Egypt, Syria, Jordan and Yemen familiar truths from the past that the cause of any water problem would be the result of external forces are safer politically than untried new truths about the water problem. Drawing attention to the real current solution involves the recognition of the role of global 'virtual water'. This concept is doubly unacceptable. It is not only an unfamiliar concept, it is also politically unwelcome because if the idea was to gain currency it would make necessary acknowledgement of the problem. Mercifully the solution of 'virtual water' operates just as effectively de-emphasised and silent as it would if it were to be publicly and loudly acknowledged.

Political leaders have experienced the same water resource history as their peoples and have had to deal with the same challenges to conventional expectations over availability. They have also been involved in, or at least been very close to, the conventions of raising a large family which accorded with the needs of their family economy and livelihoods and matched the social expectations of the past. Governments and peoples have together experienced millennia of water security despite changes in demographic and economic circumstances. Such lack of awareness contributes mightily to the absence of anxiety.

The temptation to avoid anxiety by slowing down the awareness of water scarcity has been elevated to a higher form of politics in the Middle East. The practice fits neatly with the way politicians judge they are most likely to stay in power. As a result changes in water truths embedded in traditional hydropolitics are unlikely to change quickly. Political leaders do not want to confront their people's strongly held water beliefs, deeply rooted in past awareness, with informed comment on current resource realities. This profound alliance of the perceptions of peoples and governments is of determining hydropolitical significance.

Diverse water politics in diverse economies

The MENA region is not characterised by uniform economies. There are oil rich and oil poor ones, which may be both large or small economies. All endure poor water endowments. The MENA oil rich governments and peoples were so preoccupied from the 1960s in absorbing oil revenues as effectively and as quickly as possible that the water problem was just one of many that could be addressed 'with unlimited supplies of capital' (Penrose 1971:3) though with problems of absorptive capacity (Attiga 1971:15). The oil economies swiftly and effectively addressed their domestic water supply problems for their increasingly urbanised populations. Water supplies were sufficient for the modest domestic water needs of any population. Where necessary desalination was also financially feasible for the small volumes of water needed for domestic use. Bulk water, the water needed for food, even the rising volumes demographically and standard of living driven, could be afforded by oil economies and easily provided by the global trading system.

Without analysing their good fortune the policy makers in the arid countries of the Middle East and North Africa tested the capacity of the global system to remedy its food deficits and found the trading system entirely effective logistically and economically. Unfortunately some of their natural concerns about strategic insecurity in the area of food were confirmed. The October War of 1973 brought a strong confrontation. Western oil consumers, which were also the sources of 'virtual water' appeared to confront Arab governments, especially those of the oil rich Gulf economies. Some unhelpful comments by US politicians, exposed in the media, suggesting that the US grain weapon could be just as effective as the Arab oil weapon which had been in evidence in 1973. An urgent regional review of the region's water and crop production capacity was stimulated.

The response of the Arab economies was to create institutions such as the Arab Organization for Agricultural Development to coordinate the combination of international Arab oil capital, Sudanese land and water and the technical skills of the region's agricultural professionals and farmers (Sabry 1995). The failure of this well intentioned and apparently rational initiative led to bitter disillusionment. The cooperating parties realised that they could not rely on regional cooperation. They had to rely on themselves. By the early 1980s response to the failed plans for projects in the Sudan led to extremely economically inappropriate water resource policies and management. Egypt allowed the price of Egyptian grown wheat to rise and levelled off the much cheaper grain imports. Saudi Arabia embarked on irrigated wheat and barley production to such an extent that by the mid-1980s it was a major wheat exporter.

Had the initiatives in the Sudan succeeded the hydropolitics of the region generally and the subsequent water policies of Egypt and Saudi Arabia in the 1980s would have been very different. Egypt would not have adopted its wheat self-sufficiency policies, embarked on in the mid-1980s. (See Figure 1.2) Saudi Arabia which also threw itself into an ill-inspired grain self-sufficiency policy in the early 1980s would have done so in a much more measured way. In the event by 1986 Saudi Arabia was able to export wheat. An achievement of the 1986 to 1995 period was the levelling off of the trend in MENA 'virtual water' imports as a result of the increased wheat production of Saudi Arabia and Egypt (Figure 1.2).

By 1997 a completely different pattern of hydropolitics had developed in Egypt from that of Saudi Arabia. The spring of 1997 saw President Mubarak announcing agricultural expansion in the difficult environments of the New Valley to the north and west of Aswan. The Egyptian agricultural minister was clear on Egypt's on-going commitment to wheat self-sufficiency, proud of the achievements of Egypt's agricultural policy in this area and confident of further progress, in May 1997 (Wali 1997).

> 'Egypt wants to withdraw from its key position in global grain markets by the next century, when it hopes to harvest the gains of an ambitious and controversial scheme to cultivate almost all the wheat it needs. Agriculture Minister Youssef Wali said in remarks published last week that Egypt, a major

purchaser whose import decisions have swayed international markets for years, aims to grow at least 75 percent of its consumption needs by the year 2000.

'Egyptians consume about 11 million tonnes of wheat annually and with a population growth rate of about two percent a year, consumption is set to rise further. The agriculture and supply ministries say they have been working on strategies to boost production and reduce consumption since the beginning of the 1980s.

'But this year, the government has taken the idea seriously, especially since President Hosni Mubarak launched a grand project to develop a New Valley in the south and relocate farm lands from their ancient site in the Nile delta.'

Reuters, Cairo quoting Yussuf Wali, Egyptian Minister of Agriculture, and commenting, Reuters Cairo 12 May 1997

The above report from Cairo is mainly significant in that it powerfully confirms the public commitment of Government to the old order of ideas implying with incomprehensible certainty for an economist that the WINER doctrine that **w**ater **i**s **n**ot an **e**conomic **r**esource. The same report included comments by international traders in grain that the Egyptian policies made no economic or market sense.

By contrast, the Government of Saudi Arabia had been forced by the Gulf crisis and war of 1990–1991 to confront the water challenge. It had to address both the need to change patterns of water use in agriculture as well as the perceptions of the Saudi people. Public awareness campaigns were legitimised by exceptional royal intervention. The need for water use economies in the cities was made public as well as concern about the availability and the value of water.

Saudi water use drops after halt to grain exports

Saudi Arabia's year-old decision to stop grain exports has drastically cut water use in the desert kingdom, [announced] its Agriculture and Water Minister Abdullah bin Abdul-Aziz bin Muammar. [He said] that the kingdom would grow wheat and barley only to meet demand and boost the country's strategic reserves, but not for export.

The decision to stop exports of the subsidised commodities had boosted water reserves. Water consumption by wheat farmers dropped to 1850 million cubic metres in 1996 from 6990 million in 1992, while water used by barley growers fell to 510 million cubic metres from 2010 million in the same period. [He added that in March 1997] … the Kingdom had cut wheat and barley production to a total of 1.7 million tonnes in 1996 from six million in 1992.

Analysts estimate Saudi Arabia produced some 1.2 million tonnes of wheat and around 0.45 million tonnes of barley last year. Local wheat consumption is about 1.8 million tonnes. They forecast barley imports by the world's single largest barley buyer of around six million tonnes this year.

[The Minister said that his] ministry had taken a number of measures to conserve underground water, including halting the execution of proposed animal feed projects and transforming them to other farming schemes that rely on less water. It [had] also banned drilling of wells. The ... Saudi government was also considering measures to force every farmer to install water metres on existing artesian wells to rationalise water use. It might also remove aid from farmers who violated the rules.

Reuters: Dubai April 1997

Confirmation that water discourses were changing in the 1990s came in July 1997 when the King of Saudi Arabia contributed to the national discourse on the subject of domestic water consumption. He appeared on television and went on record on a programme on water urging awareness of the scarcity and the value of water.

Saudis battle to save water after subsidy
The advertisement shows a veiled woman bathing her baby while water gushes out of taps into an overflowing basin. Suddenly the water stops and the child is left with soap all over its face.

'We made from water every living thing' an old man in a flowing white robe says, reciting Islam's holy Qur'an.

His message is clear: 'Please don't waste water.' The advert being screened on Saudi television is part of a campaign launched in July [1997] to get the 18 million people living in the desert state to cut profligate water use. King Fahd is behind the drive. In July he urged citizens to save water and said his government was revising laws regulating consumption. 'It is a religious as well as a national and development duty,' he said.

But economists and water experts say the authorities have a tough task ahead. For years there has been no incentive to save water, which is cheap at around 0.3 Riyals ($0.08) per cubic metre, thanks to generous state subsidies. Water use per person in the Kingdom is said to be higher than in other countries at the same stage along the development road. Some officials have put per capita daily consumption at 400 litres [140 m3 ppy], versus an international average of about 200 litres [70 m3 ppy]. Residents say most house and apartment rents are inclusive of water charges, or a fixed fee is levied by the building owner once a year. Others say they never receive water bills. 'The water is included in my rent so I don't even think about how much I use,' said a western businessman. A Saudi academic said he had not received a bill at home for 18 months. 'I want to pay because water should not be free.'

Farmers are the main culprits
Domestic users alone are not to blame for excess water use, as they consume only 10 per cent of total annual demand of around 18 billion cubic metres. The main culprits are farmers.

'Agriculture continues to consume almost 90 per cent of water from all sources, and the expansion in grain production has been heavily dependent on non-renewable groundwater in most areas' says the Kingdom's five-year development plan to the year 2000, The plan targets a cut in water use to 17.5 billion cubic metres by the end of the century from 1994's 18.2 billion. It also calls for a decline in the use of non-renewable groundwater to 13 billion cubic metres from about 14.8.

'These targets are to be achieved by reducing the rate of water consumption in agriculture at an average annual rate of 2.2 per cent and achieving balanced consumption rates for other purposes,' the plan says. More than 80 per cent of the Kingdom's water is non-renewable groundwater. Around 14 per cent is surface and shallow ground water and four per cent is desalinated in the 24 plants. Less than one per cent is reclaimed or treated water.

'Since the 1980s Saudi Arabia has been growing many crops not suitable to its arid conditions, economists in Riyadh say. Government help in drilling wells and cheap rates, until 1995, for diesel, used to drive water pumps, were an incentive to grow crops that would otherwise have been imported. In the early 1980's Saudi Arabia was growing four million tonnes of wheat a year. This has fallen to about 1.3 million now, because of cuts in state grain prices. But this has led, ironically, to higher output of some even more water-intensive crops.

'Production of alfalfa, which consumes more water than wheat or barley, has increased,' Hussein Mousa of the U.S. Agricultural Trade Office in Riyadh told Reuters. 'The government is planning to reduce forage production because Saudi Arabia is exporting a lot of alfalfa to neighbouring countries, which is like exporting water,' he said.

Compiled by Hilary Gosh (Reuters, Dubai, 28 July 1997)

Domestic water consumption has been a relatively minor problem both in economic and environmental terms in Saudi Arabia because the volumes of water involved are relatively so small compared with agricultural water. Saudi Arabia was, however, forced to abandon its wheat self-sufficiency policy as a consequence of its changed economic circumstances following the Gulf Crisis and Gulf War of 1990–1991. Economic deficits and obligations to Arab and Western allies reduced the economic options available to the Saudi Government. No longer was it possible to argue that it was viable to produce wheat at six times the world price for reasons of national strategic economic security. It had been comprehensively demonstrated that Saudi Arabian economic and military security were dependent on its relations with global and regional partners and not upon its capacity to respond to economic or military challenges. Its strategic security in wheat was equally vulnerable because of its dependence on non-indigenous irrigation technology, the widespread adoption of which had provided the means to address the dangerous fantasy of self-sufficiency. The post-1995 economic regime rendered it impossibly expensive for the Saudi economy experiencing the constraints of running an unfamiliar deficit economy.

Innovation: a regional pattern?

> Innovation involves change and is political.
>
> Communities need to *know about* and *want* innovations before they *have* and *operate* them. *Knowing about, wanting and allocating resources* are profoundly political processes as is the *operation* of new systems.
>
> Water is a social and economic resource—makes the acronym WISER.
>
> Turton 1999

Until the 1990s all economies, oil rich or oil poor, with the exception of Libya and Israel, had water policy strategies which were based on past expectations of water availability rather than on current realities. There has been a small but progressive shift in perceptions since 1990 which reflects a radical change in direction even if the shifts in perception are incipient rather than substantive and the consequences as yet invisible.

As with all innovation water resource wisdom has appeared across the region in widely separated locations. Israel is a well documented case where the process took a disturbingly protracted three decades before the cycle of evidence, awareness and socio-economic development had been sufficiently digested by the interested parties for politically feasible water policy reforms to be introduced. The transformation in the economy which enabled non-farming interests to gain more political influence than farming interests and dominate water policy formation took thirty years.

The transformation was characterised by industrialisation, shifts in sectoral contributions to the national economy with services and industry becoming overwhelmingly dominant by the 1980s (Allan 1996a). Feitelson has traced the transformations in water policy in Israel highlighting political economy factors in the process of the adoption of economy wide, rather than sector specific, sustainable water policies. He identified the reduction of the influence of the irrigation sector in national water politics which paralleled its decline in economic significance (Feitelson and Allan 1996). Just as important were attendant circumstances which enabled policy change such as the 'windows of opportunity' (Kingdon 1984) noted by Feitelson (1996) which occurred when there were years of serious drought—1986 and 1991–1992 (Allan 1996a).

Israel is the only MENA political economy so far which has endured the political stress of running out of water, with an outcome which is both economically sound and politically accepted. The three decade transformation in the political economy of water in Israel is a telling example of how the impact of long run economic pressures are subject to political moderation. It is also a vivid justification for those who advocate the explanations of political economy rather than the narrow and simplistic message of economics alone.

The rest of the region is at the foot of the S-shaped curve of innovation in terms of adopting the WIER—**w**ater **i**s an **e**conomic **r**esource—economic ideology. But the trend is evident if obscured by the pronouncements of some prominent politicians such as those in Egypt and officials in the Agricultural Organization for Agriculture and Development (AOAD 1997a & 1997b). However in more and more corridors and offices of the Ministries of Water and of Agriculture of a number of MENA governments there is WIER talk. The shift is more evident in informal exchanges than in official statements and published policy. But it is real in countries such as Tunisia and Morocco (Jellali and Jebali 1994) as well as in Jordan (Haddedin 1996, Al-Kloub and Shemmeri 1996, Al Kloub 1997, Al-Kloub et al. 1997), and from the above evidence also in Saudi Arabia.

Another way of capturing the contention between those that argue that water *is* and *is not* an economic resource is in the concept that *water is a social and an economic resource* (Turton 1999). These words make the singularly expressive acronym WISER. In practice it was the contentious insider WIER discourse that prevailed over the outsider WINER ideas at the turn of the twenty-first century. The innovative process of 'wiser' convergence was only just beginning in the second half of the 1990s.

Belief systems: water in the culture and society of the Middle East: the significance and meaning of water and their impact on the management and politics of water

'We made from water every living thing.' *(Qur'an, 21:30)*

Strong vested interests in using water aligned with widely held fundamental beliefs in God's existence, and in His unity, power and care, and in the resurrection, all of which are symbolised in Islam by water, appears to be a combination of beliefs much too powerful for mere authoritarian governments to confront.

Belief systems tend to endure in poor non-diverse economies.

The salience of water in Middle Eastern cultures is clear from the books of the region's major religions. The prominence of references to water in the *Qur'an* has been much cited with respect to the social and economic significance of water (Abdel Haleem 1989), to its relevance to legal principles in managing and using water (Mallat 1995b) and to the importance of treating water as an environmentally significant resource (Mubarak 1891). A *sura* cited by the mid-nineteenth century Egyptian engineer and administrator Ali Mubarak is of especial clarity and succinctness in conveying a seventh century grasp of the green principles rediscovered and re-advocated thirteen centuries later in the 1980s. 'Cultivate your world as if you would live for ever, and prepare for the hereafter as if you would die tomorrow.' (Mubarak 1891 quoting the *Hadith.*)

There is, however, some misleading emphasis introduced in interpreting the recommendations of the *Qur'an* and the *Hadith* of Muslims. For example the obligation to give water to visitors and those who request water applies to the very small quantities of drinkable water needed by humans and livestock. Such water may be life saving if the individual needing the water is in an extreme predicament on a desert journey. The obligation does not apply to water for economic and livelihood use. Meeting the needs of an irrigator are completely different in terms of volume and quality from giving a drink of water to a visitor. It is the difference between a few litres for drinking (1000 litres equals one cubic metre) and the ten thousand cubic metres needed for a hectare of crop production. Long term livelihood water is not much addressed by the *Qur'an*; life preserving water for the individual is.

The *Qur'an* and the *Hadith* are important because Arab countries, as well as Turkey and Iran, which together include about 95 per cent of the region's population, are predominantly Muslim. Their peoples are imbued with the precepts of Islam as articulated in the holy *Qur'an*. The concise and powerful statement at the beginning of this section, quoted from the *Qur'an,* draws attention to the vital importance of water (Abdel Haleem 1989:34). '*The Qur'an* … is a book for the guidance of mankind. … it treats the theme of water in its own way and for its own objectives and [identifies water] not merely as an essential and useful element, *but one of profound significance and far-reaching effect in the life and thinking of individual Muslims and of Islamic society and civilization.* On the theme of water, Qur'anic material and the way it is treated, is lively, exciting and particularly intimate to man.' (Abdel Haleem 1989:33–34).

'Qur'anic statements about fresh water constantly remind the reader that its origin is with God and not with man.' Water is sent down to man and the provenance of 'water is from the sky where it is held by His power and at will He brings it down. … The *Qur'an* never says that it falls.' (Abdel Haleem 1989:36). 'We send down pure water from the sky, that We may thereby give life to a dead land and provide drink for what We have created—cattle and men in great numbers.' (*Qur'an* 25:48–49). 'And you see the earth barren and lifeless but when We send down water upon it, it thrills and swells and puts forth every joyous kind of growth.' (22:50). There are in addition numerous references in the *Qur'an* to the bounty that derives from water (*Qur'an* 6:99, 7:57, 35:27, 22:63, 50:9–11, 80:24–32).

There is also reference to the distribution of water on the surface and beneath it. 'He leads it through springs in the earth' (39:21) He also 'lodged it in the earth' (23:18) and challenged the reader to contemplate the value of soil moisture and groundwater in the comment, 'Think: if all the water that you have were to sink down into the Earth, who would give running water in its place.' (67:30). In this passage there is recognition that naturally occurring soil moisture, well distributed through the seasons, is the most important water of all, only exceeded in importance by that water 'which … [makes] … every living thing' (21:30). The value of water is given a remarkably early citation in the economic history of water in the Prophet's urging that in the practice of cleansing themselves for the Friday prayer they should do so *'even if a glass of water would cost a dinar'* (Abdel Haleem 1989:39).

In addition to the sixty references to water in the *Qur'an* there are over fifty to rivers (Abdel Haleem 1989:40). Rivers are seen to be cooling and to be providers of irrigation water and to be environmentally enhancing. River waters are also frequently connected with references to Paradise which makes water an important element in the after life as well as in contemporary livelihoods and ritual (Abdel Haleem 1989:40). The same author argues that water is probably more significant to Muslims than to any other people in the world. 'It is a subject of profound significance, and man's senses, emotions and reason are constantly brought into play in discussing it.' (Abdel Haleem,1989:45).

The extent to which water has 'infiltrated' the language, society and religion of Muslim peoples is evident. That water is also an element in livelihoods and national economies will never be the only explaining variable in the behaviour and policy making vis-à-vis water at all levels. The meaning of water to farmers, agricultural and irrigation officials and politicians in ministries of irrigation and agriculture together have influence over the vast majority of water being used in the region's national economies.

Agricultural water accounts for between 70 and 90 per cent of water use in most of the countries of the Middle East and Northern Africa. The 'meaning' of water in the minds of Muslim peoples carries the ideas, first of 'water being the proof of God's existence, unity and power', secondly of 'the proof of God's care', and thirdly as proof of the Resurrection through the everyday evidence that water can restore life. It behoves those who attempt to understand and afterwards possibly modify water using policies and practices to be particularly careful. They must give close attention to cultural factors especially where the traditional values and 'meaning' attributed to water reinforce the interests of elements of the agricultural communities, élites and bureaucracies who together influence water allocation and management in the region.

Strong vested interests in traditional water using practices are aligned with widely held fundamental beliefs in God's existence, and in His unity, power and care, and in the resurrection. All of these ideas are symbolised in Islam by water. The alignments are much too powerful for mere authoritarian governments to confront. Such is the strength of the existing beliefs that they easily

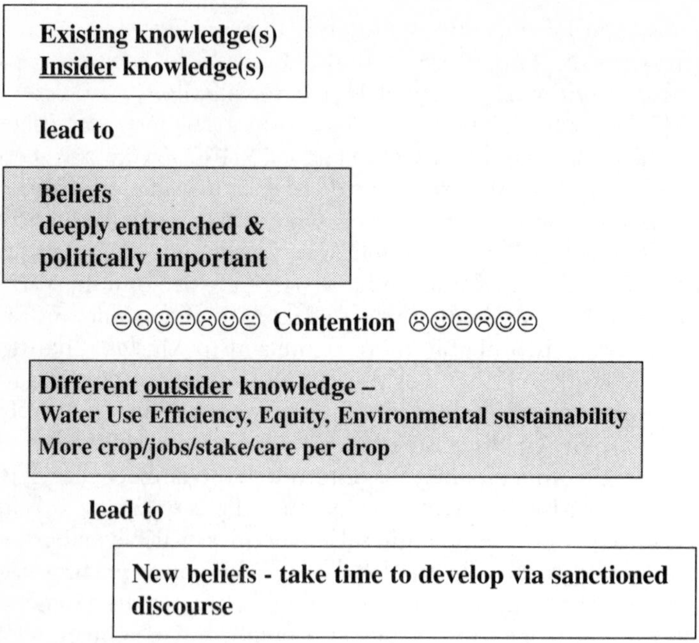

Existing knowledge(s)
Insider knowledge(s)

lead to

Beliefs
deeply entrenched &
politically important

😐☹☺😐☹☺😐 **Contention** ☹☺😐☹☺😐

Different outsider knowledge –
Water Use Efficiency, Equity, Environmental sustainability
More crop/jobs/stake/care per drop

lead to

New beliefs - take time to develop via sanctioned discourse

Figure 4.1 MENA water discourse: insider and outsider knowledge, belief and contention

withstand the assaults of new knowledge brought by outsider professionals and scientists. In these circumstances the contention between insiders and outsiders is protracted.

The main feature of water belief systems which influence societies as well as communal and national politics is the constant reinforcement of perceptions. These perceptions of water region-wide are integral to the religions of the region as well as to the secular tradition created through a long history of water resource sufficiency. Perceptions affect both water use and water policy insofar as they are relevant first, to the availability and secondly, to the rights to whatever water is available. Regarding the former there tends to be a general uniformity.

Those depending on vast quantities of water in livelihoods involving irrigation are just as optimistic of future water as those depending on small volumes of water used in urban livelihoods, with the obvious exception of users in some exceptional cities such as Amman, Ta'iz and Sana'a. Even the water short tail-ender on an irrigation channel in the river lowlands of the region does not blame the availability of Nile water to Egypt on the hydrological system. They have been assured there is sufficient water by the political leadership. They blame the Ministry of Public Works and Water Resources (Radwan 1994 and 1996). With a starting assumption that there is enough water the agricultural users of water, whether receiving water from a surface water system or lifting water from the ground will tend to believe that water has a very low value and should be available at negligible cost. The starting assumption is that water should be free or at least very low cost.

The second issue, the rights of the user to low cost water, follows naturally from the first. With the starting assumption that there is enough low cost water in the system to meet all national needs, an assumption which discourses everywhere reinforce, then beliefs on rights to water take the form of claims to water that force the communities involved into resource managing behaviour described as the 'tragedy of the commons' (Hardin 1968). Unfortunately the sound communal systems, such as the *qanat* based systems in Iran and south-east Arabia are patchy and certainly not universal. Also there are no widely agreed legal principles, religiously inspired or secular, and certainly no operational international legal regime on shared waters. There are local practices which have operated for centuries at the community level (Trottier 1999). Finally there is no operational market mechanism to contribute to the prioritisation of water allocation in the prevalent belief systems. On the contrary there is a fierce resistance to the idea of water being an economic resource. All these perceptions are environmentally unsound as well as inimical to the adoption of economic principles.

Outside the special economic circumstances of the oil enriched there are two regions which are extremely water challenged. The first is the group of economies centred on the River Jordan—Israel, Jordan and Palestine. The second is Yemen. What is happening in these political economies where adjustments are unavoidable?

In the Jordan basin economies belief systems have shifted for a major player in that catchment—in Israel. The shift was enabled by the transformation of the Israeli economy. Jordan is also shifting relatively fast (Haddedin 1996, Kloub 1997). Meanwhile the water officials of the Palestinian Water Authority are amongst the most progressive in the MENA region with respect to recognising a role for economics in water allocation and management (Haddad and Mizyed 1996, Issac 1996).

In Yemen, by contrast, the role of economics is not even on the agenda of the National Water Resources Authority. To address with environmentalist inspired and economist drafted policies the challenge faced by the political leadership of Yemen is not politically feasible. The chances of shifting perceptions, in the way they have shifted in the Jordan catchment economies, in the current economic circumstances of Yemen are very low indeed. Shifts in perception are doubly impossible insofar as the Yemen economy continues to benefit from advantageously priced and readily available virtual water in cereal imports, which de-emphasise the water crisis.

Belief systems tend to endure in poor non-diverse economies. They can be transformed in strong and diverse economies. But it would be naïve to assume that socially embedded ideas are susceptible to changes in economic circumstances. The enduring gun culture in the United States is a perverse example of the unsafeness of assuming that wiser socio-economic outcomes will be determined by prosperity.

Water use and water law in Islamic countries

> 'He created everything and ordained it in due proportion.'
>
> *(Qur'an 25:2)*

The meaning of water as expressed in the rituals and traditions of Islam is not only important in framing the approach of Muslim peoples to water, it also has a number of important influences on water management practices and has an important place in the *shari'a* (Islamic law). Ibn Manzur (d. 711AH/1311 AD), the most famous Arab lexicographer, mentions in his dictionary *Lisan al-'Arab* under the root 'sh r' that 'shari'a is the place from which one descends to water … and shari'a … [for] … Arabs is the law of water (shur'at al-ma') which is the source for drinking which is regulated by people who drink, and allow others to drink, from.' (Ibn Manzur 1959:175ff.) Mallat (1993:3) points out that a later classical dictionary is even more specific: *'Ash-shari'a'*, writes Zubaydi,

> 'is the descent (*munhadar*) of water for which has also been called what God has decreed (*sharra'a:* legislate, decree) for the people in terms of fasting, prayer, pilgrimage, marriage etc … Some say it has been called *shari'a* but comparison with the *shari'a* of water in that the one who legislates, in truth and in all probability, quenches [his thirst] and purifies himself, and I mean by quenching

what some wise men have said: I used to drink and remained thirsty, but when I knew God I quenched my thirst without drinking.'

Mallat points out that 'the connection between *shari'a* as a generic term for Islamic law, and *shari'a* as the path as well as the law of water, is not a coincidence, and the centrality of water in Islam is obvious in the economic as well as a ritualistic sense (Mallat 1993:3). The jurists who expounded the shari'a were inspired by this centrality to develop a highly sophisticated system of rules covering the whole field of law, initially purely religious law, which developed into the common law of the Muslim world. The unity of religion, law and the state which exists in Islam means that exhortations in the *Qur'an* concerning the distribution and use of water are important. A principle stated in the *Qur'an* is that the vital resource of water should not be monopolised by the powerful and privileged against the poor. The references in the *Qur'an* to water distribution 'provide the basis upon which much legal thought was formulated' in the *shari'a* (Maktari 1971:28). That water should be divided is an important principle as indicated in the *Qur'an* in the exhortation, 'tell them that water is to be divided between them' (54:28 and Yusf 'Ali 1971).

Abdel Haleem (1989:47) points out that the Prophet said that 'people are co-owners in three things: water, fire and pastures', and 'God does not look with favour upon certain kinds of people 'one of these is a man who has surplus water near a path and denies the use of it to a wayfarer', and secondly 'he who withholds water in order to deny the use of pasture, God withholds from him His mercy on the day of the resurrection'. He also notes that 'a man who is thirsty is permitted to fight another, though without the use of any weapon, if the other has water and denies him the right to quench his thirst' (Maktari 1971:21).

The role of law, both local, national and international will be more fully elaborated in chapter 7.

Natural resources as national fantasies

It has been observed that natural resources can be constructed into prominent national fantasies of a nation or community (Allan 1983b). The evidence in the Middle East and North Africa is strong. National leaders as different as President Nasser of Egypt in the 1960s (Allan 1983a), Mua'amer Qadaffi of Libya in the 1970s (Allan 1983b) and Prime Minister Sherrett of Israel in the 1950s (Allan 1996a:92) called upon their respective peoples to regard the challenge of combining water and land resources to secure their nation as an inspiring priority. Outside the region the same notions drove the former Soviet Union to reclaim its virgin lands in the 1950s and 1960s. Prominent leaders of the USSR, Secretaries Breznhev (1978) and Gorbachev made their early careers overseeing the land reclamation projects of Central Asia.

Attributing to marginal resources the capacity to solve national security issues united peoples and governments. Such common endeavours have been a normal feature of MENA political economies. The danger of the phenomenon has been that it tends to encourage the idea that natural resources, such as water, were sufficient provided that enough effort and technology were devoted to managing them. The approach is associated with the idea of the hydraulic mission (Swyngedouw 1999a and 1999b) which dominated the thinking of governments, engineering institutions and irrigation departments in the century up to the 1970s in the North. The approach was introduced to the South by the colonial interventions of the nineteenth and twentieth century especially in India but also in Egypt and to a lesser extent in Iraq.

The governments of newly independent economies embraced the technology with enthusiasm and attributed to the development of land and water a romantic and extraordinary potential. The leaderships of the MENA region remain loyal to the hydraulic mission. The recent New Valley Project in Egypt was launched in 1998. The Iraq leadership devoted resources to substantial hydraulic works in the early 1990s in the southern marshes at the point where the rivers Euphrates and Tigris meet, Turkey was planning in the late 1990s to construct major works on the Tigris river as soon as it could raise the necessary finance (Brown 2000). Iran has the ambition to further control its water resources with civil structures.

Water politics and explanations from political analysis

> Those purveying the economic and environmental facts of life which contradict the deeply held belief systems of whole populations will be ignored if they do not shape their message and pace its delivery to accord with political realities.

Interests

Political science theory has made a major contribution by identifying the self-interest of individuals, families, communities and nations as a major factor in explaining why these entities operate as they do in the way they manage natural resources. The explanation is especially useful in explaining the way they react with others to organise access water.

Water is not like other commodities in that because it flows both on the surface and underground it is difficult to own. And water rights are difficult to arrange precisely because water ownership is difficult to vest, and usually, impossible to operationalise. Unfortunately for those arguing for water markets the monitoring of water flows is technically challenging. It is for this reason that so many traditional systems of water sharing are time determined rather than volume determined (Wolf 2000b).

Water is often a common resource. The regulation of a common resource can be highly organised such as the *qanats* and *falaj* of Iran and Oman (McLachlan 1988, Wilkinson 1977) or it may be completely unregulated. Coastal tracts where the water resources have been seriously degraded because unregulated tube wells and pumps have been installed, for example in Libya and Oman, have endured increased rates of water withdrawal commonly to five times the rate of recharge (Allan 1971, Pallas 1976, Dutton 1998). Coastal Tunisia, Gaza, Israel, Syria and all the Gulf States and islands such as Bahrein have also been affected by the phenomenon of the unregulated over-pumping of coastal aquifers. Sea water intrusion has been experienced in all these areas with serious consequences for both traditional agriculture as well as for the modern irrigated farms that have generally replaced them.

Governments struggle to quantify and fathom the extent of the problem of the reduction in coastal groundwater levels by commissioning studies and plans (Latham et al. 1981), often three or more times until they obtain analyses and predictions of the position which is optimistic and fits their political needs. Water Master Plans are just as numerous for the same reason (WRA, Oman 1998). Meanwhile surveys and plans are easily and generally ignored.

Nowhere have either the conventions of traditional water management or the, albeit unenthusiastic, regulatory will of governments been effective in limiting the short term interests of the individual pump owner in achieving a pattern of groundwater utilisation which limits withdrawal so that it matches recharge. The history of groundwater use in the second half of the twentieth century in the MENA region exemplifies Hardin's (1968) thesis of the 'tragedy of the commons' disturbingly and comprehensively.

The primacy of individual interests in communal hydropolitics is observed across the region. Only Israel has imposed a regulatory regime which has levelled off and reduced national levels of groundwater use (Allan 1996a). The development and introduction of the policy was a long, two decade struggle, and one which is by no means complete as evidenced by the reception of the tough 1997 report of Arlosoroff (1997) on the Israeli water sector. But Israel was not able satisfactorily to regulate the use of water in Gaza between 1967 and 1996, and the Palestinian Authority has been even less able to achieve a balance of rates of use and recharge since it took over Gaza in 1996 (World Bank 1998a). Israel also reversed its reflexive approach to groundwater management after 1992 following the high rainfall event of December 1992.

Governments and would-be regulators have been singularly unsuccessful in confronting the primacy of individual interests in the use and management of water in the MENA region. Governments are more likely to rely on the exhaustion of the resource to be the evidence that persuades water using communities that patterns of water use have to change. The state and regional government institutions are not strong enough to implement economically and environmentally sound water policies to anticipate and avoid catastrophes. The helpful explanatory notion of the strong and weak state (Migdal 1988) discussed in a later section of this chapter provides insights into why it takes so long to innovate in the face of deeply held beliefs.

Discourse and sanctioned discourse

> Many of the statements on water resource use and management by Middle Eastern leaders are determined by the national 'discourse' and not by the recommendations of government servants and certainly not by the advice of alien visitors.

The explanations of Foucault (1969, 1971 and 1980) concerning the lack of free will for individuals and the communities 'discoursing' in political entities also bring useful theory to a discussion of the political economy of MENA water. Foucault argued that normal politics involves the interaction of interests. The resulting tension creates outcomes, which he termed 'networks of consensus' for which he used the word 'discourse'. 'Discourse' in this usage is a form of power, discursive power. He used the term discourse to conceptualise discursive power within a political entity to distinguish it from coercive power. The outcome of contention in formal and informal political processes is discursive power or discourse. Such discursive power is not just concentrated in the hands of a few. It is dispersed in the networks through which the contending parties communicate and share knowledge. The process of contention signals to those with formal power what is politically feasible. Insiders in a major way, and outsiders in a minor way, contribute to the discursive process and to the discursive outcomes.

Participants contribute to the tension of the process in ways that reflect their interests but then have their free will limited by the outcome of the 'discourse'. Those with apparent hierarchical power experience limits to their free will as well as those they 'govern'. For anyone who has tried to understand the enigma of MENA water politics Foucaultian insights are very helpful. They explain why politicians speak with many voices and fail to implement sound economic and political policies of which they are aware. Such politicians are not 'free' to initiate water reforms because they are part of a system in which their inputs are just some of many. Their main role is often to legitimise that which is determined by the 'discourse'. Many of the statements on water resource use and management by Middle Eastern leaders are determined by the national 'discourse' and not by the recommendations of government servants and certainly not by the advice of outsider visitors.

Political scientists have moved on from the explanations of Foucault (Baudrillard 1987). They are no longer comfortable with the idea that there is no free will and they have argued that new forms of communication have reduced the value of the discourse/ power idea. As with all such theory they have found that it is impossible to operationalise that of Foucault. Nevertheless, the notion of 'discourse' is very helpful in the analysis of national water politics in the Middle Eastern region. And perhaps even better is the concept of the 'sanctioned discourse' (Tripp 1996). The addition of the word sanctioned, though tautological, 'emphasises' even more the limitations on those speaking publicly about water policy. Though the two words make a tautology they make the concept much more accessible to those deterred by what they sense is obscure philosophy.

A comparison of recommended policy priorities of outsider scientists recommending the principles of economics—economic efficiency and long term economic security, with the actual prioritisation of economic policies by the region's politicians, exposes the discourse determined approach of political leaders to water policy. See Figure 4.2. It is commonplace for scientists

Economically and environmentally logical policy priorities	Comments	Politically feasible policy priorities
1 Achieving strategic water security Securing supplies of **'virtual water'** by international **regional cooperation** in the international food trade	The idea of **food insecurity** is not part of the 'sanctioned discourse' in most Mid-East countries and therefore the relationship between water and food deficits cannot be debated.	**1 Achieving improved water use efficiency** 1 Implementing **measures** of **productive efficiency** to improve **returns to water:** • **farm level - improving water** distribution, drainage & technologies • **irrigation - as for farm but** emphasising institutions • **urban & industrial water re-use**
2 & 3 Achieving improved water use efficiency 2 Applying principles of **demand management** to improve **allocative efficiency and returns to water:** • farm **level - raising** water efficient crops • inter - sectoral re-allocation • international re-allocation	Policies promoting **improved water use efficiency** by means **of investment** in technologies, **civil works** & institutions **which achieve** such are welcomed **by Mid-East** governments **and officials**	2 Applying principles of **demand management** to improve **allocative efficiency and returns to water:** • **farm level - raising water** efficient crops • **inter - sectoral re-allocation** • international **re-allocation**
3 Implementing measures of productive efficiency to improve **returns to water:** • farm **level - improving water distribution, drainage & technologies** • irrigation - **as for farm but emphasising institutions, pricing etc** • **urban & industry water re-use**	Policies promoting **improved water use efficiency** by means **of allocatively efficient measures**, are unacceptable **to Mid-East** governments **and officials** because they **are politically** stressful.	**3 Achieving strategic water security** Securing supplies of 'virtual water' by international regional cooperation in the international food trade

Comment

The above analysis identifies water policies which would provide remedies to ameliorate the water predicament of Middle Eastern and North African economies.

The analysis shows that economically and environmentally urgent policies are not the 'politically logical' way to approach the amelioration of the region's water problems.

It is concluded that measures to advance the adoption of water re-use practices are 'politically logical and certainly more 'politically feasible'.

Figure 4.2 Sanctioned discourse as an explanation of the prioritisation of water management policies as recommended by water professionals and as viewed by political leaderships in the MENA region guided by awareness of political feasibility

and professionals who are not part of the Middle East's national water discourses to have very clear views on what is economically and environmentally sound. They are certain of the fundamental principles which they claim should guide the development of policy.

An analysis of the ways that MENA Governments and outsider professionals address and prioritise three key aspects of water management policy which impact efficiency and security is very revealing. The insider reverses the ranking of the policy options of the outsider.

Current and future water security in the MENA region requires that attention be given to a number of securitising options. The outsider sees the issue as being susceptible to three major remedies. The first feature of the political economy of water in the Middle East over which there can be a principled economic approach concerns virtual water. The MENA region perforce relates to the global hydrological system via international trade in 'virtual water'. Virtual water contributes to all economies in the region and its future role in the region will be as the dominant securitising source of water. Virtual water is the crucial strategic water for the region in future. Addressing its secure availability is an unavoidable challenge. Or so argue the outsider professionals including the author.

The achievement of water use efficiency is another essential route to water security. Water use efficiency can be achieved by two groups of approaches. First, technical efficiency can be achieved by using more efficient technologies—drip irrigation instead of water spreading for irrigation, or by more efficient scheduling of water applications both by improvements in the water distribution systems and the use of crop and soil sensing systems linked to computer controlled water distribution. Domestic water use can also be improved by installing more water efficient equipment to reduce the volumes of water used per person. These practices are sometimes called 'productive efficiency'.

A second form of water use efficiency is achieved via the application of principles known as 'allocative efficiency'. Economists would be happy with the term economic efficiency, but in the case of water the term 'allocative efficiency' has a great deal to recommend it. The term captures what it is farmers, industries, services, communities, national water departments and national governments have to consider in order to achieve improvements in water use efficiency. The principle involved is to do with allocation. Applied to water, which in the Middle East is a resource which is almost everywhere limited, the principle is particularly important and very political. The idea is simply expressed in the words of a question. Which activity brings the best return to water? The approach is relevant at the level of the farm when the return to a high value crop such as fruit and out of season vegetables for an international market would be much more than to a crop such as wheat or rice. In the agricultural sector the returns gained from one crop compared with another can range from two times to scores of times. The differences between wheat and barley would be small. Between wheat and flowers for export the difference could be hundreds of times. At the sectoral level the outcomes are much more dramatic. In services and industry economic returns to a cubic metre of water and livelihood generation can be thousands of times more than they would be if the water were to be allocated to agriculture. In an ideal world at the international level the same principles, operationalised via markets, could be used to decide which national economy could make the best use of a shared scarce water resource. Insights of this sort could be basic to the development of market mechanisms which would facilitate the movement of water between willing buyers and willing sellers to everyone's advantage.

The third water policy option as ranked by the outsider professional is productive or technical efficiency. Technical measures, such as drip and sprinkler irrigation, can improve water use efficiency, possibly doubling such efficiency returns to water. Technical measures are a third priority of outsiders because the productivity gains are modest.

Figure 4.2 sets out the comparative position for the outsider professionals and scientists who sing from the sheet called 'science'. The politicians operating in live and stressful discourses sing a song called 'political feasibility'.

The professionals and scientists, including the author, can simply and quickly order the priorities in terms of their economic significance. Strategic virtual water is the region's current and future solution to its water deficit and it ought to be accorded due prominence. Allocative efficiency comes second because its wider use would bring potentially massive improvements in returns to scarce water. Measures to improve the productive efficiency of water are also important but they will never bring the economic benefits which allocatively efficient policies would.

The contrast with the priorities articulated by politicians and professionals from the MENA region is stark. The 'sanctioned discourses' which take place in the political economies of water in the region steer politicians to exactly the reverse ranking from the scientists and the international agencies whose staff are inspired by economic principles.

Productive efficiency is the popular and everywhere lauded solution to the region's water deficit. It is the prime priority for a number of reasons with most of them aligning with the national water discourse. To achieve productive efficiency all you need to do is mobilise investment and increase capacity via training. Existing interests in agriculture benefit from any investments and will welcome water managing initiatives which improve productive efficiency even if they would not be able to articulate the concept of productive efficiency. Other interests are also well served by initiatives which promote productive efficiency. Manufacturers of the technical equipment, both in the region and especially those from outside it with intimate relations with the investing agencies, are also keen to achieve productively efficient outcomes. Manufacturers of pipes, pumps, regulatory equipment and information technology hardware and software would all gain from policies based on technically/productively efficient principles. Those providing the financial services, from international agencies to merchant banks, that ultimately determine whether a project goes ahead also prefer to invest in projects that fall into the productive efficiency category. The forces allied to promote the productive efficiency approach in the MENA region are massive and almost everywhere prevail. In this reinforcing discourse there are few political stresses associated with the introduction of technically efficient methods. There are apparently no political prices to be paid (World Bank 1997:146).

The second option, involving principles of allocative efficiency, is viewed very differently by insider water professionals and those deeply involved in real political economies of water. Re-allocation of water is a profoundly political act. It disadvantages some and benefits others. Disadvantaging those that contend effectively in the sanctioned discourse is not a political option. There is far too much political stress associated with water re-allocation; too many political prices to pay. Regional politicians have a powerful intuition that economic principles and the allocative measures which follow logically from them, as with any simple, but dangerous, truth must be avoided at all costs. Speeches by regional leaders on water reflect these intuitions.

The third option is even more politically problematic than policies based on principles of allocative efficiency. Addressing publicly the challenge of understanding and monitoring global factors affecting future supplies of virtual water, and understanding the overall economic strength which alone will secure further supplies of virtual water, are challenges to be avoided at all costs. Communities believe that the indigenous water will be sufficient for current and future needs. There are so many with such unshakeable assumptions in almost all the political economies of the region that politicians dare not gainsay the contribution to the discourse coming from these interests. The options of the politicians are sanctioned. They have limited free will. They take the only option which has no political price in the short and medium term at least. They contribute to the national adjustment to the simple truth by reinforcing the politically vital lie.

Political prices

Another useful idea which deserves emphasis in relation to the understanding water policy reform in the MENA region is the notion of the political price. The concept of the political price is in wide currency in that by 1997 it figured prominently in the World Bank's (1997:146) *World Development Report*. The authors of that report had taken on board the conclusion that despite the logical virtues of economic principles they did not explain why political economies and those leading them took the directions they did. They argued that if a political price is anticipated following a controversial initiative then that price may be the determining negative factor in the policy making process.

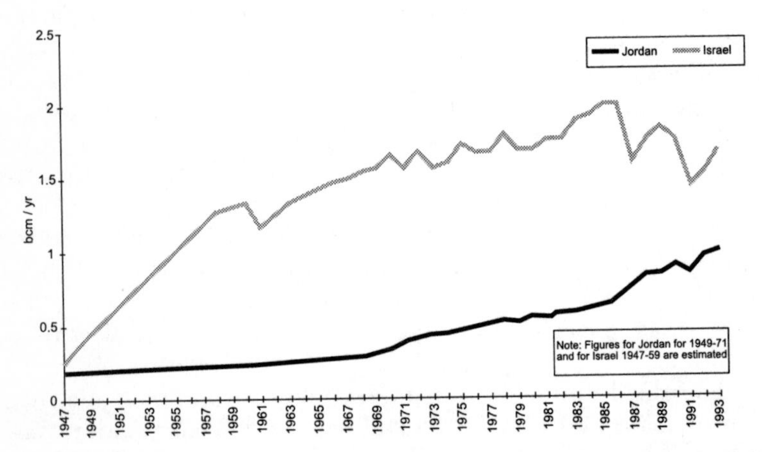

Figure 4.3 Comparison of Israel and Jordan use of water in agriculture: 1947–1995

That the World Bank had publicly and formally recognised the pre-eminence of political factors in the application of economic principles in the transformation of political economies is welcome. It is helpful to have the warning of the need for a politically informed mode of applying incentives and disincentives with a view to promoting innovation. It is certainly essential for outsiders wanting to make a contribution to water sector planning and policy in the MENA region to think beyond 'carrots and sticks'. Focusing on the economically and environmentally sound policies, which are also politically feasible, will distort and limit the projects that can be implemented. But it will not impede the process of developing water completely which would be the case if sound science were to be the only criterion for selecting the policies to promote.

Strong societies and weak states

The concept of *strong societies and weak states* (Migdal 1988) reinforces the explanation derived from discursive analysis. Migdal drew attention in the MENA region to a feature of their political economies which explains why they operate as they do. He observed that the capacity of state institutions to introduce change was limited. The power that resided in communities and in society tended to determine the way the political economy worked and the pace of change. In the MENA political economies where the size of rural water using communities is large and vested with very long traditions of water management the views of society prevail over those of officials acting on behalf of the state. The state institutions are not strong enough to ensure that there is institutional capacity to respond to national water shortages. Ideas and actions at the local level are determining. Capacity at the level of society to cope with national level challenges are limited. Yet so powerful are the views of civil society that the state is forced to seek solutions which are politically acceptable to society even where these solutions can only be sustained by denying the solution. The most obvious case of the denied solution is virtual water.

An illustration of the comparative capacity to respond to a drought challenge is shown in Figure 4.3. The droughts of 1986 and 1991 affected both Israel and Jordan. Israel responded by cutting water to the agricultural sector. Jordan was equally challenged but did not implement cuts to agriculture. The reductions in use in Jordan reflected the reduced levels of availability of water.

Windows of opportunity

> 'But being right too soon is a fatal flaw in a politician.' (Reisner 1993) Middle Eastern leaders show no sign of making the mistake of anticipating the politically feasible in their water sector policies.

The transformation of water policy at all levels is a socio-political process, subject to individual and communal psychologies, themselves rooted in the history of those societies and policies. Like other policy making transformations the rate of change is not constant. Some elements of the process may be relatively steady. For example innovative ideas which influence policy have been shown to spread through social and political space in phases; a slow initial phase of adoption, followed by a middle phase of more rapid adoption and a third phase when the final few adopters are slow to participate.

The S-shaped curve of adoption is a useful model, but it has to be augmented by other models of adoption Feitelson (1996). has drawn the ideas of Kingdon (1984) which are especially helpful in this regard. Kingdon noted that policy change, involving the adoption and implementation of new ideas, was not a steady process.

Those adopting the new ideas might grow in numbers steadily but the implementation of policies based on the new ideas tended to come as the result of particular events. Extreme climatic events, droughts and floods, for example, inevitably affect regional hydrologies and create political circumstances which are conducive to change. The drought affects everyone, farmers, urban consumers and politicians. The subject of water, or rather its absence, becomes a common topic of discussion. The media give it prominence. If the innovative ideas, which in normal politics would contradict the sanctioned discourse, are reinforced by the temporary environmental circumstances, then the extreme event can very significantly accelerate the pace of policy change. The extreme event has been helpfully referred to as an 'emblematic event' (Hajer 1996). These major events have the clout to command attention. They can also be used by those with an ideological cause and needing a foundation on which to construct alternative approaches to resource security. The 1990s have witnessed a shift from a monolithic view that security is the concern of the defence community and military priorities to a

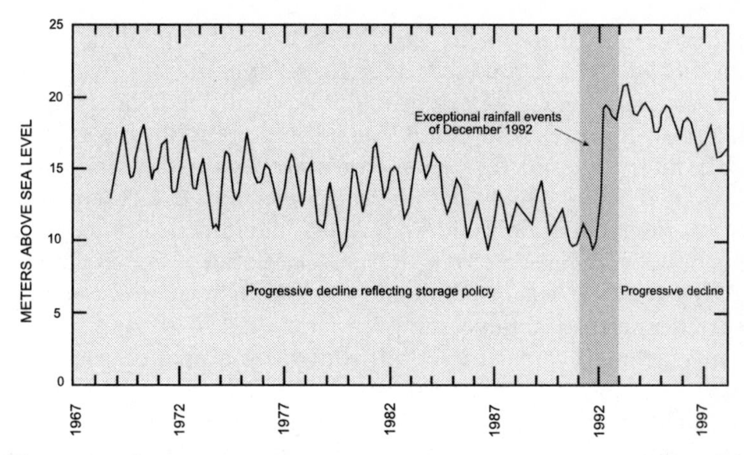

Figure 4.4 Groundwater levels in a group of wells in the West bank—1968–1998—showing the progressive declines in the 1970s and 1980s, the dramatic recovery following the rainfall event of 1992 and the results of Israeli storage policy at all stages.

wider definition that securitisation is also about the economy, society and the environment (Buzan et al. 1998).

The windows of opportunity provided by the 'emblematic' droughts of 1986 and 1991–1992 in Israel are well documented (Allan 1996a and see Figure 3.7) and provide clear examples of the Kingdon (1984) model. The profile of water use in Israel during the 1947–1992 period (see Figure 4.3) reflects both the impact of the drought on immediate water allocation policy as well as on the longer term decision making on how much water to use in the economy as a whole (Voice of Israel 1991).

That the same extreme event impacting a neighbouring economy with even worse water endowments did not have any such impact indicates that the window of opportunity does not necessarily open in all political economies when an extreme event occurs. Awareness and development of innovative water management approaches are necessary precursors of the opening of a window of opportunity provided by, for example, an extreme climatic event. Figure 4.3 compares the trends in water use in Israel and Jordan from the late 1940s to the mid-1990s and show that different pressures were operating in these two political economies. It is also the case that Jordan has a resource at its southern border with Saudi Arabia which is remote from, and 600 metres lower than the main alternative user, the population and industries of the metropolitan region of Amman.

Other windows of opportunity that have affected water in the last half of the twentieth century are the revolution in Egypt in 1952 which invested the new government with the chance to address an issue which decades of uncertain leadership and overseas intervention in its internal affairs had impeded. After 1952 the national commitment to the harnessing of Egypt's main natural resource the Nile was a politically feasible focus and one which attracted universal national support.

An extreme event which contributed to a change in water policy was the major rainfall event of 1992 in Israel/Palestine. The declines in groundwater levels in the West Bank over ten years were reversed by a few major storms. Israel reverted immediately to a storage policy which drew down the groundwater levels (See Figure 4.4).

Because the Middle East is at a very early stage in adopting Northern principles of economic efficiency and environmental sustainability it is not possible to find numerous examples to confirm the relevance of the windows of opportunity model. We have to look outside the region for confirmation, to other industrialised economies located in arid regions, to Australia and in the United States to California and Arizona (See Figure 3.9). Here there are examples of how the interdependent forces of scientifically based innovation have been focused by an extreme event. Economic principles and green principles have recently begun to contribute to the evolution of socio-economic contexts which increase the range of water policy options. In semi-arid regions droughts have proved to be very important in initiating water policy reform in Australia and the United States.

The US case is very helpful in demonstrating how difficult it is to achieve sound economic and green inspired water policy initiatives if a leadership tries to force change when there is no window of opportunity. Anyone who believes that a sense of urgency about economic and environmental issues on the part of political leadership is sufficient to bring about a major change in water policy should note the experience of President Jimmy Carter. After he became President in 1976 he wanted to achieve two goals by reforming the water industry. He wanted to reduce investment in the sector so that the monies released could be devoted to social and economic projects on which there would be a better return. Secondly he wanted to promote environmental concerns. Reducing investment in water projects was consistent with both aims. President Carter failed in his water initiatives, not because he was badly inspired scientifically; only his political timing was wrong. His ideas were adopted by the least likely implementing agent of all, the Reagan administration in the 1980s. But being right too soon is a fatal flaw in a politician (Reisner 1993). Middle Eastern leaders show no sign of making the mistake of anticipating the politically feasible in their water sector policies.

For the most part the region's economies, with the exception of that of Israel, have not yet reached the stage that they have the widespread awareness of the need for economic and environmental principles to guide policy. As a result occurrences of windows of opportunity are potential rather than real chances to shift water allocation and management policies.

The background global economy, of which the Middle East's water is a part, is, however, another important factor in affecting the rate of policy change. The global economic context is especially important in shaping water politics in the region and in determining water policy priorities. The ready availability of 'virtual water' in imported commodities has prevented serious climatic events from becoming economic and political catastrophes and windows of policy opportunity, since the 1970s. This global factor will play the same role for the foreseeable future. The availability of the alternative virtual water has made it possible to ignore the strategic significance of extreme events. As a result potential 'windows of opportunity' which might have concentrated the minds of communities and political leaders on the need to re-evaluate water policy pass unrecognised.

Global contexts: global grain-trade dinosaurs and regional benefits

> For the water short MENA economies the USDA-EU wheat trade folly of the second half of the twentieth century has brought a wonderfully silent water bonus.

> Mercifully the solution of 'virtual water' operates just as effectively de-emphasised and silent as it would if it were to be publicly and loudly acknowledged.

The role of the global grain trade in the political economy of water in the MENA region has been sufficiently established in the preceding chapters for its determining role in Middle Eastern water affairs to be understood. The phenomenon is as complex as it is significant. The professionals who have been most motivated to understand the nature of the global grain trade are based in Australia. This Australian preoccupation is not because Australia is dependent on grain imports. Rather it is because it is a grain producer and exporter. Fathoming the international grain trade is of the utmost strategic importance to the Australian economy because Australian farmers and their trading institutions have to compete with the grain producing giants, coordinated and subsidised by the United States Department of Agriculture and the European Community.

The Australian studies reveal that the MENA water economy is inflexible and ill equipped to make the adjustments dictated by economic and environmental principles; the studies also show that the USDA and the EU are extraordinary entities driven by priorities which even more defy economic analysis (ABARE 1989 and 1995, LeHeron 1993, LeHeron and Roche 1995). Only a Foucaultian explanation, and not an economic science explanation, would provide insights into the five decade fall in (constant) world wheat prices since the middle of the twentieth century. The trend has been in place not just for the last half of the twentieth century. It has been operating since the dawn of the agricultural revolution.

There have been occasional (price) spikes, but the period since 1970 is the important period for the Middle East. Since 1970, when the MENA economies became particularly exposed (see Figures 1.2 and 1.3) with respect to their growing water deficits, the grain producing economic dinosaurs of the USDA and the EU have been making water available in grain which was delivered world-wide at prices about half their production cost.

The grain was and remains subsidised and therefore the embedded virtual water was and is also subsidised. The World Trade Agreement of 1995 intended in its provisions on the international trade in cereals to bring about a progressive reduction in production and export subsidies in the United States and the EU. The aim was to achieve a closer relationship between world prices and the costs of production and distribution. The predictions of a modest increase in price (Harmsen 1995) were contradicted initially by the trading events of 1995–1996. By the spring of 1996 prices had reached US$ 240 per tonne compared with about half this level twelve months earlier. By May 1997 they had fallen back to US$ 140 per tonne (World Grain Council 1997). The capacity of the producers to respond to international demand was very powerfully demonstrated in the 1995–1996 agricultural year.

It is not surprising that visitors to the Middle East bringing advice inspired by economic principles learned in the United Sates and Europe should be treated at best with polite scepticism and at worst as being irrelevant. To re-coin a phrase 'Economist heal the economic ills and distortions of your own USDA and EU agricultural sectors before deploying "economist" remedies in the much less diverse and resilient MENA economies.'

The USDA and the EU agricultural systems are monuments to irrational economics and expedient politics. Middle Eastern water policy makers can gain no inspiration from the palpable macro-economic folly of US-EU agricultural policy. In the area of water policy Marc Reisner's (1993) withering analysis of US experience in mismanaging water by ignoring economic principles, would make them doubly puzzled. Reisner's account is enlivened by such conclusions as 'water runs up-hill to money and power', a thought inspired by such ventures as the Central Arizona Project. After examining the economic rational which underpinned most of the schemes of the 1970s in the west of the United States Reisner concluded that in water affairs 'economics are an illusion, politics are real'.

A concluding message

In the light of the non-economic approach vis-à-vis water policy adopted by the successful industrial economies in North America and Europe it is not surprising that Middle Eastern politicians and professionals remain wedded to their WINER belief that Water Is Not an Economic Resource. They believe that their political antennae are sound in detecting the negative social and political impacts of WIER (Water Is an Economic Resource) based policies and certain that there will be political costs for them if they adopt WIER policies. They know that water is a social and an economic resource (WISER) and at the phase of socio-economic development over which they are presiding society has the major voice in their national water discourses (See chapter 8 and Figures 8.1 and 8.2).

A strong conclusion, based on MENA experience and some other evidence from industrialised economies located in arid regions, is that the achievement of economic strength and diversity should be the prime priority. Achieving this status will enable a range of economic policy options to open up, including the adoption of WIER policies as well as of environmentally sound policies.

CHAPTER 5

Water, development and the environment

'Cultivate your world as if you would live for ever and prepare for your hereafter as if you would die tomorrow'.

The Hadith

Bringing risk to consciousness so that it becomes an actual threat is a demanding social process with possible associated political prices.

Not all waters are equal: some are more multi-functional than others

What is meant by water being multi-functional and how are 'some waters more multi-functional than others'? As with all the preceding analyses of the status and management of MENA water resources it is possible to consider the role of water in economic and social development and the impacts of these activities on the environment in relation to the different types of water involved—soil water, surface flows and storage and groundwater flows and storage.

With respect to multi-functionality it has already been shown that water can be used in different sectors and within these sectors it can be used for different purposes and for different functions. Water can be devoted to the agricultural sector, to the industrial and services sector, to the municipal sector or to the domestic sector for household as well as drinking uses.

The extent to which water is multi-functional depends to a major extent on the level of development of the economy. A poorly developed economy in an arid environment has few options but to devote water to the agricultural sector. And within that sector the extent to which water can be allocated to one or a number of different products or activities will depend on the diversity and capacity of the economy including the knowledge and technology which can be deployed. An industrialised economy, through its greater economic competence, will be able to allocate water to a wider range of crop production options than a non-industrialised and less diverse economy; likewise its water allocating options in the industrial and services sector will also be more numerous. Further, diverse and strong industrialised economies will also be able to address new priorities. They will be able to give priority to the state of the environment. In the context of the discussion in this chapter they will be able to develop policy to address the sustainability of a nation's environmental water capital in terms of volume as well as quality.

This second social dimension of water scarcity has led Ohlsson and Turton (1999) to argue that this second component of water shortage is as important as the physical shortage of water in determining the impact of water shortage on a political economy.

Parallel discourses

> Operationally effective, and water deficit ameliorating, global trading systems were shown by outsiders to exist. But awareness of them was of such destabilising potential that the social and political systems of the MENA region were trapped in a 'sanctioned discourse' of non-awareness.

> 'Thus the politics of risk is intrinsically a *politics of knowledge*, expertise and counter-expertise.'
>
> (Goldblatt 1996)

The last three decades of the twentieth century in the North has been a period when perceptions of water resources have been transformed. This transformation was part of the reflexive response to awareness that environments generally were being put at risk in both the North and the South (Pearce 1992, McCulley 1996, Roy 1999) by policies underpinned by the assumptions of the 'hydraulic mission' (Swyngedouw 1999a and 1999b). In the North in the 1960–1980 period, which marked what has come to be known as the end of 'industrial modernity' (Beck 1995, 1996, Beck et al. 1996, 1999), environmental activists and activist scientists influenced communities, constituencies and politicians to operationalise a different evaluation of water environments.

The reflexive response of Northern water science and the communities and polities which manage water in the North has not been taken up by an equivalent suite of activists, activist scientists, persuadable officials and politicians in the MENA region. There is no equivalent of Medha Patkar or Arundhati Roy in the Middle East.

The factor which drives communities into a reflexively modern mode has been identified as globalisation. Peoples, economies and nations have been affected by global markets, global financial flows and the global media. (Castels 1996) These global processes mean that civil society throughout the reflexively modern world is risk aware and dances to daily media-tuned anxieties. As a result of these global processes the reflexively modern political communities are hyper-risk aware. Daily media are full of stories about environmental pollution and degradation. A complicating feature is that the events covered are often not 'natural' but 'manufactured' (Beck 1999) by the same global technical and economic systems of modernity.

The communities of the MENA region are only partially involved in these systems and processes. As a result they are less prone to the construction of environmental risk by the media including awareness of the risk of regional water shortages. Circumstances in the Middle Eastern and North African region are paradoxical with respect to the water environment. The MENA region is the first region to encounter what has been argued should be strategically and economically damaging water deficits. But its communities, professionals and politicians are able to resist the outsider wisdom of reflexive modernity because civil society is, with a few exceptions, not involved in the global information system.

A range of analytical tools from economic, social and political theory can be deployed in an interdisciplinary mode to explain why the impact of the anticipated MENA water deficit hazard has not materialised. The complex interplay of unrecognised economic solutions, belief systems and political processes explains the range of perceptions that determine as yet inflexible MENA approaches to water use and allocation. Operationally effective, and water deficit ameliorating, global trading systems have been shown by outsiders to exist. But awareness of them was of such destabilising potential that the social and political systems of the MENA region were trapped in a 'sanctioned discourse' of non-awareness.

Risk society theory is helpful in providing an overarching interdisciplinary framework providing relevant analytical categories for the technical and the social aspects of knowledge [awareness and non-awareness] and impact. The water stressed MENA communities and political economies have most reason of all the Southern regions to move from water policies associated with industrial modernity to those of reflexive modernity. In practice the era akin to industrial modernity is being extended in the MENA region by the manipulation of awareness of risk by politicians. They have a natural inclination to remain in harmony with the belief systems of their peoples. Belief systems about the fundamental place of water in livelihoods are best left uncontested.

The MENA region is possibly an exceptional example of how the perception of risk, in this case the risk of water shortages, can be manipulated if socio-political circumstances allow such manipulation. MENA water resources are more perceived through a lens of cultivated non-knowledge than through a lens informed by environmental and economic knowledge. Silent solutions to the water resource risk are de-emphasised. As a result the disjuncture between knowledge and impact is extreme. There are 'specific rules, institutions and capacities that structure the identification and assessment of risk in a specific cultural context. They are the legal, epistemological and cultural power-matrix in which risk politics is conducted.' (Beck 1999:83).

The MENA region is a risk society in waiting. It has the major risk of water shortages hanging over it but it does not yet have the capacity, or at least the political will, to interrogate the problem. Even more than risk societies in the North the MENA political economies are 'trapped in a vocabulary that benefits the risks and hazards interrogated by the ... definition[s] of first industrial modernity. They are singularly inappropriate not only for modern (screening) catastrophes, but also for the challenges of manufactured insecurities. Consequently we face the paradox that at the very time when threats and hazards are seen to become more dangerous and more obvious, they simultaneously slip through the net of proofs, attributions and compensation with which the legal and political systems attempt to capture them.' (Beck 1999:83).

Policy makers and those who advise them at local to global levels can be informed by risk society theory. Some axioms of application relevance encountered in the theory and discussed in relation to the MENA region and its water resources are helpful. First, perceptions are local while the industrial way of life is spatially and temporarily open to extend across the globe and the universe. Secondly, risk may be socially visible or invisible. Bringing risk to consciousness so that it becomes an actual threat is a demanding social process with possible associated political prices. Third, and most important, points where risks impact are not obviously tied to points where the risk origins are measurable; there is a disjuncture between knowledge and impact. For example the link between the actual water deficit and the apparent adequacy of the MENA region's water resources is a non-issue because virtual water makes it possible for the link to be invisible.

The MENA case is especially powerful in exemplifying how risk, in this case water shortages, are socially invisible in the region's industrial phase of modernity. They are also invisible at the national level during the transition to post-industrial modernity. Making societies conscious of the risk requires scientific argument and cultural contestation. 'Thus the politics of risk is intrinsically a *politics of knowledge*, expertise and counter-expertise.' (Goldblatt 1996.) There is no better example of this process than the contest of ideas being played out between the insider and outsider professionals over how to value water, including water in the environment, in the MENA region.

New priorities: the environmental value of water

> Time therefore is not on the side of those with a hydraulic mission for the Jonglei wetlands.

Since the mid-1980s environmentalists in industrialised economies have persuaded governments and agencies in these industrialised economies to consider another important use of water, that is the use referred to as environmental use (Lopez-Gunn 1997). This last concept involves the nomination of minimum and maximum desirable levels of timely water in wetlands, in surface flows and in groundwater reservoirs to sustain valued ecologies. The approach rests on the assumption that past ecosystems are desirable and that the precautionary principle be applied.

The precautionary principle has been coined by a group of economists (Pearce et al. 1989, Pearce and Turner 1990). Green economists deploy normal economic principles to determine the cost-benefit of an enterprise, but they reach for different assumptions from those used in normal economic analysis. Green assumptions include intangible costs and benefits which have not normally in the past been considered to have economic value (Burrill 1996). They have persuaded a number of governments and international agencies to generate what are called 'satellite national accounts' (Lutz and Munasinghe 1991).

In brief these satellite national accounts attempt to estimate the value of the degradation and enhancement of environmental capital, for example the degradation or enhancement of a water resource. They argue that a better evaluation of the real costs and benefits present in economies and societies can be gained by taking a more comprehensive view of what the inputs to and outputs of an economic enterprise are (Green 1998). In satellite national accounts negative externalities such as pollution consequent on a resource using activity are estimated as positive impacts such as eco-sensitive improvements of wetland environments. Central to the approach is the notion that future generations should be considered and that the present generation has no entitlement to pass on an environmental resource in a worse state than that in which they found it. The approach is captured in the term intergenerational equity.

The history of the approaches to wetlands in the region reflects the above shifts in ecological wisdom. Until the mid-1980s the automatic assumption vis-à-vis wetland tracts in the Middle East and North Africa was that they should be drained and 'developed' for agricultural production.

Huleh

The history of the Huleh marshes in the northern part of the Jordan Valley is exemplary of the shift of perceptions that have taken place as a result of the transition from industrial modernity to reflexive modernity. The swamps were drained in the 1950s and transformed into what turned out to be not very productive agricultural land. The thinking and practice was part of Israel's hydraulic mission at the time. By the second half of the 1980s the 'productive' approach to water allocation and use was being questioned not only by ecologists but also by economists. By the mid-1990s the environmental value of the Huleh area had become the dominant paradigm. Some of the 'reclaimed' tracts were turned back into wetlands in accord with the shift in the political ecology of such resources. As such they had a new perceived value which counted in national politics, albeit not one that yet entered the national accounts of the Israeli economy. The Huleh development narrative accords closely in form and in timing with the adoption of a green agenda which values wetlands and the environment generally in the reflexively modern North. It is not being argued that the Israeli experience is representative of other political economies in the region. As an industrially modern economy by the mid-1980s the Israeli Huleh story accords with the predictions of those who foresee environmentally reflexive policies. The narrative accords with the argument that industrial modernity has been followed by reflexive modernity.

Other wetlands

Other major wetland development projects, the Jonglei and Iraq Marshes projects, have taken place in very different political economies. They have also been overtaken by the same global shift in perspective on the value of environmental capital. In these cases, however, contested approaches were still very much in play at the turn of the century. The '*politics of knowledge*, expertise and counter-expertise' (Goldblatt 1996) were much in evidence.

The Jonglei wetlands

The perspective on the Jonglei Project on the White Nile changed very greatly between the mid-1970s and the mid-1990s. The project to drain the second biggest wetland in Africa in the 1970s was initiated when there was general accord on wetland drainage. Wetlands contained water that could be used productively somewhere as an economic input and tracts reclaimed from inundation could be utilised for agriculture. This view was held by the government of Egypt and the northern oriented administration of the Sudan. The perspective of the international community, both that of the international funding bodies and the activist NGOs shifted in the 1980s. Later these new paradigms were confirmed by the deliberations of UNCED in 1993 in Rio (UNCED 1992). That wetlands have intrinsic ecological and environmental value for the Sudan and the world community was still by the turn of the twenty-first century not a favoured interpretation by the water resource planners in downstream Egypt and northern Sudan. Nor were the governments and water resource professionals of Egypt and the Sudan sensitive to the idea that for the local Nuer and Dinka peoples the Jonglei natural wetland regime was of economic value as an element in their traditional livelihoods (Howell, Lock and Cobb 1988).

It could be argued that there has been a major assertion of the value of environmental water in the MENA region, or at least on its southern margins, by the communities of the southern Sudan. Armed units of southern Sudanese opposition groups stopped the construction of the Jonglei Canal in 1983 (Collins 1990 and 1995). The peoples of the south of Sudan valued the water and they especially opposed the works which were to send volumes of water to the north without any apparent compensation. The environmental value of the water had also been evaluated by scientists who drew attention to the economic and ecological importance of the Sudd swamps (Howell, Lock and Cobb 1988, Howell and Lock 1995:244).

The political ecology of the Jonglei region has three major players. First, the peoples of the Jonglei region with their local self-sufficiency and sovereignty agenda. Secondly, the governments of

Egypt and Northern Sudan with their hydraulic mission to divert water for use by the populations downstream. Thirdly, the international funding agencies recently converted to a green agenda which value wetlands for their biodiversity, wildlife and as stop-overs for migrant birds. Since the 1980s the international community has aligned much more closely with interests of indigenous communities depending on wetlands. The discourse is very much alive. The longer the Jonglei Canal remains unfinished the less likely is it that the project will be completed.

The Jonglei case is important because it demonstrates graphically the significance of the political factor in the use of water resources. Politics determines whether resources are devoted to draining the wetland if Egyptian and northern Sudanese interests are pivotal. Politics determines whether they are conserved if local interests prevail. After a period in which northern irrigators' interests were being addressed via the drainage project the interests of the southern communities were expressed via armed confrontation. The circumstances of the Jonglei region are special. The wetlands and surrounding inundated tracts are remote with very few people and a low input and low output economy. The logistics of gaining the compliance of the determined local communities of southern Sudan have defied the resources of the would be dominant for centuries. The Anglo-Egyptian and then the British imperial intent was mainly frustrated by opposition groups in the last decades of the nineteenth century and through the first half of the twentieth century up to Sudanese independence in 1956. The pattern has continued since independence.

The future political ecology of the Sudd will be formed by the outcomes of the contention of the three interested parties. The military and political price of subduing the southern communities has proved to be too high for the central Sudanese Government. Meanwhile Egypt is gaining in economic strength and in some circumstances might be persuaded to mobilise the financial and engineering investments to complete the drainage project.

The political ecological context is not stable. An unfathomable factor is the rate at which Egypt will adopt the Northern ecological agenda which inspires international agencies and the international community more generally. Egypt and all developing countries have complex relationships with Northern overseas development assistance agencies and influential governments, notably the United States. Egypt is well able to wrestle with these bodies and assert its priorities. But the global process of constructing the ideas of the value of natural resources and of biodiversity was well established by the year 2000. The widespread adoption of the paradigm will be slow but the evidence is that adoption will take place. Time therefore is not on the side of those with a hydraulic mission for the Jonglei wetlands.

Iraq Marshes

There is no equivalent example of the Jonglei narrative in the heart of the MENA region itself. On the contrary there is an early 1990s example of the reverse process with the drainage of the marshes at the confluence of Tigris and Euphrates in southern Iraq (CIA 1994). In this case arguable economic priorities and strategic concerns of central government in Baghdad had precedence over the interests of the community which has for centuries lived in the marshes (Thesiger 1967, CIA 1994, USGS 2000). Meanwhile hydrological studies show that there was some doubt about the sustainability of the marshes after the completion of all of the Turkish reservoirs on the Euphrates and the Tigris in the first decades of the twenty-first century (Manley and Robson 1994).

The drainage of the marshes reflected the thinking of the 1940s and 1950s when the proposal to drain the marshes was originally suggested by British engineering consultants. It would have been in tune with the international thinking even in the 1970s when the Jonglei Scheme was being constructed. However, by the 1980s wetlands had become significant to international environmentalists. Campaigning against their drainage had become a prominent cause (Amer Project 1994). That the Jonglei was stopped by local dissident groups reflected the incapacity of the central government of the Sudan to impose its will on southern Sudan. In Iraq, however, the Baghdad Government was able to devote military and economic resources to the goal of draining the marshes. The policy reflects the non-environmental priorities of the Baghdad Government as well as its military capacity to impose its will on the marsh communities. Any economic benefit from the drainage is as yet unrealised.

The political ecology of the Iraq Marshes has had only two prominent contending parties and one minor one. The latter is the international NGO, the Amer Project, which has the mission to raise awareness of the interests of the displaced marsh peoples and the reconstruction of at least part of the traditional environment (Amer 1994). The interests of one of these parties, the Iraq Government, had been achieved by the early 1990s. The communities that lived in the marshes were dispersed, mainly to refugee camps in Iran, which bordered the marshes before they were drained. These

150,000 refugees appear to be impotent and their NGO allies though determined are not influential. The discourse of rehabilitation of the marshes has inputs from ecologists and from humanitarian pressure groups on behalf of the marsh communities. The Northern reflexive discourse that wetlands should be conserved would be advocated by Northern interests. But whereas the voice of this third party is increasingly prominent vis-à-vis the Jonglei wetlands it has very little impact in the Iraq Marshes case because the Baghdad Government has the necessary coercive power, albeit probably for a limited period, to impose its non-ecological and non-humanitarian policies. Whereas the Nile basin countries have constructive relationships with the international community that of Iraq faced total alienation and boycott after the Gulf Crisis and War of 1990–1991. The Iraq Marshes are an awful example of how ecological management is subject to politics and the interests of the wise or misguided interests of the temporarily powerful.

The Jonglei wetlands and the former wetlands of southern Iraq exemplify the status of the parallel discourses over water in the South. Outsiders have newly won wisdom on how the wetland should be valued and managed. Insiders from the region vary in their perceptions of what should be done. Insider water professionals and politicians with national priorities in mind want to reduce evaporation of standing water so that water otherwise lost to the atmosphere can be deployed as an economic input. Sometimes they want to reclaim the land for cultivation; sometimes for strategic and security reasons. Insider communities actually living in and around the wetlands who for centuries have derived a living from them resist the drainage proposals. In the case of the marshes of southern Iraq the people were forced to leave. In southern Sudan one of the consequences of the civil war is that the forces of the southern separatist movement stopped the work on the canal excavation and destroyed the excavating machinery. Since the 1980s the insider local communities have found allies in the globalised Northern, environmentally aware, international agencies and activist green non-governmental organisations. The discourses are essentially parallel with the southern professionals driven by a version of the hydraulic mission and the Northern professionals and scientists newly inspired by environmental considerations.

Environmental values

The concept of the 'environmental values' of water had not gained widespread recognition in the MENA region by the end of the twentieth century. There can be no doubt, however, that they will be recognised. Explanations discussed in this chapter will show that political economies will have to evolve if principles of the economic value of the environment are to gain currency. It is concluded that the evolution will be protracted but that it will occur.

At present an economic value is not attributed to environmental water in the MENA region. As the Middle East and North Africa is an arid region most of its managed surface and groundwaters are allocated to agricultural use. Such allocations range from about 65 per cent of water use in Israel to over 85 per cent of such use in Egypt. With populations rising so fast (see chapter 1) the need to expand crop production and therefore to increase the use of water in irrigated farming is everywhere seen as a pressing priority, except in Israel. Other waters devoted to productive sectors of the economy, in for example industry and services, are everywhere minor, rarely amounting to more than five per cent of total national use. Domestic users demand as much as 25 per cent of total use as in Israel but they generally only call upon about five per cent of such use or less.

No national water budgets in the region take account of environmental water although attention is beginning to be paid to such water. Where waters are shared, as they are so often in the MENA countries, the estimation of the desirable volumes of environmental water that should be protected and regulated, not to speak of how it should be valued, are particularly difficult and potentially contentious. It is difficult enough to gain mutual agreement on the actual volumes of water in the systems and to attribute economic values to water actually being used in economic systems when these waters are shared. To take incalculable volumes of environmental water into account is conceptually and practically challenging. To include concepts of the value of environmental water and its significance to future generations is for the moment also impossible to lodge amongst political priorities. There has been little progress in cooperative water management (see chapter 6), with the exception of the relations between Jordan and Israel, and in scientific discussions about future possible management of the West Bank/ Mountain Aquifer (Feitelson and Haddad 1995, 1996, 1997 and 1998). Without such discussions and negotiations there can be little progress in the development of policies which take into account the value of environmental water because so many of the water resources of the region are shared.

The environmental resource, water, is clearly multi-functional and it was established in chapter 3, where the economic features of water were discussed, that these different functions bring to water different perceived values and different exchange values. In users' perceptions water ranges in value from being seen as a free entitlement at one extreme to being a valued designer consumable, as branded bottled water, commanding a price of about US$ 3000 per cubic metre. The prices paid by consumers can be as little as about one US cent per cubic metre, which would cost a farmer in the agricultural sector a possibly deterring US$ 100 to 150 per hectare per year to irrigate for two season cropping. In the industrial and services sector the consumer might have to pay US$ 10000 for the same 10000 cubic metres of water. But the jobs and services provided by industry and services would be 3000 times as great and the turnover might be 10000 times that of the farmer using the same volume of water.

The political ecology of water as environmental capital

> The experience of the last fifteen years of the twentieth century demonstrates that the reflexiveness which industrially modern political economies have through their social adaptive capacity will be deployed to meet whatever risk is constructed to be the most salient.

> ... it is an example of a common circumstance where communities prefer to do the wrong thing extremely well rather than the right thing a little badly.

In order to analyse water as a factor in economic development it is helpful to regard water as a form of environmental capital. As such it will on occasions have a perceived value based on its role in traditional uses. In others it will be a commodity in a market where willing buyers and willing sellers signal value via prices transacted, albeit prices which do not take into account the economic impact of externalities. Economists with an eye to the real economic value of water attempt to attribute costs to normally unrecognised consequences of the use of a resource such as water. They try to estimate the cost of the negative impact of the degradation of quality of a water resource through use. They also estimate the diminution of the resource by over-use or the reduction, or sometimes enhancement, of the amenity value of the water resource through interventions such the construction of reservoirs and the drainage of wetlands. An example would be the loss of a wetland and its amenities through the competitive use of the water resource for example irrigation. Lundquist's concept of the 'triple squeeze', which is a feature in the management of water, helps in the analysis of the value of water (Lundquist 1996). He emphasises the progressive

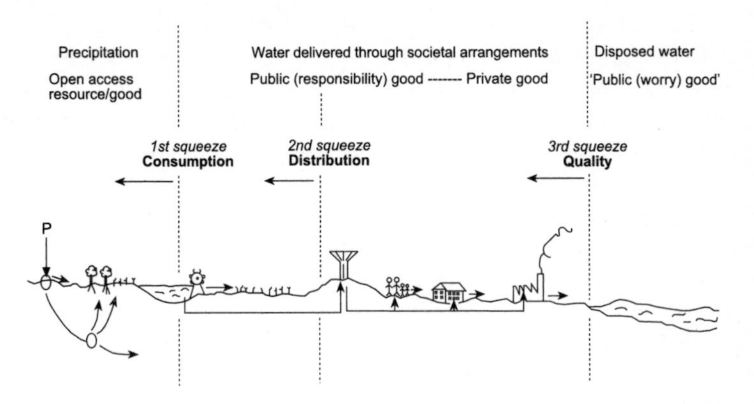

Figure 5.1 The triple squeeze on water resources: consumption, quality and distribution issues

pressure exerted on water resources first through over-use, secondly through the reduction in quality by the intensification of use and thirdly through pollution.

Water is also a public good with distribution and regulation subject to government or local government supervision and pricing. Water is often also a public good subject to no supervision. Users of groundwater and of some river waters are often unsupervised competitors engaged in a version of the tragedy of the commons (Hardin 1968). The region's coastal aquifers of the water short countries of the region have been seriously depleted, in Libya. And there are numerous examples of unsustainable withdrawals from inland aquifers in Syria, Jordan, Yemen (Handley and Dotteridge 1997, Lichtenthaler 1999 and 2000, Mohieldeen 1999), Bahrein, Abu Dhabi and Oman (Oman 1995).

The tendency of communities to regard water differently depending on the level of social and economic development was discussed in chapter 3, (See Figure 3.8) Karshenas (1994) without any apparent awareness that he was describing a trajectory predicted by the social theorists who had discovered industrial modernity (Giddens 1990) argued that it was economic development that enabled new environment and economics aware resource managing options to be contemplated and implemented. Resource reconstructing practices could replace resource degrading practices. The same trajectory is explained by Ohlsson and Turton (1999) by the presence or the absence of social adaptive capacity.

In the Middle East and North Africa the explanatory power of the environment and development trajectory can be explored in diverse political economy contexts. One political economy, that of Israel, expresses the economist inspired trajectory of environment and development Karshenas (1994). It also conforms to the risk theory (Beck 1999) explanations of a reflexively modern political economy in the heart of the MENA region responding to the awareness of environmental risk, namely the adequacy of its water. In a reflexive mode Israel reduced its water withdrawals for agriculture by 30 per cent in the 1986–1992 period. Risk theory also explains the departure from the environmental risk aware trajectory of the 1986–1992 period. After 1992 the trajectory of water use was driven by awareness of a more salient risk, that of the negotiations with Palestine over water. The risk associated with negotiation drove the level of Israeli withdrawal of West Bank groundwater back to mid-1980s levels. As always the single explanation of a change in policy is too simplistic. The major rainfall event of 1992 (see Figure 5.2) was very important in transforming perceptions of groundwater storage policy options vis-à-vis the West Bank aquifers. The rainfall event also allowed the farming sector to re-assert its interests.

The Israeli case is important for any analyst wanting to understand the water policy options and policy emphases available to a reflexively modern economy. Such insights are especially important for those involved in negotiations over water. The experience of the last fifteen years of the twentieth century demonstrates that the reflexiveness which industrially modern political economies have through their social adaptive capacity will be deployed to meet whatever risk is constructed to be the most salient. Sometimes it will be an environmental issue such as water shortages that will be constructed into prominence. But it has been possible to construct a more important risk which has reduced the prominence of the environmental aspect of the water resource sharing issue.

The foregoing discussion is consistent with another body of theory that of political ecology. Political ecologists argue that the interests of those who mainly influence the discursive outcome will as far as they are able steer first, the widely held perceptions of the nature of water, secondly, the knowledge in currency on the volumes and qualities available and thirdly, the value systems that users and policy makers hold. It is in this theory that we find the analytical tools with which to explain current water policies. These analytical tools will also enable reviews of the priorities and options that might engage professionals and government officials in future. They will not, however, enable precise prediction of water resource using trajectories. The degrees of freedom available to a reflexively modern political economy are unprecedented. They enable dramatic changes of priority according to the perceived political expediency of the politically influential.

For the political economies of the MENA region that do not enjoy the options of the reflexively modern the position has been much less comfortable. Pre-industrial and industrially modern economies cannot substitute for deficient water by freely combining other factors of production with scarce water. The politically influential, in the short term at least, have no incentive to construct the idea that water is scarce or that water should be treated as an economic good. They also have no incentive to promote the idea that water policy should be inspired by ideas of sustainability.

A first way forward could be relevant where water resources are shared with a neighbour. Here the option is to argue for an increased entitlement to water. This explanation is widely exemplified in the Jordan, the Nile and the Tigris-Euphrates valleys. The second way forward is to deny the existence of the shortage. This unlikely remedy happens to be currently feasible for the MENA region because of the capacity of the global trading system in grain to meet the region's water needs in virtual form. The third way forward is to invest in productively efficient means of water use. This approach has the virtue of meeting the political interests of both the politically influential and the water using communities they serve. Trapped in this almost optionless policy box policy makers in the pre-modern and industrially modern political economies have few ways forward.

In practice the predicament is an example of common circumstance where communities prefer to do the wrong thing extremely well rather than the right thing a little badly.

PART 3: WATER: INTERNATIONAL RELATIONS AND LAW

CHAPTER 6

Water in MENA regional relations: trading their way to water entitlement

> Not all waters are equal: some are more conflictual than others.

> Clearly hydrological factors, that is the nature and volume of water in the river systems, are not determining of political and economic outcomes. Nor is upstream and downstream location of the riparian in the catchment determining.

> Because the necessity to negotiate over water has everywhere been reduced by the availability of substitute extra-MENA virtual water there has been no imperative to de-link water from the higher order regional rivalries.

National water entitlements: the pursuit of national water rights

> Water entitlements may be achieved by gaining water rights or via secure economic access to high volumes of water? Most likely it will be both.

> Water rights: assertion easy, recognition difficult, attainment impossible. Not very digestible politics.

Conventional wisdom in the Middle East and North African region is that economies facing potential water deficits need to protect their current levels of access to surface and groundwaters. Water deficits are assumed to be potential rather than real in the public fiction of MENA national water politics.

Meanwhile the peoples of the region assume, and their governments accept, that it is the latter's responsibility to ensure entitlements to water. In chapters 2 and 3 it has been shown that Middle East and North African governments have since 1970 been singularly successful in meeting this impossible hydrological challenge in a very Senian fashion (Sen 1981). They discovered that they were able, and can continue to be able, to ensure national water entitlements by developing their economic competence to purchase that entitlement. In this they have incidentally reinforced the myth of national water self-sufficiency. Access to water entitlement has been achieved by politically invisible economic processes rather than by very stressful riparian hydropolitics; and certainly not by finding new water or establishing water rights.

In the real political economy governments have gained water entitlements via economic instruments, such as trade in food. As a consequence the internal water politics remain calm because there are no evident water crises and it appears that everyone's expectations have been met. The capacity of polities to de-emphasise uncomfortable truths by adopting essential lies was a theme of chapter 4.

In the constructed politics surrounding national water resources, (Beaumont 1994) it is the first duty of governments to *assert* water rights, gain *recognition* for them and then attempt their *attainment*. The assertion of water rights is easy; gaining recognition for them is difficult; attaining the goals announced in the political rhetoric of water rights is impossible. In politics this absence of symmetry between assertion and attainment is not crucially important. Imprecisely defined ideas are the currency of politics. One of the functions of politics is to allow such imprecisely defined ideas to be contended and held simultaneously by contending parties. The inevitable attenuation of gaining recognition for water rights and the non-attainment of the asserted national goals over water would make very indigestible national politics. As a result it has only been the occasional assertion of riparian rights, that is the first and unstressful stage in the pursuit of water rights, that has ruffled the hydropolitical waters of the region since 1970. It will be thus for the foreseeable future.

It will, however, be emphasised here that there has been a shift in the MENA international relations regime. This shift has affected regional hydropolitics. The end of the Former Soviet Union's active participation in the region's affairs in 1989/1990 has accelerated the glacial pace of innovation in hydropolitical relations from zero to a perceptible and promising level of activity. A stranger to regional relations would not register the shift but for the cognoscenti the movements in the seriously water deficit Jordan and Nile basins are seismic, unprecedented and encouraging.

Despite the changed regional international relations regime of the 1990s the contending national entities were at the end of the twentieth century still mainly involved in the first stage of asserting water rights over shared waters. In this first stage few institutions exist to achieve a convergence of views. In the water rich Tigris-Euphrates basin, for example, the Turkish government was still at the stage of making facts on the ground on the Euphrates at the turn of the twentieth century. As Turkey is the upstream riparian the impact of such facts are particularly emphatic. The engineering works have had negative impacts in reducing the average, albeit unreliable, flow at the Syrian border by almost 45 per cent, from approaching 30 cubic kilometres per year to under 16 cubic kilometres per year. The positive impact is that the flow has become a reliable average flow.

The second of the three stages in the politics of attaining the recognition of national water rights is politically unrewarding. As a result non-agreed water sharing is an unavoidable reality in present Middle Eastern international relations. The future will be similar because the sum of water demands is greater than available supplies. The third stage, attainment of self-defined water rights, is almost everywhere impossible.

The sequence—assertion, recognition and attainment (of water rights)—should be modified to assertion, *compromised* recognition and *partial* attainment. The disappointments of the second and third stages are wisely avoided by experienced politicians if they have that option. And the MENA economies have had the option, and will continue to have it, because of their ability to trade their way to water entitlements. In chapter 3 it was explained why governments have been able to assert claims to shared water extravagantly. Their economic imperatives have been achievable, not by attaining the extravagantly asserted water rights, but by quietly accessing water in the global trading system.

Assertion, *compromised* recognition and *partial* attainment of international water rights

> There is a record of much assertion, very little recognition and minimal or partial attainment.

The three significant river basins of the region, the Jordan, the Nile and the Tigris-Euphrates, have diverse hydrological endowments and very diverse patterns of individual national demography, resource utilisation and economic performance. The profiles of riparian rhetoric of water rights assertion, recognition and attainment have also evolved in very different ways. No determinism, whether environmental, economic or political appears to provide explanations for the hydropolitical dynamics of these three basins. In this chapter the capacity of international relations theory to explain and predict current and future international water relations in the region will be explored.

The state of hydropolitics amongst Nile riparians is substantially explained by the sequence of three phase political intent identified above. There is a record of much assertion, very little recognition and minimal or partial attainment. Until Jordan and Israel signed up to the water clauses in their Peace Treaty of September 1994, and Israel and the Palestinians agreed to the unsatisfactory water clauses in the Oslo Accord of September 1995, there were very few international agreements over water in the region. There were the Nile Waters Agreements of 1929 and 1959. Also the 1947 relic of the European imperial past, the agreement between Uganda, then still a colony of the United Kingdom, and Egypt relating to power generation on the Nile at Owen Falls. The 1929 Nile Agreement also reflected the interests of the United Kingdom, already something of a waning regional hegemon at the time. At that moment in its imperial vision for north-east and eastern Africa the London government gave prime place to ensuring the water security of Egypt (Collins 1990 and 1995).

Thirty years later, the 1959 Nile Agreement between Egypt and the Sudan, both by then totally independent of the United Kingdom, only determined water sharing arrangements between these two most downstream riparians. The rights asserted by Egypt and the Sudan, and given substance in the 1959 agreement were not, and have never been, recognised by the other seven, after 1993 eight, upstream riparians. The Sudan, though by the 1990s questioning of the 1959 Agreement, found its terms an immense improvement on those agreed on its behalf by the United Kingdom in 1929. The Sudan share of the Nile flow increased from four per cent in 1929 to 25 per cent in 1959.

Partial attainment of water rights had been achieved by Egypt and the Sudan by the end of the twentieth century for the last 3000 kilometres of the Nile and its tributaries. But the 1959 Agreement had not been blessed with the recognition of the other eight headwater riparian partners. As a result the status of the arrangement is vulnerable. The apparent attainments of Nile water rights by Egypt and the Sudan are being questioned, albeit as non-agenda issues, at every international meeting attended by Nile riparian representatives (Nile 2002 1993–2000). In summary water rights in the Nile basin are still at the stage of lively assertion.

Comprehensive recognition of rights, that is stage two, is for the distant future, and the compromised attainment of them, stage three, will be even further in the future. The exchange of diplomatic notes between Cairo and Addis Ababa on Nile waters in May 1997, the first since 1959, was a very positive sign that the distance of a future accord could be significantly brought forward. The evident interest of the international agencies, especially the World Bank, in promoting the pace of discussion between all riparian and especially between key players such as Egypt and Ethiopia, has been proven. The establishment of the World Bank/UNDP sponsored Committee of Experts in February 1997 marked the commitment. The purpose of the Committee was to develop ideas, improve riparian hydropolitics and especially build a capacity to resolve conflicts over water.

Meanwhile the Tigris-Euphrates riparians have not got further than the assertion stage, although Turkey has gone a substantial way to attaining what it regards as its water rights by its construction programmes on the Euphrates without having them recognised by downstream Syria and Iraq.

Only the Jordan-Israel Agreement has seen the culmination of the process—assertion, recognition and compromised attainment—with some certainty that the terms will be durable. Since the recognition and attainment process only started with any transparency in 1993 the process is extremely recent. In the Jordan-Israel Treaty of September 1994 (Allan 1996b:207) the water clauses by no means reflected the asserted water rights of Jordan because the water rights issue in the overall Jordan-Israel relationship was subordinate to other issues. The water compromise took the form it did because contention over international borders, military security and above all peace were higher than water amongst the negotiated priorities for Jordan. The water element of the Jordan-Israel Agreement, was politically pragmatic—it is ever thus—and legally unprincipled, is being implemented painfully but with reasonable expedition. The absence of principle on the relatively tangible issue of water availability has proved costly to Jordan. No provision was made for the inevitable future drought years. The first major drought of 1998–2000 exposed the weakness for the downstream and weaker party, Jordan, of sharing a volume which turns out in drought years to be insecure.

The inadequacy of the agreement was played out in the press.

A dry Israel must cut water flow to Jordan

Water Commissioner Meir Ben Meir yesterday told his Jordanian counterpart that the regional drought will force Israel to cut the amount of water Israel sends to Jordan from Lake Kinneret and the Yarmuk and Jordan rivers by 60 per cent.

The Jordanian commissioner, Dureid Muhsana, rejected the requested cutbacks. Ben Meir said yesterday that water reserves had dropped to their lowest since 1980; he said that if the drought continues, with Israel getting only 40 per cent of its average annual rainfall by winter's end, Lake Kinneret will shrink from 400 million cubic metres to 160 million. As things now stand, Ben Meir said, he has already cut 25 per cent of the water supply to farmers, and he plans further cuts as the drought continues.

Last year, Israel sent Jordan 110 million cubic metres of water, including a one-time delivery of 25 million cubic metres sent at the request of the late King Hussein. Ben Meir said that the drought is affecting the entire region and that Israel, Jordan and the Palestinian Authority would all have to cut back on water supplies. He noted that in the peace treaty with Jordan, no consideration was given to what happens in case of a drought. 'My first duty, and I want to fulfill it completely, is to the Israeli citizen,' said Ben Meir.

(Cohen, *Ha'aretz* 15 March 1999)

Jordan 'strongly' rejects Israeli plan to reduce water supplies

Jordan on Sunday 'strongly rejected an Israeli request to cut 60 per cent of this year's water supplies to the Kingdom promised under the 1994 peace treaty' in order to fend off a drought in the Jewish state, a senior official said. The official added that during a meeting in Jerusalem, the Jordanian delegation dismissed the Israeli argument and demanded full implementation of the peace deal to secure the Kingdom's rightful water share.

'Jordan wants to make it clear that the peace accord has set up our legitimate water shares, and it has nothing to do with a drought in Israel,' the official ??

According to Annex II of the peace treaty, Israel extracts 12 million cubic metres (mcm) of water from the Yarmuk River in summer 'May 15 to Oct. 15 of each year' and Jordan extracts the rest of the river's flow. In winter 'Oct. 16 to May 14' Israel pumps out 13mcm and Jordan allows Israel to pump an additional 20mcm from the river. In return for the additional water that Jordan grants to Israel in winter, Israel agrees to transfer 20mcm to the Kingdom from the Jordan River directly upstream from Deganya gates on the river in summer. In line with the treaty, the two countries last year began construction of a JD1.65 million diversion dam to more effectively utilise 40mcm of the Yarmuk's overflow per year and regulate year-round inflow of water to the 110-kilometre-long King Abdullah Canal. Also in accordance with the treaty, the Kingdom is currently storing winter water in Lake Tiberias, which it reclaims in the summer, supplying 60–80mcm.

'Israel will stand by all its commitments in the peace accord with Jordan and shall continue cooperation with the Kingdom on various issues, including water, as outlined in the agreement between the two countries,' Israeli Minister Ariel Sharon said then in an official clarification after protested Eitan's statement.

(Khatib, Jordan Times, 16 March 1999)

Cabinet meets to discuss water situation, possible shortage. Government maintains demand for full water share from Israel

The government Monday stood firm in demanding Jordan's full water rights from Israel, while drafting contingency measures to deal with any possible water crisis.

'Jordan and Israel have signed a peace treaty, which we insist on implementing as is, and we have officially informed [the Israeli side] of our stand,' Prime Minister Abdur-Ra'uf S. Rawabdeh told reporters after a Cabinet session at the Water Ministry.

Israeli Ambassador to Jordan Oded Eran said his country was not able this winter to collect the 20 mcm from Yarmuk because of a regional drought. Therefore, he said, the situation should be dealt with on annual basis depending on the circumstances. 'The two sides have to understand that when there is shortage, it should be shared between the two,' Eran told the Jordan Times.

(Khatib, Jordan Times, 16 March 1999)

Watering peace

Israel's intention not to provide Jordan this summer with agreed-upon supplies of water, in violation of the peace treaty, has rekindled thoughts of a repeat of last year's water crisis. Last summer, despite the fact that we drew water from Israel, we found ourselves without water for weeks at a time. No running water in homes, no bottled water in the stores. This year, thanks to the Israeli decision, 4.5 million people will have to contend with some 25 million cubic metres less.

While we share the conviction that a real solution to the excruciating water problem can only be regional and comprehensive, we also urge the Likud-led government of Prime Minister Benyamin Netanyahu not only to read the treaty, but to also read peace.

Drought, an 'act of God,' could not possibly be mentioned in the treaty to constitute a 'reason' for lack of full implementation. But the 'spirit of peace' should prompt partners to share whatever resources are available, because satisfying one party's needs at the expense of the other contradicts the spirit of peace.

(Jordan Times editorial 16 March 1999)

The terms of the 1995 Israel-Palestinian Authority 'accord' (Allan 1996:233) over water are also subordinate to a number of even more contentious issues—borders, sovereignty over occupied territory, settlements, refugees, the sharing of a multiply sacred capital city and statehood. Any agreement over water is even more likely, than in the case of the 1994 Jordan-Israel water agreement, to be an unprincipled accommodation to the need to reach agreement on whatever will be the major political and economic issues at the time.

Generalising about whether riparians will achieve the second level of hydropolitical contention, the achievement of recognised or unrecognised riparian water rights is everywhere difficult. This is certainly the case in the Middle East. Where riparians are weak economically there is little chance of them taking water using initiatives which would threaten the waters of another riparian. Where one economy is significantly stronger and more diverse than the others and no other has significant hegemonic power then the economically competent entity has taken the opportunity to assert its perceived water rights without them being recognised by the other riparians. This is true whether the basin is in serious deficit, the case of Israel in the Jordan basin; in significant deficit, the case of Egypt in the Nile basin, or not in deficit, the case of Turkey in the Tigris-Euphrates basin. Israel is the example in the case of the Jordan basin, Egypt in the Nile basin and Turkey in the Tigris-Euphrates basin.

The attainment of recognition of asserted riparian water rights has been frustrated for all the riparians in Middle East and North Africa river basins. Those that have been achieved unrecognised riparian water rights, Israel, Egypt and Turkey, are likely to have to accept reductions in the waters taken by them from their respective systems as the economies of the basins develop. Tables 6.1 and 6.2 provide basic analyses first of the comparative position of each riparian at the end of the 1990s and secondly of the role of the four main factors which influence the attainment of recognised and unrecognised water rights.

Table 6.1 analyses the comparative positions of the riparians of the three river basins in contention. The analysis confirms the diverse pattern of international relations and affirms the arguments of the realist school of international relations (Waterbury 1994, Lowi 1990). Economic strength combined more or less with hegemonic advantage explains the privileged outcome for a mid-stream riparian Israel, a downstream riparian, Egypt and an upstream riparian, Turkey. They have all achieved their unrecognised water rights by dint of being economically competent in the mobilisation of

Table 6.1 The pursuit of national water rights: summary of the international relations of water rights in the MENA river basins

Basins/countries	**Assertion** of water rights	Gaining **Recognition** of water rights	**Attaining** water rights
Jordan basin countries— including Yarmuk			
Lebanon	Secure —geographical (upstream)	Not challenged	**Attained**
Syria [Jordan & Yarmuk only]	Secure—geographical (upstream)	Challenged by Jordan	**Attained**
Israel	Secure—hegemonic	Challenged by Syria & Palestine	**Attained**
Jordan	Insecure—geographical (downstream of Syria)	Denied by Syria on the Yarmuk	Very partial attainment
Palestine [West Bank & Gaza]	Insecure—Dominance by Israel	Denied by Israel	**Attainment very conditional**
Nile basin countries— including Blue and White Niles			
Countries upstream of the Sudan	Secure but unaddressed	Challenged by Egypt with Sudan's 'support'	Latent
The Sudan	Nile Agreement waters only. Challenged by upstreamers	Challenged by African riparians	Achieved but being challenged
Egypt	Nile Agreement waters only. Challenged by upstreamers	Challenged by African riparians	**Achieved but being challenged**
Tigris-Euphrates countries— including Tigris tributaries			
Turkey	Secure	Challenged	**Being achieved**
Iran	Secure	Minor challenges	**Achievable**
Syria	Insecure	Denied by Turkey	No attainment on Tigris/Euph'. [Totally achieved on Orontes & Yarmuk.]
Iraq	Insecure [But is hydrologically well endowed and has major water resources.]	Denied by Turkey	No attainment

Source: the author

Table 6.2 The pursuit of national water rights: factors affecting international relations over water. An analysis of some major and minor national entities

Country/ economy	**Hydrological capacity [Approx. water self-sufficiency %]**	**Economic capacity [Strong & diverse to very weak]**	**Hegemonic power [Strong to very weak]**	**Access to global support [Very significant to very little]**
Jordan basin—extremely water short				
[*Upstream*}				
Syria	70%	**Moderate** [some oil]	Weak	Little
Lebanon	100%	**Moderate**	Very weak	Little
Israel MIDSTREAM	25%	**Strong & diverse**	**Strong**	**V. significant**
Jordan	25%	Weak	Weak	Little
Palestine	20%	Very weak	Very weak	V. little
[*Downstream*}				
Nile basin—significantly water short				
[*Upstream*}				
Entities south of the Sudan	100%	Weak & Vw	Very weak	V. little
The Sudan	100%	Very Weak	Very weak	V. little
Egypt DOWNSTREAM	70%	**Moderate**	**Moderate** [some oil]	**Significant**
[*Downstream*}				
Tigris-Euphrates—water surplus				
[*Upstream*}				
Turkey UPSTREAM	100%	**Strengthening**	**Strong**	**Significant**
Iran	90%	Weak	**Moderate**	V. little
Syria	100%	**Moderate** [some oil]	**Moderate**	V. little
Iraq	100%	Weak—temporary [much oil]	Weak—temporary	None
[*Downstream*}				

Source: the author

accessible water. At the same time they have been sufficiently strategically secure so that neither the threat, nor the actuality, of armed response to their water using initiatives have deterred them. Clearly hydrological factors, that is the nature and volume of water in the river systems, are not determining of political and economic outcomes. Nor is upstream and downstream location of the riparian in the catchment determining.

Table 6.2 examines the major factors which affect the internationally recognised and unrecognised attainments of water rights. It emphasises again that hydrological endowment is not determining, and possibly should not be assumed to be such a prime factor in the debates on water rights and the introduction of international water law.

That a mid-stream riparian in the most water stressed of the MENA river basins has been most successful in its attainment of control over water of any state in the region suggests that non-geographical and non-environmental factors are prime determinants. Economic strength and hegemonic power are the most influential factors. In the absence of economic strength and hegemonic power entities in all basins have not gained the water to which they have claimed entitlement. Egypt, a downstream riparian, and Turkey, an upstream riparian, have both achieved access to what they regard as their water rights, and neither has been as successful as the economically strongest, most militarily secure and the most super-power befriended, Israel, which is a mid-stream state.

The third level of hydropolitical contention, namely the attainment of water rights, has with the exception of the limited agreement on the Nile between Egypt and the Sudan, and the Jordan-Israel agreement, not been achieved in the anarchic hydropolitics of the MENA region. This third level of hydropolitical contention will be discussed later in the chapter after a discussion of some international relations theory.

National politics and riparian water rights

> Winning the water battle at the expense of losing the socio-economic development war would be a negotiating disaster. Winning the water battle will involve negotiating resources to develop a strong and diverse economy rather than gaining too hard won entitlements to small volumes of freshwater.

> 'When nations negotiate, often the toughest bargaining is not between nations but within them. The reason is simple: international agreements, no matter how much in the national "interest", inevitably have differential effects on the

> factional concerns [...] experienced negotiators almost invariably insist that the more difficult part of their job consists not in dealing with the adversary across the table but in handling interest group, bureaucrats and politicians at home.'
> Mayer 1992:793

Strategic national water resources have been shown to be subject to intense political discourse within individual political economies (Waterbury 1979, Lowi 1993). These introverted discourses have resulted in extraordinarily monolithic outcomes in terms of the perceptions of water availability at all levels in a national polity (See chapters 2 and 3 and Feitelson and Allan 1996). These dominating and misleading perceptions have had a profound impact on the way international water issues are approached by riparians and the pace at which they gain any, or a high, place amongst other contended political priorities.

Water is economically vital and at the same time politically symbolic. As a political symbol it is not as significant as threats to, or actual occupation of, territory. Nor is water as significant as versions of territorial sensitivity such as building settlements on disputed territory or sharing a capital city, or as the refugee aftermath of wars. Water's symbolic significance can, however, on occasions be given a high priority in inter-state relations, if it is thought that international relations goals can be achieved through making it a prominent issue. This process is called linkage.

The history of the riparian relations in the Jordan basin testify both to the enduring concern about water and especially to its periodic assumption of a prominent conflictual, or at least an emphatic precautionary, role (Lowi 1993, Wolf 1996a and b, Medzini 2000b). The Nile riparians are the next most concerned about water, reflecting the deficit position of the main Nile waters user, Egypt. But because there has been no significant utilisation of water by the riparians to the south of the Sudan, and the Sudan itself has not used its Nile share agreed with Egypt, there has been no reason to turn occasional strong rhetoric into hot conflict. The Tigris-Euphrates riparians are conscious that there are potential water shortage problems but these will be some decades in the future.

Cognitive mapping: revealing shifting scenarios

Other analysts have addressed issues of the water resource itself and the needs of the economies (Naff and Matson 1984) and the way that political entities view their water resources at different periods (Frey and Naff 1985, Frey 1995, Medzini 2000a). The very useful approach called cognitive mapping has been used to good effect by both Frey and Naff (1985) and Medzini (2000a:223–230).

The important contribution of the cognitive mappers is their revelation of very different perceptions of water that communities and nations have at different periods. Medzini's (2000a:223ff) analysis of the three Euphrates riparians shows how each of them had very different concerns and priorities in three periods, namely before 1960, between 1960 and 1990 and finally after 1990.

Figures 6.1, 6.2 and 6.3, for Turkey, Syria and Iraq respectively, reveal the enormous shifts in perception that have taken place. A number of general points also become clear in the three periods. First neither Turkey nor Syria was much concerned about Euphrates waters as a domestic national issue before 1960 (Figure 6.1). The waters were not used significantly by the upstream riparians and they certainly perceived no water deficits in relation to their water needs. The plans to develop Euphrates water on major irrigation projects were in the future. Iraq was much more concerned about the Euphrates as a domestic issue because it had been using the waters for thousands of years and had rural populations and their livelihoods dependent on the river's unreliable flow.

Before 1960 (Figure 6.1) Euphrates water was counted as a foreign policy issue by all three riparians. But it was not a major concern. The waters touched ideology and unquantifiable perceptions of entitlement but did not influence the approach of governments to the international relations of the Euphrates basin countries.

Between 1960 and 1990 (Figure 6.2) the Euphrates gained a very high domestic profile in all countries. The Euphrates and the development of its waters were significant to Turkey in the relations of the Ankara Government to the poorly developed region of the upper Euphrates. It was here that the alienated Kurdish population lived. The issue of security made relations with downstream Syria of international significance for Turkey because Syria, which benefited from Euphrates flow from Turkey, chose to be an ally of the Kurdish groups opposing the Ankara Government.

Meanwhile the Euphrates assumed a very high profile in Turkish power generation. The Keban and the Karakoya Dams were planned and completed and the construction of the Ataturk Dam was started. In this period Syria also had ambitious plans to command Euphrates waters for irrigation but encountered insurmountable difficulties. Syria also achieved some useful hydro-power generation gains.

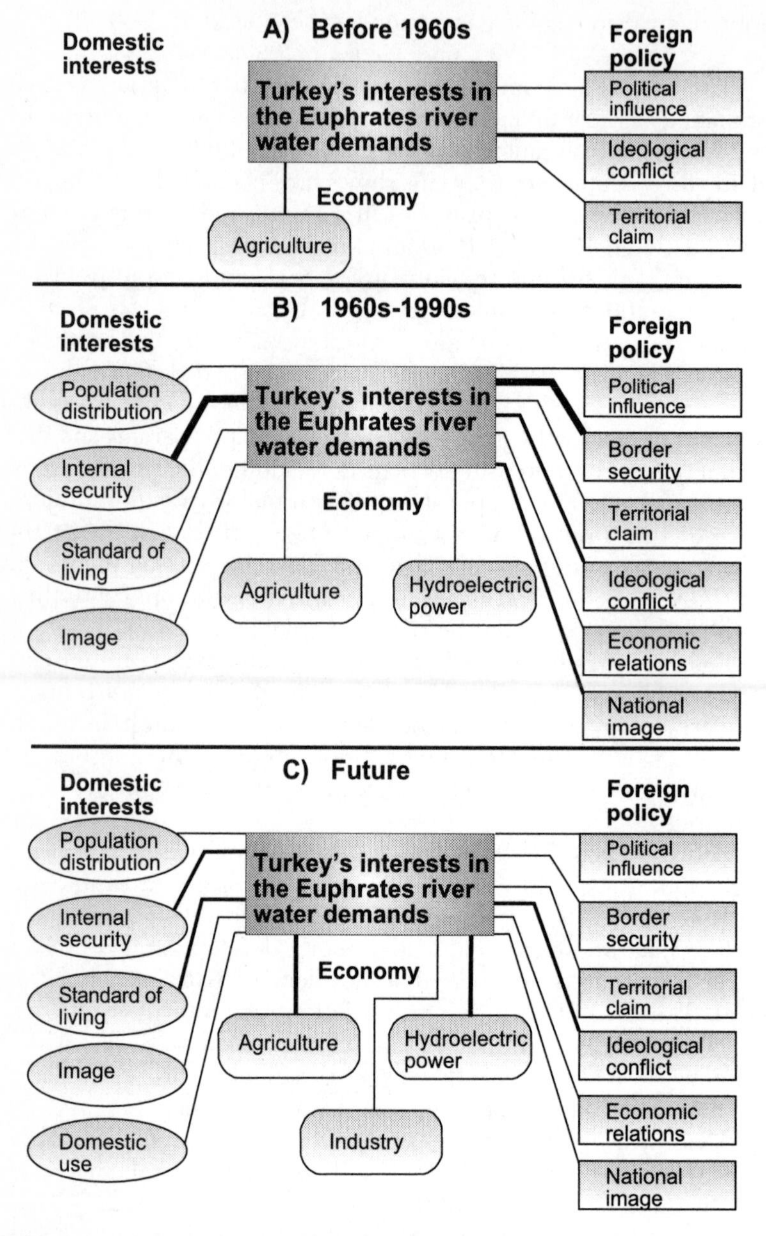

Figure 6.1 The cognitive mapping of the Tigris-Euphrates showing the different cognitive scenarios of the pre-1960 period, the 1960–1990 period and the post-1990 period—Turkey
Source: Medzini 2000a

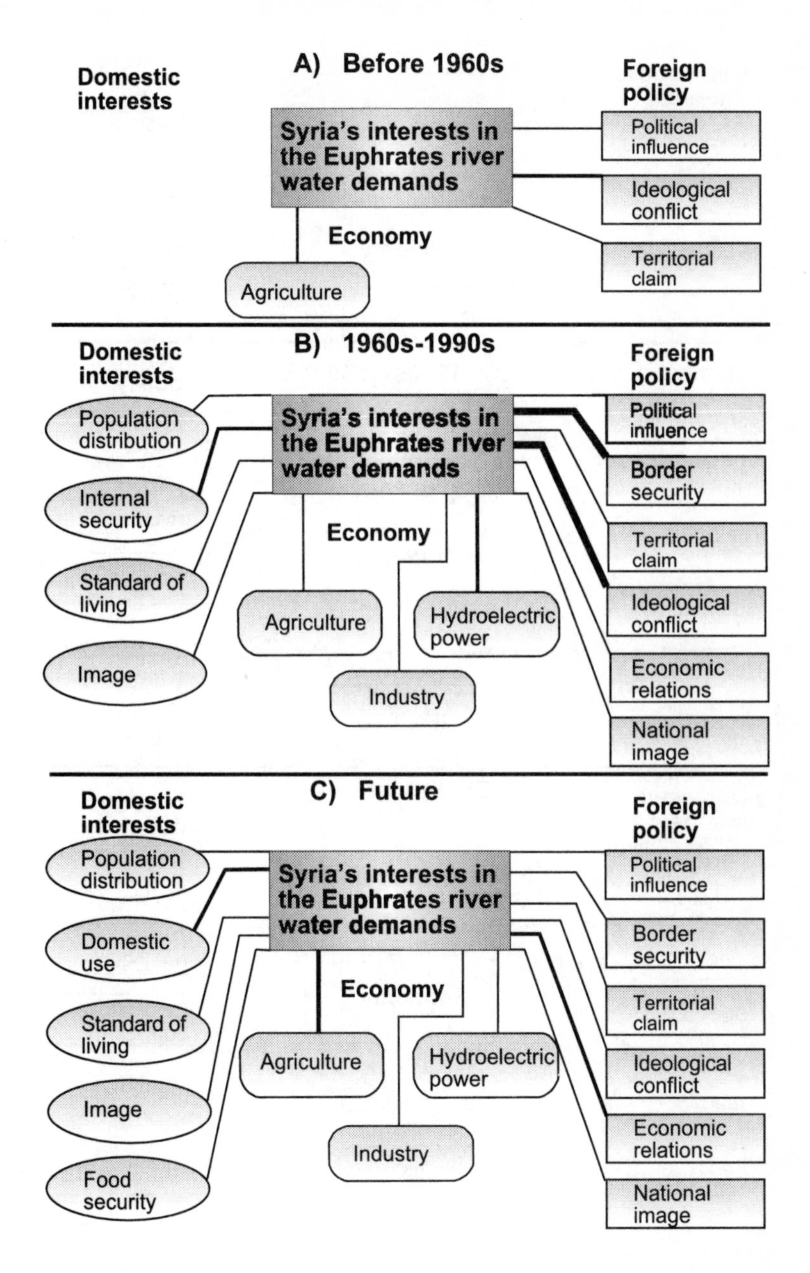

Figure 6.2 The cognitive mapping of the Tigris-Euphrates showing the different cognitive scenarios of the pre-1960 period, the 1960–1990 period and the post-1990 period—Syria
Source: Medzini 2000a

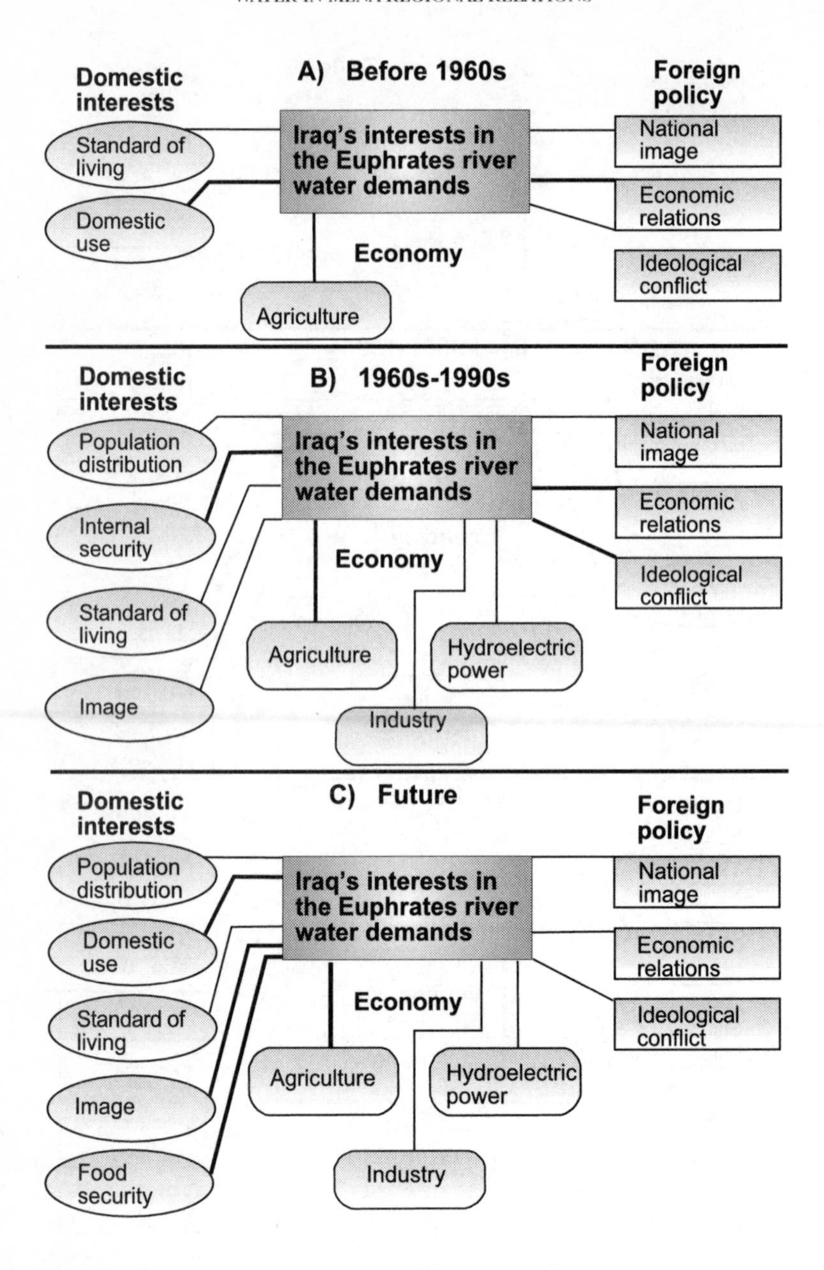

Figure 6.3 The cognitive mapping of the Tigris-Euphrates showing the different cognitive scenarios of the pre-1960 period, the 1960–1990 period and the post-1990 period—Iraq
Source: Medzini 2000a

The main change in perception on the strategic importance of the Euphrates by Syria was the result of Turkey's initiatives in constructing three major dams in the 1970s, the 1980s and the 1990s with other dams, both major and minor, planned for construction up to 2020 (Bagis 1989, Bilen and Uksay 1991, Kolars 1990, 1992b, 1993). The impacts of the construction were observable in the reduction in the flow of the Euphrates. The filling of the Turkish dams required over 100 cubic kilometres of water plus increasing evaporative losses of over three cubic kilometres per year by 1990 from an annual flow regime of 29 cubic kilometres.

From the fifteen years of flow from the mid-1970s to the early 1990s Turkey had to find an average of seven cubic kilometres annually to fill the new storage capacity of the Keban, Karakoya and Ataturk reservoirs plus the increasing evaporative losses. These numbers, adding up to a total of about 10 cubic kilometres annually, are lower than the actual reductions in flow which by the early 1990s had fallen by an average of about 13 cubic kilometres per year to a nominal 15.6 cubic kilometres per year, equivalent to 500 cubic metres per second. The 15.6 cubic kilometres level is predicted and guaranteed by Turkey in future, when the reduction in flow will be the result of the consumptive use of water, mainly in irrigated agriculture rather than the filling of reservoirs. In this future phase the quality of the Euphrates waters crossing into Syria will be impaired as a result of the return flow of drained irrigation water which will have passed through Turkish soil profiles at least once.

Engineering water 'facts on the ground' according to opportunity and economic and military strength has shaped the hydropolitics of the region. Internationally recognised regulatory frameworks on sharing surface and groundwaters have not. Chapter 7 will argue that the absence of recognised international water law occurs because international relations have not evolved to a point where initiatives to establish formal legal frameworks can be articulated, discussed, and finally negotiated. Before legal principles appear as addressable agenda items at inter-riparian meetings, especially with respect to the elusive concept of *'equitable use'* of shared waters, there is a deal of inter-riparian (political) relations foundation building to be done.

Achieving the acceptance and successful implementation of international water law is a very challenging process of innovation. Innovation is a political process. A number of stages are involved. They can be summarised as the phases of knowing, wanting, having, operating and finally operating effectively. In the case of international water law all governments, and especially those in the MENA region know about the principles of international water law, and they are aware of the United Nations International Law Convention of May 1997 (McCaffrey 1998). But with the exception of Syria, which is one of the handful of states that has signed the Convention, only Jordan, Lebanon and Yemen had agreed to sign the draft Convention by the middle of the year 2000. The politics associated with acceptance would be too uncertain. Even if the states did accept the Convention they would find it difficult to mobilise the political will to allocate the resources to implement their obligations. The principles of international law are important, if prone to ambiguous application. The politics of introducing them are monumental.

In the absence of an accessible and acceptable legal framework with which to address contention over shared water, the next section will comment on attempts by international relations theorists to explain and predict international water relations. The purpose will be to throw light on current outcomes and possible future scenarios. The analysis is considered relevant to the inescapable process of building foundations of riparian relations.

International relations theory and MENA waters

Realist, functionalist/idealist and regime theory contributions

Middle East and North African water issues have generated a steady stream of analysis from political scientists and international relations analysts. Those of the 'realist' school (Waterbury 1979, Lowi 1990 and 1993) argue that creating facts on the ground according to economic capacity and strategic strength substantially explain the nature of international water relations in MENA river basins. Waterbury has looked mainly at the Nile but has also made reference to the Euphrates. Lowi's studies have focused on the Jordan but she has also looked beyond the region to South Asia and Africa to develop her realist theoretical framework. Frey (1993 and Frey and Naff 1985) has made valuable contributions by establishing the changing perceptions of communities and national entities on water and food security and on the national symbolic role of water resources. Regime theory has also been deployed by a number of regional relations specialists since 1990 to draw attention to the context in which international water politics operate (Ergil 1991, Kibaroglu 1998). Again their theory emphasises the dynamic, albeit slow moving, internal and external factors that are integral to the process of evolving hydropolitics.

The international relations approach is particularly useful in being sensitive to the overall political context and to many, although crucially as we shall see not all, of the diverse pressures endured and opportunities enjoyed by foreign ministries and political leaders. Water interests are just one concern jostled by many others. The Syria-Turkey confrontation is used by all the analysts to show that Syria used its protection of Kurdish dissidents from Turkey as a lever to gain attention from the Ankara Government for Syrian water interests. Syria was signalling that Turkey's single-minded self-interest assertion over water, a shared economically strategic resource, could have serious political and security costs in another area. (Kibaroglu 1998)

The problem with international relations theoreticians is that their commendable lateral thinking is nonetheless limited. They can only identify politically challenging initiatives and responses within the MENA regional political economy. They appear to assume a closed political economy, closed that is except to privileged superpower intervention. Such initiatives and responses contribute to the conflictual scenarios, for example the current construction of upstream works in Turkey or the initiative to develop Nile water for the New Valley irrigation and industry. They are also very helpful in identifying behaviour, again indigenous to the region, which might reduce conflict, for example an offer by Egypt to invest in Ethiopia's water conservation programme.

To date international relations theorists have, with a few exceptions (Turton 1997 and 1999) presumed that apart from the influence of the international community beyond the region, and particularly the United States since 1990, the only types of behaviour which would contribute to enhancement or reduction of conflict have their provenance in the region. That the effective damping of water conflict has been possible in the region through the effective participation of MENA economies in the global trade in food is not an idea that has been taken on board by the international relations community (Buzan et al. 2000, Homer-Dixon all dates). The problem of using an idea which is invisible to a discipline is compounded by the conspiracy of silence which surrounds the practice in the economies themselves. These blind spots and the deficiencies in the analytical tools have seriously impaired the role played by international relations theorists.

Turton (1997) has attempted to deploy international relations theory in analysing the potential for cooperation over water in the Zambesi River basin. His analysis is different from other international relations theorists in that he has incorporated the concept of 'virtual water'. His approach is of general relevance to water short regions because he attempts to marshal theory as it might affect water relations. He identified three relevant approaches of international relations theorists and some theory from eclectic environmental scientists such as Falkenmark (1979, 1984, 1986a and b, 1987), She has been attempting to draw the water problems of water short region to international attention for over a decade (Falkenmark 1989, 1990).

The first group of approaches is that of the 'realist school' of international relations. The approach is flawed by what a geographer would regard as rather banal environmental determinism especially that of Homer-Dixon and others (Homer-Dixon 1991, Homer-Dixon and Percival 1996, Hudson 1996, McNicholl 1984, Myers 1989, Renner, Pianta and Franchi 1991). Another realist contribution is that of Frey (1993:59) who points out that political relations are much more complex than scarcity of resources alone would suggest. Frey argued that it was necessary to analyse a number of key characteristics of political significance. The following were identified as being of key importance—first, the degree of scarcity, secondly the degree of maldistribution (between political entities), thirdly the extent to which waters are shared and fourthly the perception of the importance of water by a community or nation.

We have already seen that Frey has provided a technique, cognitive mapping, which enables changes in the real and perceived importance of water in domestic affairs. Water was identified as an economic resource for agriculture, industry and livelihood generation and as a factor under-pinning the material and social standard of living of a population. Water was also recognised as a factor in foreign relations through its relevance to national security and as a symbol of national identity and self-image. The aggregate of these factors changes through time and especially as the demographic and water demand circumstances evolve (Medzini 2000a:223ff).

The third exponent of the realist approach who has taken a close interest in the region is Lowi (1993:13ff). In looking at the Jordan basin and researching comparative information in the other parts of Asia and Africa she identified four groups of variables which appear in all the cases she had analysed. First there was the issue of *resource need/dependence*, secondly the issue of the relative power of the respective riparians. The latter is sometimes referred to as the *'rapport de forces'* in the French literature. The term captures the hegemonic as well as the dynamic features of this variable. States within a river basin interact and the nature of participation or non-participation in the rapport de forces varies according to their status as well as their perceptions of their status. Thirdly, Lowi addresses the nature of the political rivalry amongst riparians, within which she notes three characteristics of great significance—first the protractedness or not of the conflict, secondly the concern for identity, recognition, legitimacy and survival and thirdly the perception of the riparian with whom there is a conflict. The fourth issue she identified is the extent to which there have been efforts at conflict resolution.

Lowi (1990:p12) emphasised that only *resource need* is unambiguously linked to riparian issues in theoretical and practical terms. The other variables are associated with a range of international politics theory. *Hegemonic stability theory* and *realism* underpin the rapport de forces; riparian relations is rooted in *neo-functionalist theory*. *Conflict resolution* has a burgeoning theoretical literature (Renner et al. 1991, Priscoli 1994).

International relations analysis of the central issue in riparian relations, *resource need*, echoes the discussion in chapter 1 where scientists were grouped as optimists and pessimists. In resource need analysis there are *realists*, who are pessimists, and *idealists*, who are optimists. In international relations analysis the pessimism is not about the resource itself but about the capacity of contending parties to cooperate. Realists are confident that they can demonstrate that national interests prevail and prevent non-zero sum *functionalist/idealist* outcomes.

The *idealist* school deploys the concepts of the *functionalists* and the *neo-functionalists*. Along with international regime theory these approaches attempt to explain 'why and how states may cooperate in areas of mutual concern such as in areas related to shared water resources.' (Turton 1997:30). Both functionalists and non-functionalists aim to demonstrate that functional cooperation, that is over water resource allocation, management, conservation and environmental regulation could be a preliminary to cooperation over other contentious issues. Such thinking has been increasingly prevalent in the Jordan basin.

Functionalism is an attractive inspiration in the field of international relations. There appear to be a number of examples where the development of a functioning institution to regulate the cross-national activity has led to bigger things and multiplied outcomes. For example the coordination of the production of iron and steel in what was the European Iron and Steel Community in the 1950s has evolved via the European Community to become the powerful and expanding European Union. Power politics contradict functionalist principles. Functionalists urge sacrifices in national sovereignty in order to achieve cooperation across national borders. In the case of shared waters in the MENA region such cooperation could facilitate small endeavours such as the security of water for water deficit communities.

At the international level such cooperation could contribute the foundation building element in riparian relations as a preliminary to grand schemes leading to a peace treaty after decades of conflict. This latter foundation building phase in inter-riparian relations is the essential preliminary to cooperation in water sharing and joint management. It finds a place in functionalist/idealist theory rather than in realist theory, and in conflict resolution theory which flows naturally from the functionalist approach. It is important to emphasise that even if idealist theory has not yet gained the ascendancy it is likely that it will. For those of us who have witnessed the rapport de force at conferences and workshops where advocates of the two schools have been in contention—for example the Harvard 1993 water in the Arab world conference (Rogers et al. 1994), it is possible to conclude that the realists could prevail until 1993 but thereafter the era of the idealists and resolution of conflict theorists could claim to be making an increasing contribution to explanation (Priscoli 1994).

The concept of 'spill-over' (Haas 1980, 1989, 1992) is central to the functionalist predictions. The EU case is supposed to exemplify the phenomenon. The progressive elaboration of the economic union was assumed by functionalists to be a foundation on which the more political aspects of European cooperation could be developed. The European experience has shown, on the contrary, that the process of progressive spill-over is by no means automatic. And in the Middle East there are no straws in the wind that would suggest that there will be a progressive movement via improved water relations towards more comprehensive regional cooperation.

The idea was actively promoted by the Turkish Ministry of Foreign Affairs in the late 1980s and early 1990s. Its spokespeople even used the example of the European Iron and Steel Community as the model for an interdependent grouping of Tigris-Euphrates riparians creating foundation interdependence through the development infrastructures to exchange and trade water, oil and electric power (Arab Research Centre 1993). The Turkish officials articulated clear functionalist principles and argued that the disruption of the infrastructures would seriously and negatively impact all the riparians. The downstream states responded with extreme and classic caution to the Turkish diplomatic initiative with statements of emphasising narrow zero-sum nationalist interest.

How to achieve willingness to cooperate over resources such as water and to create institutions that involve the reduction of national sovereignty is the crux of the hydropolitical challenge. It is difficult to achieve partly because such cooperation is unfamiliar. Those which existed in the region until 1994, the Nile Waters Agreements of 1929 and 1959, were arranged in water surplus circumstances. The 1959 Agreement has been tested by the Gods in the protracted 1981–1987 drought when there would not have enough water in the Blue Nile/Atbara systems to supply the agreed allocations of Egypt and the Sudan. Live storage in Lake Nasser/Nubia was all utilised by the summer of 1987. But in periods of more normal flow, which always include some extremely useful high flows—for example in 1988—the agreed allocation of 84 cubic kilometres annually is available. The Sudan still does not draw, by the end of the 1990s, all its 18.5 cubic kilometres annual 'share' and so there is no reason for there to be any conflict according to the Agreement.

The only expression of willingness to deal over water and water rights is the Peace Agreement between Jordan and Israel. The history of this international water relationship is scarcely stereotypical in that low profile cooperation had been in place for 25 years despite the official state of war. Also the intense involvement of the third party super-power, the United States, was a determining factor in achieving willingness to discuss and then finally sign the agreement.

Haas (1989 and 1992) had already noted that the active involvement of third parties, the super-power, international agencies and NGOs, can be catalysts in promoting willingness to negotiate first steps and especially in creating a context of confidence to take the last step to make an agreement. Such factors as Nye's 'elite value complementarity' (Lowi 1990:21) are also considered to be potential facilitators of willingness to deal. However, in the case of the Middle East it has been comprehensively shown in chapter 4 that the 'value complementarity', that is the perception of the value and availability of water in this case, brings about adamant opposition to having a transparent treatment of water issues. Elites, officials and peoples argue that there is no shortage of water.

Hegemonic cooperation: a critique

Lowi concluded that at the end of the 1980s that circumstances had not yet reached the point that co operation explained by functionalist developments were feasible in the Jordan basin or in the other river basins of the region. She suggested that the hydropolitical dynamics in the Jordan catchment have features which are clearly realist, and based on hegemonic stability theory. At the same time she recognised others which were inspired by neo-functionalist tendencies. The outcome she called a theory of hegemonic cooperation (1990:23). In this she recognised the transitional state of MENA hydropolitics in which both realist behaviour and functionalist tendencies were intertwined.

Her arguments rest too much, however, on a perceived link between water and state survival. (Lowi 1990:364) Such ideas have become part of pervasive and essential water lies for the peoples and governments of the region (See chapter 3). That state survival depends on water security based on indigenous water is a starting point in the argument. Water security is either underwritten by existing access to adequate shared waters or if there is inadequacy then water security is absent. If hydrologically based water security exists then hydropolitical tension is unnecessary. If hydrologically based water security is impossible as it is for all except Turkey, Lebanon and arguably for Syria and Iraq, state security based on such water security is unattainable. In these water insecure circumstances it is argued in the chapters of this book that governments have a choice, either to announce the insecurity or to hide it by sanctioning the topic and preventing it from entering the national discourse. They can only do the latter if there is substitute water, albeit not indigenous water.

The subject of water insecurity has occasionally been cited by national leaders, Sadat in 1976, and by King Hussein and Boutros Boutros Gali in the 1980s, prompting anticipation of water wars. But their anxiety was always about the water insecure future and not about the actual water insecure present that has existed since the mid-1970s. The water deficits have grown in line with the population increases region-wide, 2.5 per cent per year; a doubling rate of thirty years approximately. Yet water wars were even less likely by the mid-1990s than they were 25 years ago at the onset of the regional water deficit.

Why has there been no link between water security and state security? The answer lies in the availability of an alternative supply of water from outside the region via a system, which is not normally seen to be a means of moving and trading water. The import of water intensive commodities such as wheat has very effectively balanced the MENA economies' water deficits. The role of this 'virtual water' is complex and is paradoxical in a number of ways.

The first paradox is that because virtual water is de-emphasised and not part of public or inter-riparian discourse it apparently does not exist. Yet it enables economies to survive and politicians to manage apparently stable, yet seriously water deficit, economies. Fortunately virtual water's contribution is as effective silent and de-emphasised as it would be if it were to be acknowledged and prominent.

One of the very important contributions of 'virtual water' in MENA hydropolitics is in enabling past assumptions to continue to be embedded in water politics far beyond the stage that they are viable. This capacity to extend the life of a protracted rhetorical conflict is the second paradox. Virtual water actually solves the water problem and enables riparians to avoid armed conflict, but at the same time it allows the familiar form of rhetorical conflict to continue. And there are always sufficient advocates in every country to keep the traditional form of the water conflict, linked to the enduring higher order regional conflicts, in currency. The region's unconfrontable belief systems see to that.

Lateral thinking is necessary in this type of analysis. It is essential not to fall into the trap of regarding the MENA region as a bounded political economy as well as a closed hydrological system. There is some justification in regarding the region as a series of shared and closed hydrological systems. It is scientifically derelict, however, to regard as closed systems the regional political economy and those of its individual national political economies.

Lowi tends to see the absence of water as a determining constraint; and thus it is also seen by those who only know about MENA waters. All the governments and peoples of the region are of the same opinion about the 'water security—state security' link. To date the temptation for international relations analysts to compound the problem by emphasising the non-existent link, at least in its banal closed MENA sense, has been irresistible. The essential lie that the individual political economies are water self-sufficient is a deeply reinforced perception and a 'political' fact which has to be taken into account in the hydropolitical analysis. To ignore the *economic* mechanisms which allow the lie to endure is basic survival *politics* for the respective governments. The political survival instinct leads to the sanctioning of discussion of the role of virtual water. For international relations analysts to ignore the phenomenon is, however, an example of the impaired behaviour of narrow disciplinary science unprepared to stray beyond familiar and very limiting epistemic territory.

The MENA water deficit constraint is seen to be modest if one regards water as an element in the overall global economy. Whether there will be sufficient water to meet the needs of water deficit regions such as the Middle East and North Africa in the global hydrologies should be a concern of Middle Eastern governments. In chapter 1 it was concluded that there was almost certainly sufficient water to meet the needs of a doubled future world population and of a more than doubled future MENA population. But the uncertainty is an urgent topic for Middle Eastern scientists to research.

In the narrow perspective of regionally focused international relations analysis the theory of hegemonic cooperation is nevertheless a useful analytical tool to apply to MENA hydropolitics. It takes into account the perceptions which underpin positions, albeit fantasy based, taken by MENA riparians.

The third level of hydropolitcal contention — the attainment of water 'rights'

We can now return to the third level of hydropolitical contention—the achievement of perceived water rights. A central and valuable element of analysis based on hegemonic cooperative arguments is the notion of 'relative gains'. In the protracted water conflicts which span the period of water surplus pre-1970 and the three water deficit decades since no contending riparian has been willing to agree to anything which would 'strengthen' the other party. Recognising the right of a contending riparian to a share of water, and then actually ceding it would enhance the adversary (Lowi 1990:368). But water is just one element in a greater political rivalry. Any gains have to be substantial if one party is to agree to them. But this benefit is precisely the reason that the other contending party would not agree to such cooperation.

Lowi (1990:372) points out that up to 1990 the only functional—that is cooperative—arrangements over water in the Jordan basin had been between Jordan and Israel, the most needy entities until the establishment of a state of Palestine. The upstream riparians do not depend on the upper Jordan waters and use it 'as a geopolitical strategic resource vis-à-vis Israel' (Lowi 1990:372). The neo-functionalist predictions that cooperation would occur had not come about by 1990. The shift in regional regime as a result of the departure of the Former Soviet Union has, however, brought about significant functionalist cooperation between Jordan and Israel, and the beginnings of it between the Palestinian Authority and Israel.

The Nile basin riparians also appeared to be readying themselves for a new international relations regime in the late 1990s. Diplomatic notes were exchanged by Egypt and Ethiopia in May 1997. The World Bank/UNDP sponsored meetings of a Committee of Experts from February 1998 were also auspicious. There were signs that a phase was being initiated at the turn of the twentieth century in the Nile basin which will see the construction, albeit protracted, of the foundations for effective riparian cooperation. These effective foundations are an essential preliminary which must precede the consideration of legally principled or politically pragmatic bi-lateral or more comprehensive water agreements. They certainly have to precede the detailed consideration of the hard to define and impossible to operationalise concept of 'equitable utilisation', coined by international lawyers.

If we leave out of the analysis of regional international relations, including those over shared waters, the removal of the need to cooperate over water through the availability of extra-MENA virtual water, the theory of hegemonic cooperation is helpful. Following this line and making the same incorrect assumptions as the actors in water conflict Lowi concludes that in the Jordan and in the other river basins she researched in the second half of the 1980s—the Nile, the Tigris-Euphrates and the Indus—the theory of hegemonic stability held. The outcomes were that cooperation had only been achieved if the dominant power accepted it, or had been coerced by an external power (Lowi 1990:386). The hegemon took the lead in establishing the regime, accepting change in it, and had the capacity to enforce compliance if the hegemon gained from the arrangement. She concluded that in the absence of external coercion the hegemon will only give priority to strategically significant water resource needs. One has to modify this conclusion to *apparently* strategic water resource needs, in that virtual water is the most strategically important water (see Table 6.2) but being politically embarrassing is therefore silent. She also suggested that to initiate and even enforce cooperation the party would not be a high order riparian that is an upstream state. This is puzzling both from a 1990 perspective when she was writing and even more from a 2000 perspective. Certainly the initiators of co-operation, Israel a mid-stream state and Egypt, a recent initiator, a downstream state, fall into such a category. But since 1990 Turkey, the main advocate of cooperation over water on the Tigris-Euphrates, is an upstream state. Lowi argues that the capacity of Syria and Iraq to create a regime to gain their interests is prevented by Turkey's hydrological advantage and its economic strength. However, the non-water inspired rivalry of the leaders of Syria and Iraq and the extraordinary recent military and economic history of Iraq since 1990 are much more powerful in explaining the international relations of the riparians than water resources.

Lowi's conclusion that hegemonic cooperation would only come about—that is in the future in that there had been no formal co-operation by 1990, except between Egypt and the Sudan, over water in a river basin (Lowi 1990:386–387) if:

- it were to be in the interests of (for example Egypt vis-à-vis the Sudan) and enforced by a hegemon (no examples of enforced co-operation only enforced and non-recognised assertion of rights—for example Turkey and Israel; and also Egypt vis-à-vis Nile riparians south of the Sudan)
- a higher order rivalry would be settled and water conflicts would be resolved as part of the deal (none by 1990 but the 1994 Jordan-Israel Peace Agreement could be an example)
- very needy but economically and militarily weak entities have no viable alternatives. (In practice even the neediest have been able to avoid such environmentally determined behaviour through the availability of global virtual water.)
- cooperative regimes would only come about if the dominant entity wants one or is induced to lead one. Cooperative water sharing arrangements are unlikely to be the basis of a more comprehensive settlement of higher order rivalry. Such co-operation had not come about by 1990. That which has since 1990—Jordan-Israel 1994, and the putative cooperation on the Nile initiated in 1997 are consistent with the argument. But the capacity of non-hegemons, Palestine, Syria and the Nile upstreamers to attenuate the process is as yet untested.

Again it has to be emphasised that these conclusions are viable according to the arguments of political players and international relations specialists who conspire to ignore the underlying water resource reality. It would be helpful if the international relations science community would adopt a more comprehensive set of assumptions which would enable their analysis to embrace both the evident as well as the non-evident factors determining the policy and negotiating options of contending riparians.

Regime theory since 1990

The impossibility of delinking the water issue from 'protracted political rivalry' in the region has been widely observed (Lowi 1990:375, Shapland 1997, Turton 1997:32). One of the dangers of the functionalist and neo-functionalist approaches is that it tends to de-emphasise that water issues are part of a much higher order conflictual scenario. The 'high politics' of the higher order protracted conflict cannot be subordinated to water disputes, and certainly not to potentially diverting cooperative initiatives anticipated by the neo-functionalists. The reverse is emphatically not the case. Water can be subordinated to higher order priorities. Further, because the necessity to negotiate over water has everywhere been reduced by the ready availability of extra-MENA virtual water, there has been no imperative to de-link water from the higher order regional rivalries. This last has proved to be vital and determining of regional international water relations for half a century and its role will endure for another 20 years and might well still be in place in some form by the middle of the twenty-first century.

Despite the circumstances of the 1990s being much more susceptible to functionalist and regime theory the success of the regime theorists to explain developments have been limited. This is partly because they have been carried out by international relations specialists from Turkey. Analysis at the beginning of the end of the era of communism (Ergil 1991) emphasised the functionalist regime theoretical arguments, and later studies by Kibaroglu (1998) and Guner (1996 and 1997) struggle to find evidence of cooperative behaviour despite the assumed potential made possible by new circumstances. The Tigris-Euphrates basin is, however, unique in the region in that its waters are not yet fully utilised and it still has the capacity to meet the needs of its dependent peoples if effectively managed. The need to cooperate is low and there is no external pressure to bring about agreement as is palpably the case over Jordan waters and was becoming the case in the 1990s over Nile waters.

Securitisation: a new approach to water security

> 'Normally the patterns of conflicts stem from factors indigenous to the region. ... [which] outside powers, even if heavily involved, cannot define.'
>
> (Buzan et al. 2000)

Security as a concept

The purpose of this section is to present frameworks of analysis that will help both insiders and outsiders to understand the MENA international relations context in which water is one of a number of contentious elements. It will be shown that water though a significant resource is a minor element in the complex of issues over which the individual states of the region contend. The discussion will be set in the post-Cold War world since 1990 when there has been a significant expansion of the concept of security following the fall of the Soviet Union.

In the event the Cold War was lost by the super-power with the weaker economy. The capacity to fund an intensified arms race of the 1980s proved beyond the political and economic capacity of the Soviet Union. The consequences for the economic, environmental, social and political security of the Soviet Union and its allies were elemental. That the Cold War was won by accelerating the arms race rather than actually using missiles and other armaments emphasised the importance of an expanded concept of security.

Conventional military security had preoccupied the security community throughout the Cold War. The focus of security professionals had been narrow. They were concerned primarily with military capacity and accounting for weapons. The 1990s required a new approach. Awareness that security had numerous dimensions first impacted the security and military services of the Western economies. Without the palpable challenge of the familiar super-power enemy old preoccupations required less attention. There was not only the need for new thinking there were human and budget resources to adopt new lines of analysis. Members of the CIA who had graduated in environmental and Earth science in their student days could make a job-securing virtue out of their environmental science insofar as it enabled insights about natural resource sustainability and therefore environmental security.

The academic international relations community also spotted the need for a shift in emphasis. The work of Buzan et al. (1998) provided a new analytical framework. On the basis of their examination of how security agendas get set they observed that there were 'normal' international relations circumstances and exceptional circumstances. Exceptional circumstances were defined as those where a political economy was faced with clear threats to its security. These circumstances required a different political response to normal international relations. It would be necessary, they argued, to construct awareness that would mobilise the social and political processes to devote new resources to remedying the insecurity. This political process of resource re-allocation was termed 'securitisation'. The politics necessary to bring securitisation about was 'security politics' which were different from 'normal politics'.

The argument is very similar to that of the social theorists who identified the process by which prominence was given to green principles in the 1960s and 1970s. These social theorists noted that 'emblematic events' were necessary to raise the profile of environmental degradation and vulnerability. Hajer (1996) coined the term with reference to water policy and Bakker (1998) found the concept very useful in analysing the way communities, water providers and governments transform their perceptions of water resources and reform water policy. The new approach to security of the international relations theorists also argued that awareness raising was necessary to generate the necessary political will to deal with newly identified insecurities stemming from environmental, social and political vulnerability.

Security complex theory and MENA water

> 'Each region is a security constellation made up of the fears and aspirations of the separate states, which in turn partly derive from domestic features and fractures. Both the security of the ? states and the process of great power intervention can be grasped only through understanding the regional security dynamics.'
>
> (Buzan et al. 2000)

Security complex theory advocates that within an anarchic global system regional security complexes and sub-complexes have established themselves. Within the MENA security complex there are three cores or sub-complexes. There are two main sub-complexes. One is in the Levant and the second in the Gulf. The third weaker sub-complex is in the Mahgreb. If a diagram were to be drawn showing positive and negative interactions between the referrent units, the individual nations, dense lines would be evident in these nodal locations.

The political economies in and around the two main nodes, the Levant and the Gulf, are different (See Figure 6.4). They have different natural resource endowments. The Levant is very deficient in water. Less than three per cent of the water of the region is found there. The sharing of water resources is a significant issue. Its mineral resources are poor. All the Gulf economies have oil. Two, Iraq and Iran, have significant water resources. Water sharing within the Gulf security complex is not an issue despite the water poverty of the Arab states on the southern shore of the Gulf.

If a MENA region diagram were to be drawn showing lines of interaction over water they would not fit the nodes recording the interactions over other factors, crude oil prices, oil negotiations, borders and military security, air traffic, trade generally and tourism.

One set of lines would reflect the hydropolitics of the Nile. Another showing the hydropolitical activity to the north of the Gulf node reflecting the minor level of diplomatic activity over water in the Tigris-Euphrates basin. Neither would coincide with the security sub-complex pattern. The Jordan basin hydropolitical node would coincide with the security node pattern, but here the water issue is so minor in comparison with the other contentious and non-contentious factors that water has to be seen as non-pivotal. It will be shown in the next section that partly because of the minor status of water in the security complex of the Levant the way that water is perceived by one party at least can be quickly changed.

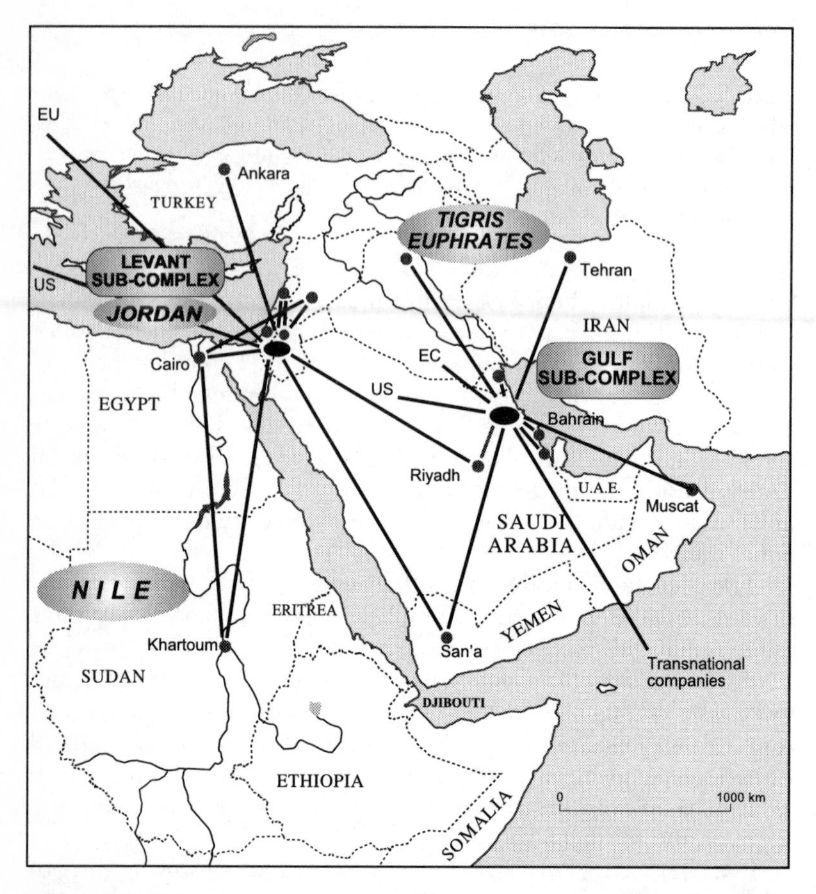

Figure 6.4 The security sub-complexes of the MENA region—the Levant and the Gulf—showing how eccentric the major hydrological systems and their related hydro-politics are to the dominant regional strategic security sub-complexes.

With respect to the MENA region's security sub-complexes, the Levant and the Gulf, interactions over water resources are either minor, as in the Levant, or their focus is separately located, as in the Gulf, from the main nodes of interaction. In these circumstances it will be difficult to make sense out of the positions adopted over water by the contending entities. In the Jordan basin negotiation over water was always overshadowed by five or more other prominent, important and symbolic issues during the peace talks of the 1990s and after. And though water was not treated as an economic resource in the peace process, despite the recommendations that it could be (Fisher 1994), in practice it is the very fungible economic qualities of virtual water that has made water into a relatively minor negotiation issue.

Water in the MENA region is a suitable candidate for analysis according to the framework of the 'security' theorists. It has been shown that the manipulation of awareness of the status of water resources is a fundamental feature of the water politics of the region. The case of Israel will be examined to determine if security theory is a useful explanatory framework.

Israel and water security theory—1950s–1990s

Water resource policy in Israel has been subject to a number of clear phases (Feitelson 1996). It has been established that while the scope and reach of the water issue in the Levant coincides with that of major security sub-complex water is not a high security priority compared with other issues such as statehood (for Palestine), Jerusalem, settlements, borders and refugees.

The Israeli case is, however, very useful because it is possible to investigate the nature of security politics with respect to water and why shifts in water policy occurred. Security theory argues that normal politics can be transformed into security politics when it is perceived that an issue in contention has increased in importance.

In discussing water and environment we have seen that the process of socio-economic development brings with it the capacity to perceive environmental resources differently (Karshenas 1994). In the early phase of socio-economic development natural resources tend to be used and degraded. Groundwater reserves tend to be depleted. In later phases developed economies have the capacity to reconstruct natural resources.

In the 1947–1986 period water resources for Israel were developed via a policy which regarded the absence of food self-sufficiency as a risk. Water was not a normal politics issue. It was a security politics issue. In the period 1970–1986 green activists, sympathetic activist environmental scientists and government officials allied to construct a different and more persuasive risk. The over-use of water in Lake Kinneret/Tiberias and from the West Bank aquifers was unsustainable and its continuation would destabilise the economy. The Israeli public was sold the idea that there was a scientifically based threshold level, a red-line, below which it was dangerous to lower the level of Lake Kinneret in the Jordan Valley. So successful was the idea of the red-line that hydro-geologists concocted one for groundwater levels in the Western Aquifer of the West Bank. So strong was the evidence of the unsustainable progressive declines in the level of storage in this aquifer that the Water Commissioner was persuaded to respond to the two drought events of 1986 and 1992. Sharp reductions in the levels of water delivered to agriculture were imposed.

The two phases of security politics based first on the basis of the perceived risk of food insecurity and secondly on the basis of environmental insecurity were both revealed to be political constructs. In the period 1992 to 2000 a third phase followed. Again the politics were not normal. With the start of the Peace Process the Israeli attitude to water resources changed. The peace negotiations presented risks. Both Jordan and Palestine would have to be accommodated. As a downstream riparian on the West Bank Aquifer Israel behaved as a normal downstreamer and began to increase withdrawals to increase its level of 'current use'. Such a statistic would have substance in negotiations. A share of a large assumed annual flow of water, some of it unreliable, is almost always safer than a share of smaller assumed flow for a downstream riparian.

The Israeli case shows that security theory explains what has happened but it does not predict what will happen. The three trajectories—see Figure 6.5—are all consistent with security theory but there is nothing in the theory that would predict when the shifts in the prioritisation of perceived risk would occur. Buzan and Waever (2000) observe in a somewhat realist tone that 'Within the structure of anarchy, the essential structure and character of security complexes are defined by two kinds of relations, power relations and patterns of amity and enmity'. Arab-Israeli contention has been played out in circumstances of unequal power and in versions of high enmity. The Israeli water policy appears to have depended on a number of factors—initially its awareness of food insecurity, next environmental sustainability and finally its awareness of the likelihood of the loss of control of negotiable water in peace

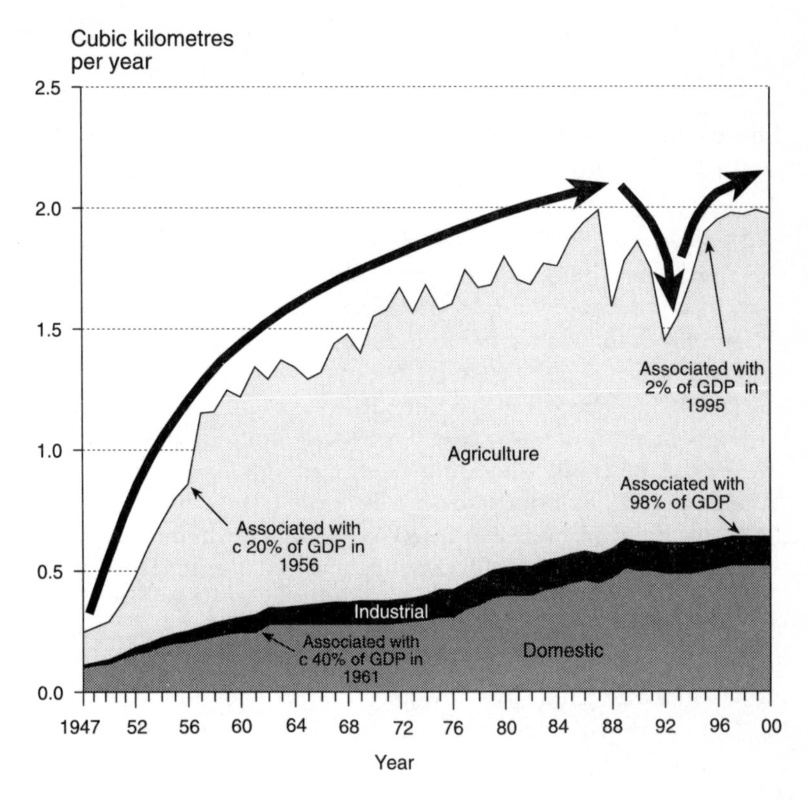

Source: Israel Bureau of Statistics

Figure 6.5 Three trajectories of water resource use in Israel: 1947–1986 when perceived food security risk determined water security; 1986–1992 when a wareness of environmental risk determined water security politics; 1992–1999 when awareness of peace process risk determined water security politics.

negotiations. Meanwhile Israeli economic power means that it has easy access to strategic virtual water from the global system and the capacity to meet its domestic and economic water needs by desalinating sea water.

An attempt has been made in this interdisciplinary analysis to show that different voices can be heard at different moments in the international relations discourse. It has been shown that the international relations over water in the MENA region security theory is saying useful but very similar things to risk theory (Beck 1996), to social theory (Hajer 1996) and to political ecology theory. For water specialists it would appear that political ecology is the most comfortable framework in that the realist dimension of politics is captured as well as the importance of the intrinsic value of the environment.

The rainfall event of December 1992—high rains as well as droughts have hydropolitical significance

As already stated no change of water policy comes about because of a single factor. The 1986 Israeli reflexive and 'sustainable' policy was partly the result of a drought, partly the result of the green discourse of the preceding 15 years and partly a response to political pressure from the United States to reform the economy. The change of policy which occurred after 1992 was also caused by more than one factor. The Peace Talks scenario led to a reappraisal of the policy on levels of current water use. There was another very important environmental event which impacted thinking. Twice the normal annual rainfall fell during and after December 1992 (See Figure 5.2). The West Bank groundwater reservoir recovered within weeks to the level of a decade earlier. Overnight there was a transformation in the approach to the management of the West Bank aquifers. A high level of storage was no longer a virtue. In the event of major rainfall events it was necessary to have substantial storage capacity to retain the storm water. Otherwise it would flow uselessly to the sea. Both the Rabin and the Netanyahu administrations pursued the same policy of allowing increases of water use in the agricultural sector. There were some modest increases of demand resulting from the million additional immigrants from the former Soviet Union from the late 1980s. Figure 4.4 shows the groundwater levels in a group of wells in the West Bank—1968–1998—and the progressive declines in the 1970s and 1980s, the dramatic recovery following the rainfall event of 1992 and the results of Israeli storage policy at all stages.

External powers and hydropolitical regimes

The capacity of external powers to influence water affairs in the region is clear. The pace of Jordan and Yarmuk waters development was significantly determined by the United States in the 1950s and especially in the early 1960s (Lowi 1993, Wolf 1996a and 1996b, Medzini 2000b). The Eisenhower government took the view that the cooperative management of Jordan water would enhance the likelihood of a comprehensive peace between the Arabs and Israelis as well as enhancing their economic capacities. The model was the US Federal Government led inter-state cooperation exemplified by the Tennessee Valley Authority. A relic of this thinking is in the title of the Jordan Valley Authority in Jordan.

The failed initiative of Eisenhower's water ambassador—Ambassador Johnston—who devised the Johnston Plan discussed by the Jordan riparians and the Arab League between 1952 and 1955, is a testament to the subordination of the water issue to the wider regional rivalry. The plan gained some technical acceptance especially by Jordan and even the Arab League engineers, but it could not be ratified because the Arab League political level officials could not cede resources which would enhance Israel (Lowi 1993).

The protracted struggle by Israel to obtain US financial support for its Water Carrier in the early 1960s and the balancing investments by USAID in the Jordan Valley of Jordan are further examples of super-power interest in water. The Former Soviet Union was also active in the Nile Valley when it picked up technical and financial obligations in connection with the construction of the Aswan Dam and in the associated land reclamation endeavours which the new Aswan water made available in the 1960s (Waterbury 1979). The Peace Agreement of 1994 between Jordan and Israel was very significantly facilitated by the United States as was the 1995 accord between Israel and what has become the Palestinian Authority.

The United States will also be a key player in supporting the initiatives of the international agencies—The World Bank and UNDP—in the 1997 initiatives to advance cooperation amongst Nile riparians.

The commitment of the remaining super-power, the United States, since 1990 has been shown to be without consistency. The US election of 1996 appeared to influence the officials responsible for Middle East policy to reduce the priority of the Peace Process to a minor level compared to the 1993–1995 period. A Syrian-Israeli deal was always going to be more difficult than a deal between Jordan and Israel or between Israel and the Palestinian Authority, but the de-emphasis by the United States of the Peace Process following the assassination of Prime Minister Rabin and the Israeli election in the spring of 1995, was very evident.

Inter-riparian relations foundation building

The importance of a grasp of international relations theory and the place of water in such theory has not been negated by their capacity to explain rather than predict. The theory is very useful in highlighting the factors which contribute to water conflicts and their alleviation in the MENA region. An understanding of the nature of the perceptions of political leaders and of the ill-founded assumptions on which these perceptions have been, and will tend to continue to be based, is also necessary if sound international relations analysis is to be achieved. There has been a tendency for international relations analysts to assume that only the public statements of regional leaders on water, conveying a traditional and politically acceptable position, needed to be taken into account. The underlying global economic systems which allowed the constructed positions to appear viable are just as important.

There is no sign yet, except in Israel, that the connection between economic diversity and strength are very closely connected with water security. But the process of spreading the good news that water is an economic resource which through trade can be accessed effectively is gaining ground against the entrenched but untenable position of those who still insist that water is a zero-sum issue.

An important message of the discussion in this chapter is that while idealist/functionalist arguments have still not achieved wide adoption they are gaining currency. The most important message is that the achievement of a cooperative mode in each of the region's river basins, and the development of institutions which promote such cooperation, will be an essential preliminary to introduction of legal principles into the discussions. Such principles are contentious and have proved elsewhere to be impossible to operationalise if parties are unwilling to deal at a preliminary political level. Politics tend to be unprincipled. Hence the problem of making contentious legal principles part of pre-negotiation foundations.

Recurring problems: some high profile water disputes; the Jordan, the Nile and the Tigris-Euphrates

Despite the solutions of socio-economic development and virtual water which have ameliorated the problems of water shortage in the Middle East and North Africa region in the last quarter of the twentieth century there are points of friction which keep recurring. The most prominent in media terms involve the countries of the Jordan catchment. The initiatives of Turkey in the upper Tigris-Euphrates also receive a deal of attention. Meanwhile the riparian relations on the Nile, which have been quiet for some decades, could in the coming years be the focus of cooperative if sometimes tense progress and will therefore receive more attention.

The Jordan: Syria, Lebanon, Israel, Jordan and Palestine

The waters of the Jordan and of the groundwater of the West Bank are strategic issues for Israel, Jordan and Palestine. (Beaumont 2000, Wolf 1996a and 1996b, 2000) In practice the 1993–1994 peace talks between Jordan and Israel culminated in September 1994 with a deal which provided a phased 200 million cubic metres per year partly out of Israel resources and partly from jointly developed resources, in 50 million cubic metre increments. This was less than Jordan wanted but was an acceptable figure in a complex deal which achieved peace and deals over land as well as water.

It has been shown that one of the unavoidable challenges for those making international deals over shared water resources is that the discussions on water are always linked with other issues. The position is made even more complicated because the extent and nature of the linkages changes during the progress of the negotiations.

The next two phases of negotiation, with Palestine and Syria, began in 1995. The Oslo Accord of 1995 addressed the five major issues facing Palestinian and Israeli leaders and negotiators; statehood, territory, Jerusalem, settlements, refugees and water. When the talks started all the issues appeared to be very significant. In the years since Oslo the hand over of the West Bank territory and the related issues of settlements have been the priority topics. Jerusalem is closely connected with both settlements and territory and will take its place at the top of the negotiating agenda in due course. The fraught last years of the century have shown that while water was both a symbol of Palestinian national interest and an important current and future national contended resource, it was also a foundation element in potential cooperative relations between Israel and Palestine. This was true both during and beyond the final stage negotiations.

The reason that water would be a source of cooperation rather than long term conflict is that the solution to the water problems of both Palestine and Israel lie beyond the water resources of the Jordan River and the West Bank shared variously by Jordan, Israel and Palestine. The major water resource in terms of volume for all these economies now and in the future lies not just outside the Jordan basin. It lies outside the MENA region. The gross water needs of the Jordan riparians can only be met by importing 'virtual water'. The capacity to afford such imports is a problem of socio-economic development, in the achievement of which water plays a part. But it is not a determining part. Other factors of production such as capital and human capital are the crucial determinants.

It is not yet clear to all Israelis that strong economies in the West Bank and Gaza would be an essential condition for a peaceful and secure Israel. A sufficient and growing minority is aware of the developmental pre-requisites for a comprehensive peace. The discourse has been moving slowly but favourably despite the high profile politics over land. The water issue would not significantly slow either the accord or the socio-economic development.

Agreement with Syria and Lebanon seemed imminent in the early part of 1996. Golan has been perceived by both Syrians and Israelis as a key element in peace negotiations. Significant resources have been devoted to gaining and retaining it and much national emotional energy has been expended in anticipating a new status for the tract and its water. The terrain is militarily strategic but the waters which flow from Golan into the upper Jordan River are minor. They comprise only about five per cent of current Israeli water use and about one per cent of those of Syria. Because of their location they could not easily be used economically in Syria. The Banias tributary provides about 130,000 cubic metres per year; other Golan streams provide minor flows into Lake Tiberias/Kinneret. It appeared that Israel had digested the discomfort of returning the Golan territory and its water to Syria by 1995. Syria was still not yet ready to make a deal as late as 2000, but water would not be a significant impediment to such a deal.

Lebanon exports a similar quantity of water to that exported by Syria to the Jordan River. Lebanon's Jordan tributary is the Hasbani tributary. Lebanon was not disposed to make a deal over water during the Israeli occupation of southern Lebanon. Nor is it disposed to cooperate over water with Israel despite enjoying a water surplus. Israel's past interest in the Litani waters is an ongoing concern for the Lebanese people and their Government. A period of adjustment of a decade at least will be required before the Government of Lebanon will feel confident about discussing water with Israel.

The diplomatic distance between Lebanon and Israel is reinforced by the close relations between Syria and Lebanon which are very much determined by Syria. Any enhancement of Israel's water resources as a result of a rapprochement between Lebanon and Israel will have to have the approval of Syria. There is a good economic case for both parties for the sale of surplus water by Lebanon to its southern neighbour but the political moment for such an initiative was not close at the turn of the twenty-first century.

In brief the Jordan riparians are amongst the most badly endowed with respect to water in the region. But the manner in which Israel has coped, and Jordan had begun to cope by the 1990s, with the problem point the way to secure strategies of appropriate allocation for all the Jordan riparians. Appropriate allocation, together with sound water re-use policies will enable the development of livelihoods for individuals and the generation of economic strength in the economy as a whole so that the strategic water to balance the national water budget can be imported.

The Tigris-Euphrates rivers

The Tigris-Euphrates system is shared by Turkey, Syria, Iraq and Iran. The system was still in surplus in the year 2000 in that use does not yet take up all the water available. Turkey supplies over 90 per cent of the Euphrates water and about half of the Tigris flow. The rest of the Tigris flow comes from the Iranian tributaries.

Dam building by Turkey during and since the 1970s has changed the regime of the Euphrates. The Keban (1976) and the Karakoya (1988) Dams, each designed to store about 30 cubic kilometres of water were part financed by loans from the World Bank. Non-controversial power generation and the control of flow were the purposes of these structures. The largest dam, the Ataturk Dam, with 46 cubic kilometres storage, was designed for consumptive water use in irrigation projects as well as for non-consumptive hydro-power. The reservoir filled in the first half of the 1990s. The project had to be financed by Turkey because the World Bank has an operating directive (OD 7.50) which requires that any structure which might 'harm' another riparian could only attract the Bank's concessionary finance if all potentially affected riparians approved the project. No such approval was forthcoming from Syria and Iraq.

The hydrological impact on the Euphrates of the new storage has been as predicted. The, albeit very variable, flow of the river at the Turkish-Syrian border had fallen from about 30 cubic kilometres per year to just under 16 cubic kilometres per year by the 1990s. Since the mid-1970s the gross reduction in flow of about 150 cubic kilometres has been the result of filling reservoir storage plus evaporation therefrom at the three major, and at other minor, structures. 150 cubic kilometres amounts to five years total Euphrates flow or nine years of the 47 per cent of flow which is the 500 cusec average rate of flow promised by Turkey to Syria. Turkey argues that the flow is now reliable and assures its southern neighbours that the lower average flow will be guaranteed even after serious consumptive use of ten or more cubic kilometres per year will be the norm when the irrigation projects are constructed and operating.

Because the annual flow was unreliable the average use of the flow by Syria and Iraq was never above 15 cubic kilometres a year in total. Had Syria's 1970s irrigation projects been more successful the consumptive use downstream of Turkey would have been greater. Meanwhile Syria was continuing to develop the irrigation potential of the Euphrates in the 1990s by diverting water to projects near Aleppo where soils are more productive than those in the Euphrates lowlands which proved to be so unrewarding in the 1970s.

Turkey was also beginning to develop its Tigris waters in the late 1990s which meant that there will be reductions in the river flow and improvements in the reliability of flow. The Tigris is mainly a Turkey-Iraq concern. Syria is only a riparian of the Tigris for a few tens of kilometres and the terrain makes the diversion of water into Syria costly.

Iraq is the main beneficiary of the Tigris and any problems which might emerge on the Euphrates as a result of Turkish upstream engineering could be made good by the development of River Tigris waters. Iraq, however, was in an extraordinary and temporary economic predicament during the 1990s. In July 1990 before the invasion of Kuwait, Iraq was importing 90 per cent of its food, implying that it was importing about 90 per cent of its water needs. In the circumstances of the prolonged embargo there have been a number of significant shifts in the Iraqi economy. High priority is now being given to food production, especially irrigated crops. However, the necessary and sound emphasis being given to the use of land and water resources is hampered by the shortage of skilled farm labour. In the years of high oil revenues of the 1970s Iraq came to rely on Egyptian expatriates to run and work its farms. In August 1990 the extent of the continued reliance on Egyptian expatriate staff and field labourers was revealed when a million Egyptians plus their families congregated at the Jordanian border to escape the 1990 Gulf Crisis.

The problems faced by Iraq in making use of its very considerable water resources—it is almost as rich as Egypt with a much smaller population—is not the water volume at its disposal, and not yet the water quality of that water. It is the intractable nature of its low lying and saline soils and the absence, as yet, of appropriate rural institutions to mobilise the exceptional levels of expertise and technology needed to develop productive irrigation in an unfriendly soil environment.

The issue of water quality has been a legitimate concern of Syria and Iraq. When the Turkish irrigation schemes north of the Syrian border come on stream they will drain saline water back into the Euphrates. Such consequences can be monitored and international precedent has demonstrated—in the United States where the US now pays to remedy the pollution of the Colorado where the river crosses into Mexico—that the costs of this type of pollution can be determined and remedies prescribed so that costs fall on the polluter. The Turkish approach to this issue will reveal whether its cooperative initiatives will be given substance by a responsible approach to monitoring and treating the flows of the Euphrates and the Khabur rivers as they cross into Syria after receiving the drainage water from irrigation schemes in southern Turkey.

The Nile

'Too much history for some: too little water for others'

The use of Nile waters has a 6000 year history. For all except the last 30 years of the twentieth century the only major user of the Nile waters, Egypt, had sufficient, safe and secure water. Such an apparently endless awareness of security of supply and a belief in an assumed entitlement to it cannot be changed by new knowledge based on a few carefully researched hydrological and demographic numbers. The Egyptian certainty about its entitlement to Nile waters is accompanied by an eternal vigilance about the state of the upstream riparians and their capacity to control and consumptively use the water upstream.

In the second half of the nineteenth and the early twentieth century the imperial presence of a very Egypt aware United Kingdom established a power structure of a sort across much of the Nile basin. The power structure and the rudimentary communications network enabled the imperial power to survey and monitor the hydrology of the Nile relatively comprehensively (Hurst et al. 1966, Collins 1990 and 1995). The terms of the 1929 Nile Basin Agreement which allocated 96 per cent of the average flow to Egypt and only four per cent to the Sudan reveal the priority given to Egypt's interests by the United Kingdom.

After independence, nominally in 1923 for Egypt, but in a concrete form in 1952 with the revolution led by President Nasser, and for the Sudan in 1956, it was possible for Egypt to take a radical approach to securing the sufficiency and quality of its future water resources. The construction of the Aswan Dam impounding 168 cubic kilometres of Nile flows when full required an agreement with the upstream riparians. In the event only the Sudan was prepared to enter into an agreement, the Nile Waters Agreement of 1959. Egypt recognised the entitlement of the Sudan to a much greater share of the Nile flows and the agreement allocated 25 per cent of the flow to the Sudan, compared with only four per cent in 1929.

Egypt's strategy in Nile basin relations has been impressive in its attention to detail and in its global scope. Egyptians were closely involved in the agreements over the Owen Falls hydro-power structure in 1954. They still have had an observer at the site ever since to monitor flows. They were also financially involved in the aborted Jonglei Canal project. The Jonglei Canal was one of the projects for which the terms of the 1959 Nile Waters Agreement had made provision. New water gained from such projects in the south of the Sudan would be jointly financed by Egypt and the Sudan and the benefits would also be shared. This expression of bi-lateral cooperation was based on an approach not otherwise initiated by Egypt. In all other dealings in the Nile involving the other seven riparians before 1993 and eight thereafter with the independence of Eritrea they had to be consulted over any water management scheme or policy. The participation of such a large group meant that no basin-wide consensus could ever be achieved. The upstream riparians were aware from the 1960s of the disadvantage of the bi-lateral option being available for downstream Egypt and the Sudan but the Egyptian orchestrated basin-wide consensus condition always blocking upstream initiatives. As a result key upstream riparians, Ethiopia, Kenya and Burundi would go no further in their commitment to the Egyptian dominated cooperative structures than attending meetings as observers.

Egypt was especially professional in its access to the main gatekeeper of investment funds which might have affected Nile river flows, the World Bank. In addition to its national representative in Washington two key Egyptian professionals headed the environmental directorate and the international law advisory position during the late 1980s and the 1990s. No professionals from the other Nile riparians held senior positions of relevance to international law, water and the environment in the Bank. The combination of two decades of political instability, war and civil war in major areas of the east African and Ethiopian highlands and the absence of World Bank support on water projects was a fiercely hostile environment for water projects.

The Hydromet organisation was a UN inspired agency put in place with the good will of most of the then nine riparians, not Ethiopia, to monitor the Nile hydrology to enable better basin-wide management. Hydromet was formed in the 1960s and foundered by the 1980s. Since the end of the Cold War a major initiative in the Nile basin has brought together the Nile basin riparians—six members and four observers—in a body to enable technical discussion to take place amongst representatives of the water resource planning and management departments of the respective governments. The body called TeccoNile was the forum for these technical exchanges. It met normally twice per year. TeccoNile was superseded in 1999 by a new organisation technical body with its offices in Entebbe. The change in name, to the Nile Basin Initiative, is intended to signify a new cooperative energy in the Nile basin and a move towards further cooperation.

At the political level a Council of [Water] Ministers was formed of the ten riparians which met normally once per year. A large international conference took place each spring since 1992, rotating around the countries of the region. The conferences were attended by senior technical staff from each riparian, international scientists and representatives of the international agencies. These meetings were symptomatic of the preliminaries needed to establish the common language with which to achieve greater understanding, confidence building and in due course, possibly after a decade or more, a 'regime' productive of agreements and joint management.

The MENA experience in general and the experience in the Nile in particular confirms the contention that progress is most likely to be made at levels below the level of the total river basin system. The sub-basin will be more likely to witness a 'successful' hydropolitical regime ahead of a regime for the whole basin. The 1994 agreement between Jordan and Israel confirms this contention. The 1959 Nile Waters agreement over the waters of the lower Nile between Egypt and the Sudan is also confirming evidence. The move initiated by the World Bank to recommend a subsidiary approach to the Nile basin with an Eastern basin consisting of Ethiopia, Eritrea, the Sudan and Egypt, and a Southern Nile consisting of the six other riparians is consistent with this theory.

Until this recent shift in approach the influential Nile basin riparian, Egypt, was committed to foundation discussions for what amounts to regime formation for the whole basin, although it was not called by this name by those engaged in the process. They referred to it as river basin cooperation. It is the lower basin riparian, Egypt with the support of the Sudan, that has been committed to whole basin scenarios involving the ten riparians. Upstream riparians regard insistence that the basin level with all riparians agreeing water projects was a deliberate delaying tactic on the part of Egypt and the Sudan. The need to get the agreement of ten riparians has always been sufficient constraint to frustrate the wishes of the upstream Nile riparians in the Ethiopian highlands—Ethiopia and Eritrea, and in the East African Lakes basin—Kenya, Uganda, Tanzania, Zaire, Rwanda and Burundi, in their planning and most importantly in their fund-raising for major civil works on their upstream tributaries.

The Nile River Basin Action Plan (TeccoNile 1996) reflected a step along the road towards the development of norms at the river basin level for hydrological monitoring, capacity building and river-basin modelling and planning. The Plan nevertheless did not meet the aspirations of the upstream members—Tanzania, Uganda, Zaire and Rwanda and the observers—Kenya, Burundi, Ethiopia and Eritrea.

The World Bank response in its 1998 reply to the 1996 Nile River basin Action Plan was delivered in January 1998 in Cairo at a meeting of the Council of Ministers of the Nile basin governments. The occasion was significant in shifting the emphasis of regime building. The Bank recommended the notion of the sub-basin and subsidiarity as likely scenarios for the acceleration of urgently needed upstream investment in regulating flow, creating upstream storage and reducing land degradation. The Bank also emphasised that there were still major water management prizes to be won at the river-basin level through cooperation at the whole basin level.

MENA hydropolitics: insider and outsider perspectives

MENA international relations over water have been shown to be the subject of a number of narratives. As usual in the MENA region there are groups of insider and outsider narratives. Insiders are usually prisoners of their history and of a suite of perceptions and expectations more determined by past water resource experience than by actual current water deficits. Past misconceptions can continue to be projected into the discourse because water security for all the countries of the region, apart from Turkey and Lebanon, is achieved by invisible and unacknowledged virtual water imports. Outsiders have been observers and analysts of as well as participants in the region's international relations. The remaining super-power and the rest of the international community have played and continue to play roles of uncertain effectiveness. In providing the invisible virtual water which stabilises the region's economies water outsider interventions are very effective but unacknowledged. The influence of outsider nations and international agencies on insider international relations is very limited. This has been as true of water just as much as it has been true of the major insider concerns within the security sub-complexes of the Levant and the Gulf. The USA did not prevail in its well inspired Johnston initiative of the 1950s. Nor have its views, nor those of other outsider perspectives, been taken into account by the negotiators in the Peace Process of the 1990s and beyond. The capacity of outsiders to impose their interests in what they regard as circumstances of extreme insecurity cannot be in doubt after the Gulf crisis and war of 1990–1991. Oil was strategically important for the Northern interests. But water is not a strategic resource of global significance. A water issue or conflict would not be the reason to move a single outsider soldier; oil mobilised briefly over half a million during the Gulf events of 1990–1991.

The MENA international relations discourse is subject to three insider contributions. One is major and there are two others which in most cases are latent. First, the region's international relations over shared waters are largely driven by the belief that the large farming communities of the economies of the region are entitled to water. The achievement of food self-sufficiency and the security of rural livelihoods are prime insider driving forces. This approach allied to the hydraulic mission prevalent in the MENA region during the twentieth century contributed to tensions over water as the region slipped into water deficit since the 1960s.

The second contribution to insider discourse on regional water relations are environmental concerns. The third is the long held and many sided patterns of enmity and amity. The second contribution has been expressed in the notion that water should be managed according to a holistic perspective. Principles of integrated water resource management give expression to this approach. These principles have been shown to be difficult to implement in the politically fractured and non-integrated political circumstances of the MENA region's river basins. They were briefly acknowledged in the 1986 to 1992 period in Israel. Despite their current low profile in the discourse they will become paradigmatic as the hydraulic mission is gradually reversed in favour of more environmentally and more economically aware policies.

The third factor driving international relations over water is the link with patterns of enmity and amity. Water is just one of many features of the region's international relations which can be overwhelmed by periods of confrontation. These periods of tension are caused by transgressions which aggravate and draw attention to underlying enmities. The reversal of Israeli management of the Western Aquifer of the West Bank is just the most recent example of such dynamics.

Outsiders are generally impotent with respect to the international relations of the region including water relations. They have little influence on awareness of the strategic water situation and the role of the global system. At the same time outsider economies do provide the unacknowledged solution to the region's water deficit via the trade in virtual water. Outsider international agencies promote, with some success, their newly discovered environmental and economic principles as well as the virtues of integrated water resources management (World Bank 1998). They also used their leverage on the weak entities in the region by insisting that water projects be handled by international private sector companies. The rehabilitation of the Amman water services and the ugrading of the Gaza water distribution system are examples of this outsider preferred mode of implementation.

The outsider risk conscious media struggle to understand the dynamics of the water relations in the MENA region. There is a constant search for the link between water deficits and armed conflict (Bulloch and Darwish 1993). They also promote the arguments of green and human rights activists and bring pressure to bear on funding bodies involved in dam projects such as the Ilisu Dam in the Kurdish region of Turkey (Brown 2000). The impact of the NGO and media campaigns are limited and indirect. They have no impact on government policies in the region where there has been no space for the principles being advocated by outsider activists. The arguments could, however, impact professional and public opinion and in turn the decision making of funding bodies such as the UK Department of Trade and Industry in the case of the Ilusu Dam in 2000. Unlike South Asia the MENA region has not developed influential environmental NGOs to protect the interests of communities negatively impacted by water projects. They will certainly emerge. When they do the international relations of the region will be affected. The extent to which the resulting tensions will be contained will depend on the extent to which the economies of the region develop into diverse and strong economies. Economic diversity and strength will enable even very seriously water deficit economies to adopt conflict reducing environmental and economic policy options.

CHAPTER 7

Water law and international water law in the Middle East and North Africa

Water law and international water law and the MENA region

Definition of the region and legal principles

> 'The regulation of international waters has assumed today a vital importance in the peaceful relations among human communities organized as states.'
>
> (Vidal 1995)

> At the turn of the twenty-first century addressing the politics of adoption is a more important priority than the arguing the quality of the legal principles.

At the national and the international levels any legal system addressing the shared waters of the MENA region has to take into account not only increasing water deficits but also the seasonal and periodic fluctuations of flows and water availability. Communities, nations and whole river basins in the MENA region endure variable water insecurity. This last feature will be a theme to which the discussion in this chapter will return frequently. The problem is how to protect and compensate those enduring periodic water deficits. Nicol (2000a) has pointed out that water can be hydrologically secure or socially secure. Rephrased water can be hydrologically secure, hydraulically secure or socio-economically secure. In the first case Nature ensures security. In the second case engineers make water secure. In the third case socio-political and economic institutions devised by communities, national entities and international bodies make water secure. Law and international law are part of a suite of transaction cost saving socio-political tools that can make a contribution to securing water. Ohlsson and Turton (1999) have emphasised that it is the possession of the social adaptive capacity to have and deploy the socio-political and economic remedies that distinguishes a secure from an insecure community or nation.

The success or not of the installation of water law and international water law will depend on the social adaptive capacity of the players at the different levels of water allocation and management—communal, national and international. As with economic instruments such as water pricing discussed in chapter 3 it is not the logic of the approach which determines adoption it is whether the adoption of the approach, in this case water law, is politically feasible. At the turn of the twenty-first century addressing the politics of adoption is a more important priority than arguing the quality of the legal principles.

The development of legal conventions and regimes has proved to be difficult partly because defining and generalising about MENA water resources is not straightforward. The core of the region is arid. Only Turkey has relatively abundant water resources deriving from rainfall falling within its borders. Lebanon, a much smaller territorial entity, also enjoys abundant water resources. Sovereignty, and the legal status of water, is complicated by the movement of the majority of the region's water across international boundaries. Both water rich Turkey and Lebanon 'export' water which falls on their territory. Other political economies 'export' water, but that water has been 'imported' from an upstream riparian. For example, Syria receives water from Turkey via the Euphrates and then exports most of it to Iraq. The Sudan also imports water from riparians to its south; from Uganda and from Ethiopia and then passes over 75 per cent of these waters on to Egypt.

Generalisation about the waters of the Middle East and North Africa is much easier if Turkey is not included in the discussion. The economies to the south of Turkey are almost uniformly rainfall scarce and endure water deficits. They depend on surface waters which come from outside the core Middle East and North Africa and on locally and by no means uniformly available groundwater. In this core region the management of water became a significant economic and environmental challenge in the second half of the twentieth century. As a result water has gained a prominent, but by no means determining, place in discussions of the international affairs of the region (See Figure 6.4).

But the deduction that serious, and even armed, conflicts would be the response to the growth of water demand beyond the capacity of the region to meet its population driven water needs (Bulloch and

Darwish 1993) have proved to be unfounded. Nor has there been any haste to install agreements and regimes of legally founded joint river basin management argued by outsiders to be the rational response to resource deficits (Brans, et al. 1997, Dellapenna, 1997, Falkenmark, M., 1997, McCaffrey 1997a, 1997b, Saeijis, et al. 1997, Wouters 1999b).

The region consists of a substantial part of the African continent, including most of the Sahara Desert. Also the arid countries of south-west Asia. The latitude of the region, between the 40 degree line of latitude in the North and the northern hemisphere tropic in the south means that the region experiences the Mediterranean regime of rainfall and temperature (see Figure 2.1). The modest rains, rarely above 600 mm per year and for vast areas less than 200 millimetres per year, occur between the less hot months of October and March. Only Yemen experiences the tropical monsoon pattern of seasonal rainfall. The uplands of Yemen receive their rainfall between May and October. The same monsoon regime causes the late summer flood regime of the Ethiopian Nile tributaries which traverse Ethiopia, the Sudan and Egypt. Elsewhere surface waters in the north of the region are subject to the winter rains of the Mediterranean region. The annual peak flow of surface water can be increased by spring snow-melt. The Tigris-Euphrates system was especially affected by the effect of snow-melt until the construction of the dams on the Euphrates and those being planned and constructed in the early twenty-first century on the Tigris.

The low and variable rainfall regime and an even more variable surface water flows have been a challenge first for insider farmers for millennia, secondly for insider ingenuity in devising durable customary legal conventions for rural water sharing communities, and thirdly for insider and outsider engineers for the past two centuries. With the inadequacy of engineering measures and of customary legal institutions the MENA water deficits are seen fourthly by outsider economists as a problem ripe for the application of economic instruments and fifthly, by the 1990s, by outsider water lawyers as susceptible to water law reform and to regulation according to principles of international water law.

In practice it has been found that once engineering solutions which provide higher than average levels of reliable water have been exhausted, then the compensation of those with livelihoods affected by drought periods have to be short term and fiscal, or long term and economic via new jobs or forms of government subsidy. Even strong economies find the politics of such compensation difficult. In the 1999–2000 agricultural year the decision of the Israeli Water Commissioner to reduce the allocation of water to agriculture by 40 per cent was frustrated by the unwillingness of the Ministry of Finance to compensate farmers for lost income. As a result the Israeli farmers ignored the directive and continued to use water at the level of the previous year. If legal principles could be deployed in these contentious circumstances then the costs of such conflict could be ameliorated for communities. But only if the economies have the capacity to support the subsidies.

If the introduction of a system based on legal principles is difficult at the national level then it has proved to be impossible in the MENA region at the international level. The outsider and insider discourses on water law are different at the international level from those at the national level.

Outsider legal professionals and academics argue for the adoption of legal principles. At the same time they recognise that the United Nations International Law Commission (ILC) process, unavoidably political, since the 1970s has not achieved swift ratification of its International Convention on the law of non-navigational uses of international watercourses (ILC 1994 and 1997, Wouters 1998, 1999b and 2000, Crawford et al. 2000). After nearly thirty years of slow but promising struggle the process had become influential. Ratification was delayed by the informed but unwilling.

> The [ILC] Convention contains rules of international law that apply regardless of whether or not it comes into force. It is not a perfect document, but contains some provisions that would help a lot of States, especially on the procedural side.
>
> The Convention was adopted in May 1997, and is open for signature until 21 May 2000. To date, it has 15 signatories and seven ratifications. Contrary to the views expressed by many noted 'experts', the Convention will not fail to enter into force if it does not obtain 35 ratifications by 21 May 2000. As with many other global international treaties, the UN Watercourses Convention will come into force once it has acquired the required number of ratifications (Article 36.) This could occur at any time in the future and in fact, is a feasible possibility. However, even if the Watercourses Convention never enters into force, it is clear that it already has generated a considerable influence on States drafting new agreements or involved in diplomatic negotiations regarding their shared watercourses. The Southern Africa Development Community Protocol on Shared Watercourses has been revamped to include the main provisions of the Convention. Moreover, the International Court of Justice underscored the importance of the Convention when it referred to a number of its provisions in its decision in the

> Gabcikovo-Nagymaros case, a dispute between Hungary and Slovakia over the Danube. In any event, many of the substantive rules contained in the Convention reflect customary international law, which binds all States, regardless of entry into force of that instrument.
>
> (Wouters 2000)

In the MENA region, engineering instruments have been the preferred first option for the regulation of the availability and allocation of water rather than institutional instruments. This is always true at the level of the state. States are characterised by a form of hierarchical governance. They have the capacity to mobilise investment, indigenous or international, to ameliorate a water deficit. To date MENA states have pursued a hydraulic mission approach to water regulation with one exception. Israel has shown that it is possible to adopt different principles when alternative national interests are identified as discussed in chapter 6.

The need to sub-optimise and for flexibility?

When the hydraulic mission solution, in reality if not in rhetoric, had ceased to provide a total solution to water problems MENA governments have still preferred to avoid the tough international politics of making water agreements and treaties with neighbours. They have resorted to the stress free option of importing virtual water from the global system. For the MENA insider virtual water works as a politically feasible ameliorative alternative. Put in simpler words this alternative is a version of the widespread phenomenon that 'the second best normally works politically'. For the differently inspired outsider, with ideological attachment to economic and environmental principles, this description captures their evaluation of current MENA approaches to water allocation and management. To insider water challenged governments and peoples in the MENA region untried international legal institutions have very little appeal. They are not seen to be intuitively risk avoiding or transaction cost saving (Coase 1960). For the MENA political and professional insider international water law is an example of the politically unworkable 'best' being the enemy of the workable 'good', namely the status quo. Unfortunately international law comes a very poor second to virtual water as a stress free and adoptable solution for coping with the regional water deficit for weak economies with low social adaptive capacity.

The notion of second best has been in wide currency in economics for half a century (Meade 1955, Lipsey et al. 1956).

Economists discovered that (Paretian) optimum outcomes could not be achieved even if one (Paretian) condition was constrained. Ideal economies do not exist. Second best outcomes are all that exist. An important insight of the economists of the 1950s was that if one economic condition is not optimum then second best optimum outcomes are achieved by departing from all other optimum conditions.

Politics determine the constraints on the achievement of optimum economic conditions as well as the operation of optimum legal principles. If law requires that interdependent conditions including beliefs, values, social norms and willingness to comply obtain then if any one of these conditions is compromised then perhaps all the others will have to operate sub-optimally even to achieve the second best legal regime. There is no space here to explore the possibility that there is a second best theory in the domain of law. The notion of the workable second best is just as uncomfortable for the legal professional as it was for the professional economist. The compromise of legal principles appears to be a contradiction of the unquestionable perfection of the rule of law and very difficult for advocates of legal regimes to accept. In circumstances when second best legal systems are the only option and the only way forward then there is evidence that the need for flexibility is being recognised (du Bois 1995). Ben Venisti (1997) has pointed out the need to sub-optimise and has gone so far as to say that 'law should only present vague standards and avoid strict rules, as the former induce negotiation while the latter induce litigation.'

The purpose of this discussion is not to add to the litanies that already exist on what would be the best allocative and management models for MENA waters based on law. Its purpose is to explain why water is allocated and managed as it is and why the introduction of water policy reform, including the introduction of water law and international water law, is slow. The pace is slow because water policy reform based on economic, environmental and equity principles, and reinforced by law, has perceived and real negative impacts on existing communities. The negative impacts cannot be assuaged by coercive legislation nor by invasive regulations which are the tools of the law makers and implementing agencies. The same applies to economic tools such as pricing and also to the regulation of environmental pollution. All such reforms would become acceptable if those impacted had options which would enable them to achieve better livelihoods and incomes in the reformed circumstances.

That water markets and customary law do exist in communities remote from central MENA governments confirms for outsiders that such tools are relevant and operable. To central government insiders customary regulation of water in for example Morocco and Sinai (Wolf 2000b) and in the West Bank are aberrations that have not yet been incorporated into a central government system with a new or soon to be installed comprehensive national water law (Trottier 1999). Embedded customary practices such as these and new local responses to water shortages, such as the water markets experienced in the city of Ta'iz in Yemen in the 1990s (Handley 1999), may appear to confirm the relevance of outsider approaches. However, the awareness, motivations and above all the transparency which brought these circumstances about cannot be replicated at a higher level of the governance hierarchy. Higher level political economies lack the social adaptive capacity to construct shared awareness of benefits and more importantly to provide the compensating resources for those who lose by the water policy reforms.

No one has been able to say with certainty whether political or economic 'development' comes first. In the case of MENA water it would seem to be the case that institutional reform tends to follow economic advances made in unreformed circumstances. Economic advances are the result of incremental improvements in productivity and reductions in transaction costs. Law is one of a number of additional transaction cost reducing measures that can be adopted if there is the social adaptive capacity to do so.

Unfortunately the contribution of water law has been much less palpably beneficial to date at the national level than technical transaction cost reducing measures. Engineering measures have been judged to be most beneficial and have attracted investment resources. If water law and water pricing are to gain support and currency in MENA communities and with MENA governments their benefits for livelihoods must be made more clear. Without such benefits being evident legal reforms will have to wait for the economies to diversify and strengthen sufficiently to provide resources which compensate the losers affected by the process of reform. The lesson that water policy reform and water management legislation comes late in the process of socio-economic development is proven in all the other semi-arid regions where the same challenges have been faced. The United States (Reisner 1986), Spain (Swyngedouw 1998), Australia (Langford 1998) and Israel (Allan 1996a).

The adoption of international water law has been even less evident than the adoption of national water law in the MENA region. The preceding preliminaries have been necessary to relate an analysis of water law to attempts in the international domain to introduce principles developed by international water lawyers. Water in the Middle East and North Africa is, in perception and in reality, different from any other region in the world. The region has less conventional water per head—that is surface waters and groundwater—than any other region. It has by a considerable margin the worst endowment in soil water. For the regions from which outsider water lawyers come it is soil water which is the dominant element in national water budgets. When outsiders from such better endowed regions argue for legal principles as if the challenge was the same in the MENA region as in humid temperate regions they have not been persuasive.

Outsider lawyers and international lawyers who advocate a more comprehensive role for water law in the MENA region operate at the national and international levels. At the international level the position over water relations the situation could be described as a version of anarchy. At the national level there is patchy adoption of water law. At the local level there exists a wide range of transparent, trusted and effective systems for managing scarce and variable water supplies in locations as far apart as Morocco and Palestine (Wolf 2000b, Trottier 2000). At the local level the political challenge of operating customary water law is possible because of the transparency of the practice and the mutual awareness of the consequences of non-compliance. There is no equivalent transparency and awareness at the national and international levels. The construction of such awareness is a massive and unavoidable preliminary to the adoption of national and international water law. It is costly both in terms of politics and professional inputs. Currently impossibly costly for many of the MENA region's economies and governments.

In concluding this review of the contexts in which water laws exist it is necessary to emphasise that there is a mosaic of old established practice at different levels of the MENA region's social organisation. Pervasive Islamic tradition has meant that common principles lie behind the practices adopted by the separate communities (Caponera 1990 and 1996). The MENA region has experienced two centuries of the ebb and flow of secularisation especially in the area of law. (Mallat 1993) Part of that ebb and flow has been the adoption of the outsider concept of the nation as the preferred model for governing territorially identified political entities. Any compatibility that the Islamic tradition made possible between the elements of the mosaic of traditional legal practices, including those relating to water, has throughout the second half of the twentieth century been weakened by the intrusion of other Northern approaches. From the late nineteenth century the hydraulic mission was very widely adopted and still prevails in the big water economies of Egypt, Iran and Turkey. In the last decade of the twentieth century the visions of the economist, environmentalist and the water lawyer have been successively pressed on the region. Adopting the institutional instruments that these approaches require has proved to very challenging, much more challenging than the adoption of the engineering instruments of the hydraulic mission.

The contention between outsider international lawyers and insider MENA politicians and professionals has been a sub-discourse of the much bigger contention over water policy reform between MENA insiders and Northern outsiders more generally. The pace of innovation in the main discourse determines the pace at which legal institutions, both national and international, can be transformed. The big issues have been addressed in the previous chapters—economic policy, environmental policy and approaches to international water relations. MENA governments' delegates played a full part in the contentious discourse which developed the ILC Convention (ILC 1997). Upstream interests were pressed by Turkey and downstream interests by Egypt. Syria supported the process throughout and was one of the 15 signatories by the end of the third year after the UN General Assembly passed the Convention in May 1997. No other MENA government had signed. Turkey and Egypt were and remain especially cautious. The third way of equitable utilisation had not by the end of the twentieth century provided a secure framework for the mediation of the contending interests of the upstream and downstream riparians.

A tendency to de-emphasise legal principles in the MENA water sector

> Untried international legal institutions do not appeal to water challenged governments and peoples in the MENA region. Unfortunately they are not intuitively risk avoiding or transaction cost saving.

Devising laws which can be readily introduced and enacted, and then operated free of contention, is rarely easy. The ease of operation of a law depends on a range of social and political factors. Whether a law will be 'workable', even in a 'strong state' (Migdal 1988), is often used by those who oppose the enactment of such a law to question its legitimacy as well as its utility (Wouters 1999). In addition to debates about legitimacy, workability and utility other debates rage around the role of law in the management and allocation of water. Very familiar is the debate between those who are inspired by legal principle and those who regard political feasibility as the sure guiding principle for the introduction of operable water legislation (Dellapenna 1995:57, McCaffrey 1998:7). The discourses are active at the national level and particularly at the international level in the Middle East and North Africa.

The debates are of particular relevance at the international level to the Middle East and North Africa because the region is palpably short of water, howsoever water scarcity is measured. The circumstances which have led to a very clear de-emphasis of international water law in the Middle East and North Africa in the settlement of international disputes over water are special. They provide rich analytical material for those interested in determining the relevance of such law at the end of the twentieth century. It is evident that methods of dispute resolution, exemplified in normal political processes and in formal and informal negotiations, have been the preferred means of addressing international water disputes. Even so the most usual situation is that water disputes have not been discussed at length by the parties; there has been little negotiation, none over water alone. Nor has there been resort to law. Even when contentions over water have gained an agenda place, such as in negotiations between Jordan and Israel (1993–1994) and between Palestine and Israel (1994–....) water has been a subordinate issue. Sequences of complicating linkages with, and de-linkage from water and other issues such as peace, territory and freedom from intimidation, are the norm.

Why is water difficult to bring into the formal embrace of law? And what are the particular circumstances of the Middle East and North Africa with respect to water law?

A diversity of water uses and water types

A first reason why water is difficult to bring within the embrace of law is that water is a resource which occurs in many forms. This is true of all regions and not just the Middle East. The utility of water can be very different in one form of storage compared with another. Some forms of storage are ideal for a particular economic activity and not for others. Low quality water in soil is available for the most demanding activity of all, crop production. Such soil water cannot be used for any other economic activity. Soil water is not a minor economic resource. It is the major water resource globally although not in the Middle East and North Africa. Globally soil water accounts for about 70 per cent of the water used in agriculture. Agriculture accounts for about 90 per cent of global water use. The rest of the water needed for agricultural production comes from surface and ground waters which can be used for other competing purposes. Thus most—that is the soil water—of the economically useful water in the global system will not be susceptible to law, except indirectly as an ephemeral quality enhancing or disenhancing a tract of land.

In the Middle East and North Africa, very little of the water available for economic use resides naturally in the region's soil profiles. Out of the approximately 240 billion cubic metres of water accessed annually in the region for economic use only about 40 billion cubic metres (17 per cent) finds its way naturally into the region's soil profiles following rainfall.

In these circumstances the Middle East is in a special position with respect to the international trend to introduce legal instruments into the water sector. With so little of its water—less than 20 per cent of total water used—occurring as soil water, the majority of the water in use is water which can be measured, albeit with difficulty, controlled and legislated on. However, there are a number of problems facing those who would attempt to introduce laws on water into the region.

Naturally occurring soil water is used in one sector, agriculture, and is beyond the reach of legislation. Surface and groundwaters on the other hand are subject to competing demands, could be commodified and be subject to legal regulation. In water markets such waters could be subject to commercial law as well as to laws sanctioning pollution and regulating water quality as well as ensuring approved entitlements and remedying unauthorised use. In the absence of such economic instruments and especially in the absence of the monitoring of water transactions the development of enhancing legal instruments was inevitably poorly developed by the end of the twentieth century.

The problem of legislating for a resource traditionally assumed to be free

Secondly, water is a natural resource which has not traditionally had a cost of delivery attached to it. As a consequence water has not been valued in the economic activities to which it has been an input, often a major input. Surface water, in for example river channels, and groundwater, can be used in all productive sectors provided the necessary diversion or pumping works are constructed. Such surface and groundwaters could be competed for and could be subject to law provided that flows and storages could be measured to the mutual satisfaction of those accessing the resource. In practice the challenge of monitoring water delivered to irrigators in the big irrigation schemes has proved to be insurmountable. It is likely to remain so (Perry 2000). The idea that water should be valued has been prominent in discussions about water in the Middle East and North African region since 1990. The discourse is unprecedented and also stormy because deeply held beliefs are being challenged. The operational problems of instrumenting vast surface water irrigation systems reinforce the resistance to such regulation.

The ignored water resource: soil water

Thirdly the hydrological cycle is not widely understood by the managing economies including the water resources of those economies. This is evident in the way that water resources are enumerated by national and international agencies. All economies use about 90 per cent of their water for agriculture. In some cases almost all that water is derived from rainfall infiltrating into soil profiles and then used for crop production. Most of the economies of north-west Europe, for example, enjoy year round well watered rainfed soil profiles. Likewise in the humid equatorial regions. In arid regions, conversely, almost 100 per cent of water used in agriculture has to be lifted from river flows or from the ground. In these circumstances there are real, if de-emphasised and non-accounted, costs of delivering water. These are the waters that are accounted as part of the national water budget. Soil water is ignored by water resource authorities, by economists and by governments in soil water rich and soil water poor economies. Until 1997 it was not mentioned in the international debate until emphasised at a number of international meetings by the author. It is possible that the issue will become an element in the hydropolitics of the Nile.

A mobile rather low value resource

> In water, we must always start from reality and move priority by priority. Often we will trip and fall at the first step— but that is better than ignoring the first priority and pretending to address the next three…
>
> (Perry 2000)

Fourthly, and most importantly water moves. The movement of a resource is not unusual. Labour moves, capital moves, technology and information move. All of these factors of production, with the possible exception of information, can, however, be relatively easily enumerated, attributed a legal identity, be paid for or bought and sold and regulated by rules and laws. Complex institutions have been set up to regulate immigration, protect copyright and to bank and transfer money. All of these are reinforced by legal arrangements. Informal activity takes place; there is smuggling, illegal immigration, illegal work practices, plagiarism and worse. But the majority of activity is regulated by legislation albeit associated with ongoing and sometimes highly contentious debate.

With water, even the water at the surface, the monitoring of volume, is difficult. The surveillance of withdrawals is also difficult and socially unpopular. The regulation of groundwater use is particularly difficult. All these problems arise because water flows and when used in large volumes is not readily measured with any precision. Other fluid resources such as petroleum are more valuable and merit much more elaborate accounting procedures. It is these accounting procedures which make it possible to install regulatory practices which enable the resources to be managed coherently. Once the resource has been commodified it can then be subject to elaborate commercial legal regimes. High value water is subject to just as rigorous monitoring and regulation as other highly valued fluids, such as petroleum. In the year 2000 chilled bottled water available to consumers in convenient locations was more costly than refined petroleum. And this chilled water did not bear the taxation to which petrol sold at the pump was subject in industrialised economies. Bottled water everywhere in industrialised economies and in much of the MENA region was as expensive as taxed petroleum and just as regulated. The regulation was necessary for both as one can explode and the other can poison.

In agriculture, however, water is a low cost resource. Where it is very low quality it poisons plants and tends not to enter economic processes. Water in agriculture has a price, often unrealised, a cost of delivery and a value. It is usual for the revenue raised from farmers for irrigation water to be less than one US cent per cubic metre. The cost of delivery often approaches 10 cents per cubic metre. The value of the water is often likely to be about US$1 per cubic metre in terms of the value which can be added by its use in farming. Value can be accounted either by the opportunity cost of such water elsewhere in the economy or by the value added to the farmer by the water (Perry 1997 and 2000).

With such vast differences in the perception of the 'value' of water the unavoidable politics of bringing about convergence in the minds of the parties involved are elemental. Those using irrigation water, those delivering it and those wanting to establish a legal framework to regulate its use have vast journeys to make to reach some common ground of awareness. The problem is compounded in the MENA region because water supplies are unreliable. There is a need for flexibility. Flexibility is not a quality commonly associated with law.

The problem of establishing ownership: national owners and local owners

> '... water law ought to focus on structuring common dicretionary water management rather than defining water rights.'
>
> (du Bois 1995:112)

> '... the principles of precaution and sustainability might be better served by using economics to determine efficient ways of delivering politically and legally defined standards of service and quality than for defining the principles in the first place'.
>
> (Morris 1996:201)

Fifthly, and following from the fourth characteristic of water, its mobility, the ownership of water is very difficult to establish except where it has been put into containers or pipelines. Because water is subject to the forces of gravity it leaves tracts where it has fallen as rainfall and flows to the lowest point regardless of international boundaries or intra-national forms of land tenure. In an engineer-free world the only water which the owner of a tract, or the peoples of a nation can count on using is the water that is held in soil profiles. Such water is then moved upwards by the incredibly powerful natural water lifting capacity which is the life support mechanism of plants. The crop element of the vegetation domain lifts vast quantities of water from the soil. Soil water provides in turn the life support system of animal populations including the human population.

Other waters than soil water have to be diverted, stored and distributed after engineering interventions. The raising of capital resources to achieve control of water is an important stage in changing the relationship of individuals, communities and national governments with a body of water. Recovering the cost of the investment, except possibly in oil-enriched circumstances which do obtain in the Middle East and North Africa, is associated with all such capital works. The forms of cost recovery put in place can sometimes be judged to be very unsound to an economist. The assumptions would be especially unsound to the environmental economist who has been much in evidence since the late 1980s taking into account the costs of environmental impact (Pearce et al. 1991, Winpenny 1991 and 1994).

Where the engineering works have been put in place by state capital, or a package of state and international capital, borrowed from an international bank such as the World Bank and associated agencies, users tend to assume that the water is similar in provenance from water in the natural system. The political resistance to recognising the capital and operational costs of maintaining the flow of water is fierce and generally successful in preventing the establishment of instruments to achieve full cost recovery. Even strong political economies find it difficult to enact cost recovery policies and implement water charges in the rural sector. Many projects have been set up in Northern economies that were not designed in the first place to achieve cost recovery in the farm sector. The costs were to be recovered in associated sectors such as hydropower (Reisner 1993). Australia is probably the first economy to achieve full cost recovery in current water delivery systems. And this only in the late 1990s after 30 years of reform (Langford 1998). In France the levels of cost recovery for water distributed to farmers was about 40 per cent in 1998 (Cheret 1998). In Israel similar figures had been achieved (Arlossoroff 1996). If this is the level of successful implementation of legally defined economic principles in agriculture in diverse and strong economies with elaborate administrative institutions, their implementation in less well institutionally founded political economies will take a number of decades.

The concept of property in the domain of water

> '... despite the asserted advantages of private property systems, such systems do not work for ambient resources like water.'
>
> (Dellapenna 1995:153)

> 'If, therefore. privatisation is viewed primarily as a means of improving productive efficiency, there are benefits which result. But if the primary problem to be addressed is that scarce water resources are being directed into the wrong activities, privatisation will be of very limited potential in the water sector.'
>
> (Storer 1995:265)

The last decade of the twentieth century has been a period when the privatisation of resources as well as productive and public services and utilities has been rapid and widespread. In the water sector the rate of privatisation is fast but in a service industry where the public sector has been regarded as the natural provider most water services remain public. World-wide private sector services in the domestic water sector only amounted to about six per cent of all provision in the year 2000 (GWP 2000). The proportion was lower in the MENA region. In the big using sector, agriculture, private sources of water provided only a tiny proportion of the water in the region. The water sector has not escaped the outsider attention of those who believe that gaining a private property status for water will be in the interests of the economies of the region.

The purpose of this section will be to discuss the relevance of a private property inspiration for those involved in water policy formulation during the first two decades of the twenty-first century. It will be shown that privatisation exists at two extremes. In remote circumstances where users are dependent on a well or spring which is owned then local arrangements to purchase water may have been in place for generations. At the other extreme where there is a highly regulated, diverse and strong economy, then privatisation is being extended. The 90–95 per cent of water used by most economies in the irrigation sector is not experiencing privatisation.

'There are just three basic types of property in flowing fresh water' (Dellapenna 1995:150). These are—common property, private property and public property (Demsetz 1968). Most groundwater in the MENA region is treated as common property. This is an approach reinforced by Islamic principle that associates the right to develop groundwater with the duty of a farmer to utilise land effectively. The act of utilising land and water can be the basis of legitimising such use. Both soil water and groundwater are embraced by this approach. Groundwater can be treated as public property if legislation is passed to give it that status. The Israeli water law for example did this in Israel soon after the establishment of the state. Other states usually treat groundwater as public property subject to regulation but it has proved impossible in practice. This is true in Jordan (Lancaster et al. 1998).

Rivers, streams and springs can also be treated as common property. As water responds to gravity upstream users are always at an advantage in common property regimes on flowing water resources. At the international level this anarchical approach is shown to exist when downstream users are denied water by upstream users with prior access. Only Turkey and Syria have carried such policies to extremes: Turkey on the Euphrates and Tigris and Syria on the much smaller Orontes. Syria claims that the Orontes is not an international river between Syria and Turkey as the territory into which it flows is disputed and should be Syrian. The assertion of upstream privilege is feared by a number of downstream riparians. Egypt is very vigilant and precautionary in this regard as is Israel. Egypt and Israel have been the 'successful' hegemons on their respective shared rivers and groundwater systems. None of this 'success' has been achieved via legal regimes or compliance with international law.

The majority freshwater in the MENA region is surface water in rivers. While such water is treated as common property at the international level, at the national level most river water is treated as public property in the irrigation sector. Such waters have been put at the disposal of farmers as a result of public investment and operation and maintenance by public sector agencies. They are subject to public supervision and regulation of variable efficiency. A consequence of the variable efficiency is that regulatory regimes and the collection of water charges are ineffective. Users refuse to pay for unreliable water and ignore the regulations of inefficient systems.

The introduction of successful and functioning legal systems which regulate water as private property require that the property to which the law refers can be clearly identified, and then monitored with respect to changes in location, volume and quality. It is further required that the ownership of the property can be registered and remain unambiguous. Similar principles would apply to water at the international level if governments had the capacity to control totally the water within their national boundaries. There are problems in operationalising such law as water is frequently difficult to contain, measure and evaluate. Water it would appear does not operate 'naturally' as private property in low value activities. Low value activities are those where the value of inputs and outputs are low. Irrigated agriculture in extensive alluvial lowlands, where water is used in large volumes, is the main example. The Nile river lowlands and those of the Tigris-Euphrates are typical of such high volume, low cost of delivery and low economic return water regimes where water is not treated as private property. The waters economically used in these lowlands accounts for most of the consumptive water use of the region, about 80 cubic kilometres annually in the late 1990s and over 90 cubic kilometres annually when the GAP agricultural projects will be functioning by 2020.

Table 7.1 shows the water volume and quality requirements of

Table 7.1 Water use by volume and quality and indicative returns to different waters

	Volumes needed		
	High	Moderate	Low
Quality			
Low volume/High cost/High return $1/m3			
Very high	None	**Some industry** **Some services**	**Drinking**
Medium volume/Medium cost/High return$1/m3			
High	None	**Domestic**	None **Some industry** **Some services**
High volume/Low cost/Low return $0.1/m3			
Low	**Agriculture**	**Some industry** **Some industry**	None

Source: the author

the major water using sectors. It also shows the sectors which are susceptible to regulation by water law.

The river lowlands of the Middle East and North Africa have been engineered for agricultural water distribution based on a number of assumptions all of which are inspired by water being a public good. The first assumption is that there would be limited technical capacity at the farm level to regulate water to achieve water use efficiency. Metering would be prohibitively expensive. Secondly, it would be difficult for a variety of practical engineering and numerous other social and cultural reasons to levy water charges. Thirdly, cost recovery would be achieved by regulating the price paid for crops. The difference between the price paid to the farmer and the world price for the commodity, or the price in a government regulated 'market' national system, would enable governments to recover something towards, or even all, of their operating and capital costs.

The advantages of such systems for governments was that they were administratively relatively simple and light in bureaucratic overheads. The disadvantages are also obvious. The water distribution agency, free of the constraint consequent on knowing the value of the water and also without the responsibility of gaining the best possible return on the water, will tend to oversupply water to minimise the aggravation of complaints from users. Users will, if they have the chance, that is when they are upstream on the distribution canal, tend to over-use water. Tail-end users will have to remedy their periodic water deficits by installing pumps to raise groundwater or access drainage water (Radwan 1994). No one owns the water or values it except in a crude presence or absence sense. At no point in such a system is there an instrument which stimulates water use efficiency, except amongst the minority and disadvantaged tail-end users who have to anticipate water deficits and pay to remedy them.

Determining whether the unvalued public good approach to water allocation and management is economically effective or not is beyond the capacity of this analysis as it involves comparing unmeasurable transaction costs of the existing system with those of an as yet undevised system. Also there are environmental circumstances which make the oversupply of water a benign management activity, as in Egypt where excess water is the supply of users downstream. Farmers in Egyptian agriculture use their water 2.3 times from the time the 40 cubic kilometres of water used annually in agriculture leaves storage at Lake Nasser/Nubia annually until ten or so cubic kilometres reach the Mediterranean over the year (Abu Zeid 1997). Egypt achieves a high level of water use efficiency because it enjoys a very favourable river geomorphology and soils which are capable of intensive use. The Tigris-Euphrates lowlands are different. The alluvial deposits used for irrigated farming are neither so deep nor so fertile. Worse they are much more prone to natural salinisation and they are easily degraded by the use of 'excess' water. The management regime which works admirably in the Nile does not in the Tigris-Euphrates lowlands.

The relevance of the above discussion to law is that the Middle East and North African region has some way to go before it will be possible to introduce a legal regime for the majority water using sector agriculture. The installation of engineering, water use monitoring and water charging infrastructures which must precede the introduction of law based water regulation will take place at a rate determined by the overall social and economic development of the involved economies. A number of decades will elapse before such infrastructures are built.

Reforming water law in the MENA region: law and privatisation

> The sectoral re-allocation of water implicit in industrialisation is invisible to most governments and to all but a few professionals and scientists.

Water law at the national level in the MENA region is one of the institutions which in the 1990s was given considerable attention by insider governments and professionals and also by outsider international agencies. The World Bank took a lead in urging governments to draft and legislate new water laws. The drafting, legislating and implementing experiences were in the event attenuated and the impacts partial.

Introducing land reform including the reorganisation of the ownership of land and a new water law are profoundly political acts. Much existing legislation at the national level has been in place for many decades or more. There are also customary laws that local communities have observed for centuries in isolated hydrologies where use does not impact on other water users. Changing the legal status of water and the relationship of water users with water resources is potentially destabilising. Such change has been perceived to create uncertainty and winners and losers and legislators do not have the stomach for the debate. The Palestinian Authority and Yemen were involved in lengthy debates on their new water laws in the 1990s.

Water law is developed in differing institutional circumstances. The legislative process in the development of water law in Jordan for example was affected by the existence or not of parliamentary procedures. Temporary laws enacted in 1977—Law number 19, the Jordan Valley Development Law, and in 1983—Law number 18, the Water Authority Law, had to wait for debate by a parliament until 1988 and 1989 and enacted as regular laws in 1989. The Jordan case is also interesting in that passing laws was integral to the establishment of water institutions. Law number 19 had its origin in the Law of the East Ghor Canal Authority of 1959, was updated in 1966 in the Natural Resources Authority Law, and updated again in Law no 18 in 1977 with the Jordan Valley Development Law. Law no 18 of 1989—the Water Authority Law—originated in 1959 as the Central Water Authority Law, was updated in 1966 in the

Natural Resources Authority Law, and in 1974 in the Domestic Water Corporation Law, and in 1983 in the Water Authority Law (Haddedin 2000).

The politics associated with such innovation in environmental policy have been shown to be different when there is a major change occurring in a political economy. Emblematic events enable unfamiliar and even unpopular innovation to take place (Hajer 1996). 'Normal politics' have to be superseded by 'security politics' if a community or nation is to achieve the necessary awareness of the need for change (Buzan et al. 1998). Israel, for example, was able to install its new and very radical water law in the extremely dislocated political circumstances immediately following the first Arab Israeli War in 1947. The law was discussed and implemented over a period of five years. The contexts of land ownership, the national vision for the economy and the role of water in that economy were unprecedented. The need for political unity was extreme and the sense of public purpose was fundamental. The political will to act was focused and effective. Water was nationalised and became a public resource supervised in a framework of law by a central government system that was for Israelis essential and legitimate. Displaced Palestinians had a different view.

Most states in the MENA region have experienced dramatic political change since the mid-twentieth century. In the 1950s and the 1960s the revolutions and reforming initiatives were usually inspired by socialist principles allied with Arab nationalism and a recognition of the essential role of Islam. Accommodating the sometimes conflicting principles was always a challenge for the leaders of the respective states. The belief that water should be a public good accorded with both Islamic and socialist principles. Water in the MENA region is almost without exception seen as a public resource and water law evolved accordingly. In some isolated circumstances water from individual wells could be treated as a private resource. The extent to which the Islamic convention that water can only be sold in containers is observed varies.

The outsider inspired campaign of the 1990s to introduce the privatisation of water services and the concomitant need to establish ownership encountered opposition for a number of reasons including privatisation's legal ramifications. The opposition was partly expressed in reluctance to pass the necessary legislation to enable the privatisation process to begin. Privatisation of water has to be considered in two parts—first the agricultural sector and secondly the domestic and industrial sectors. In irrigated agriculture privatisation was opposed first, because it was unfamiliar and raised uncertainties about livelihoods and the cost of water. Secondly it was opposed in some quarters because it contradicted the precepts of Islam concerning the sale of flowing water. Thirdly it was opposed because it would be difficult to the point of being impossible to install metering and regulation systems. Fourthly, it would be very costly indeed to install metering. Fifthly, privatisation was regarded as alien and inappropriate for a sector with an ancient history and a public sector tradition. Private sector type transactions for water for farms are rare. They do occur in isolated villages. These arrangements are long standing and not the result of recent outsider inspired initiatives.

Water is being delivered to farms by public sector organisations at prices related to the cost of delivery in only a small number of MENA countries. The practice is comprehensive in Israel and Cyprus, increasing in Tunisia and Morocco and incipient in Jordan. None of these countries welcomed or adopted the privatisation of irrigation sector water provision. The pattern of resistance is evident in all the other MENA economies. Outsiders misguidedly think the opposition is merely an ill-informed response to the unfamiliar. In practice, in the agricultural sector it is mainly because pricing water, the main instrument of privatisation to achieve cost recovery, is a very blunt and ineffective instrument (Perry 1996 and 2000).

Opposition to the privatisation of domestic and industrial water. The reasons for the opposition are similar to those for resisting the privatisation of irrigation water supply. However, circumstances are very different in the domestic and industrial sectors. First, the technical problems of metering are surmountable in these sectors. Secondly, the volumes of water are about ten per cent of those in irrigated agriculture. Thirdly, the livelihood dimensions of water in the domestic sector are complicated and misunderstood. Water in the domestic sector is not generally considered to be relevant to livelihoods. In practice the impact of the availability of a close-by water supply instead of at some kilometres distance can release a number of family members for other social and economic activities, previously devoted to daily water collection. Fourthly water in the domestic and industrial sector can be charged at full cost of delivery prices because users do not use much water in these sectors and the sums paid for water are affordable by all except the poorest consumers. Fifthly and perversely, the role of water as an input to industry and the job creativeness of the industrial sector does not register. The transformation of the diversity of a political economy by industrialisation and the institutional change associated with it can be spectacular. The impacts on the economy and on its social adaptive capacity are registered in new demographic trajectories, improving economic indicators, high levels of employment, and a suite of favourable social indicators. That water plays a role in such transformation is rarely discussed and the users of high volumes of water in agriculture are unaware of the process, and if aware, such awareness would not be acknowledged. In practice the movement of farming family members to jobs in industry effects the necessary change in employment structure to enable sectoral re-allocation. The sectoral water re-allocation of water implicit in industrialisation is invisible to most governments and to all but a few professionals and scientists. The re-allocation is essential, but, if it were to be implemented via a prominent and very public reform programme, it would be prevented by political opposition.

The commitment of the leading international agency, the World Bank, to privatisation of the domestic and industrial water sectors and related water law reform in the MENA region has been very high during and since the 1990s. The World Bank has limited influence in the major economies of the region, in Egypt, Turkey and Israel and is only called on in an advisory capacity in the oil economies. It has an office in Riyadh and has been invited a number of times by the Libyan leader to evaluate projects. In less rich economies, however, the World Bank has succeeded in making the participation of a private company part of the assistance package in the rehabilitation and improvement of water supply and sanitation services. Jordan had to accept that the rehabilitation of the Amman metropolitan area would be carried out by a French private company. The same multinational company was given the contract to repair the network and improve its efficiency in Gaza. At the same time the World Bank sponsored the writing of the new water law of the Palestinian Authority.

For the outsider the logic is clear. Two decades of private sector ascendancy in the North had brought the deregulation of many public services in North America and Europe. In practice there was only one thoroughgoing privatisation of a national water sector, in the United Kingdom. (Kinnersley 1994) The public sector model for water provision was retained in the metropolitan heart of the capitalist project, New York. The Netherlands emphatically announced that it was to retain the public sector model early in the year 2000. Nevertheless the World Bank and the international agencies embraced privatisation and related legal reform as tools to address the challenge of providing water for domestic and industrial users in the MENA region.

The World Bank inspired Palestinian Water Law was being drafted and legislated in 1999 (Palestinian National Authority 1999). It is a potential tool not only to regulate water supplies and the raising of revenues to sustain the sector. It is also a centralising tool which will bring Palestine Authority government to the previously institutionally remote parts of the West Bank. The new Palestinian water law has already been shown to be susceptible to opposition from traditional priorities in Jericho. The arguments of those using the centuries old water sharing customary practices at the Jericho spring prevailed in 1999 over the principles of the new water law when the Palestinian leader was called on to adjudicate (Trottier 2000). Political forces favouring the local *status quo* prevailed over legal principles formalised in the 1999 draft Palestinian water law.

The customary practices in Jericho were in 1999 an example of the very complex local arrangements which successfully achieved the challenging task of sharing a common property resource. Such complex arrangements are durable in circumstances where the geographical scale of the water sharing is limited and the numbers of people and families are small enabling mutual awareness across the community. In these circumstances there can be transparency and trust in the processes of supervision and appeal. Informal information sharing is adequate to secure the system at this local scale. Compliance is achieved via social mechanisms rather than legal instruments. Replicating transparency and trust over larger geographical areas and larger user groups requires extravagant institutions including water laws and compliance mechanisms. Wolf (2000b) has drawn attention to the success of local arrangements when dealing with complex common property circumstances. He identifies the following five principles and practices as fundamental—the allocation of time not volume, the prioritisation of 'demand sectors', the protection of downstream minority rights, mechanisms of dispute resolution, and ritual forgiveness. Unfortunately the financial and social opportunity costs of installing and implementing new water laws and compliance mechanisms at a higher level governance are deterringly high. Making them effective when they contradict long standing social arrangements in diverse geographies is very difficult.

It has to be concluded that it is appropriate for outsider international agencies to support the development of water law and related legal institutions at levels higher than those where customary practice still works. The separate customary arrangements may be ideal for isolated communities living in traditional, pre-modern, political and economic circumstances. As individuals and families participate more intensely in the wider political economy, and thereby with global systems, the old livelihood and water distribution systems on which families depend will gradually become redundant. There will be inevitable conflict between those preferring long established and previously unmolested local customary regimes who want to continue with these traditional institutions and those who want to install principles relevant at higher levels of the governance system. Mediation will be needed.

It will be the transformation of the political economy which will provide a new context where the contention between customary and constitutional provision will disappear because livelihoods will not be dependent on water for irrigation. Jobs will be in non-agricultural sectors. The re-allocation of water will be achieved invisibly by an increase in jobs in industry and services, as demonstrated in Israel between 1955 and 1985 (Allan and Karshenas 1996). The move from customary to constitutional water law will be similarly achieved because the demand for water for irrigation will tend to decline as the farming families find livelihoods in other sectors. The two systems will work alongside each other for as long as local communities can demonstrate the effectiveness of their customary institutions.

International water law and the Middle East and North Africa

> MENA hydropolitical hegemons and non-influential MENA states both selectively reject the well meaning principles of international water law developed by the end of the twentieth century.

The Middle East and North African region is the most river free area on Earth. It would be even more river free if some tectonic accidents in the geological past had prevented, rather than enabled, the major international river systems, the Nile and the Tigris-Euphrates to flow into the region from high rainfall highlands in north-east Africa and Anatolia. Until the late twentieth century the waters of the Nile (85 billion cubic metres average annual flow) were sufficient for all would-be downstream users. The Tigris-Euphrates provides sufficient water (75 billion cubic metres average annual flow) for would-be users for the foreseeable future. The minor system—the Jordan-Yarmuk provides only about 1.4 billion cubic of metres of annual flow on average. Unlike the major regional river systems this minor system is heavily over-used and the upper catchment users reduce flows to the lower riparians. Both Israel and Syria seriously deplete the flow of the Jordan and the Yarmuk respectively.

Self-constructed entitlements to water are non-contentious when the economic systems of users do not demand water beyond sustainable yield. In water surplus situations perceptions of the resource itself, the likelihood of water stress, and international relations are relaxed. Anxiety about the adequacy of a resource are heightened when the demand approaches a high proportion of the supply, especially when no obvious augmentation of water resources can be mobilised.

In the domain of shared international river basins and aquifers the terminology differs from that in the national and customary domains. The concept of 'ownership' used about water within nations and communities becomes an element of national 'sovereignty'. At the international level the concepts of common property, public property and private property have little relevance. Although the notion of 'unregulated common property' does convey a sense of the anarchic situation that prevails in the capture and management of water in international hydrological systems.

Legal principles are undoubtedly of great potential significance. It is not the purpose of this section to argue otherwise. The problem, however, with alien legal principles evolved in alien outsider institutions, is that they have little appeal to MENA politicians, professionals and communities when they will disrupt existing practice and are not founded on the cultural and religious conventions of the region.

At the international level it is not just pre-modern and modernising MENA countries that reject, and/or are unwilling to adopt, international water law. The post-industrial political economy of Israel also neither observes the incipient international law to be found in a comprehensive interpretation of the articles of the International Law Commission (ILC 1997), nor recognises ILC principles that contradict its interests. It has no enthusiasm for any version of water rights that compromises the principle of prior use. Its lack of enthusiasm for the constructive concept of 'equitable utilisation' is overwhelming. MENA hydropolitical hegemons and non-influential MENA states both selectively reject the well meaning principles of international water law developed by the end of the twentieth century. The rapidly developing hegemon in the Tigris-Euphrates basin, Turkey, is the most trenchant in its rejection. Egypt, the hegemon in the Nile basin, is also very cool indeed on the subject of the adoption of ILC principles. Israel, alone in the region, with the economic capacity to adjust to any impacts that the 1997 ILC Convention might bring abstained at the General Assembly vote on 21 May 1997.

A sequence associated with the role of legal principles and practice in international water affairs at different phases of water shortage has been simplified in Table 7.2. Such an environmentally determined analysis is not recommended by the author. However, it is helpful to note that the status of the water deficit is not determining of how a state would reach for international water law in shaping its riparian relations. A water deficit is not a forcing factor in getting governments to adopt either the principles of international water law nor the articles of the as yet unratified 1997 International Law Convention. The analysis of the legal dimension of international relations in Table 7.2 follows that of an analysis of responses in the economic and political domains. Socio-economic development reduces the probability of armed conflict. It also makes less urgent resort to legally based processes to assuage the conflictual situation between riparians.

At the turn of the twenty-first century we are still at the very untidy political stage of establishing water law as an internationally acceptable instrument. The outcome of the political contention by the year 2000 was a set of UN International Law Convention articles which show very clearly where the interests of the contending parties have shaped both the content and the language of the Convention. The absence of compromise and absence of accord on the prioritisation of principles has had to be left for a later stage of the political process. The narrative is clear for those that assume that politics will determine the pace of evolution and adoption of international water law. Those who expect worthy principles to be adopted because they make sense to professionals find the pace of innovation frustrating.

The Articles of the 1997 Convention are afflicted by politically determined contradictions. The role of the principles of international water law in currency at the end of the twentieth century in the MENA river basins has been to provide conflicting legal principles to serve the arguments of contending riparians. First, the principles are 'prior use' and 'no harm', favour the long standing downstream user. Secondly, versions of 'sovereignty', all of which have an intuitive appeal, are favoured by the upstream user and especially the new upstream user. Both sets of principles lend themselves to popular and very selective chauvinist advocacy. Thirdly, the concept of 'equitable utilisation', a consensus converging notion

Table 7.2 International relations and water law during the transition to water deficit in the Middle East and North Africa

No water stress	Approaching water deficit	Water deficit	Serious water deficit
Perceptions			
No problem	Precautionary approach to other riparians	a) Deficit de-emphasised b) Deficit selectively emphasised	a) Deficit de-emphasised b) Deficit selectively emphasised
Actions			
In the political economy			
None	Seek new water Mobilise substitute water—virtual water	Continue to mobilise virtual water. Increase allocative and productive efficiency	Virtual water imported
Significance in international relations			
None	Initiate or respond to basin management initiatives. Identify principles appropriate to the 'interests' of the riparian in question. Identify a water volume/share for & political negotiation . purposes. Participate in or observe at International Joint River Basin Technical Commissions – bi-lateral or basin-wide	As for 'approaching water deficit' Possible establishment of political relations sufficient to achieve bi-lateral and/or multilateral agreement. This is entirely a political process.	As for 'approaching water deficit'
In the domain of law			
Actions	Participate in UN international water commission discussions—eg the ILC	as in 'No water stress'	as in 'No water stress'

Source: the author

developed by the International Law Association in Helsinki in 1966 has only gained support amongst legal professionals and some outsider scientists. The concept was refined by those meeting at the three decade long deliberations of the United Nations International Law Commission. The ILC Convention articles produced in May 1997 cannot be precisely defined nor effectively operationalised. They lead to impasse rather than resolution. They are not considered as tools with which to achieve agreement. They are not yet viewed as a language to achieve mutual understanding, nor as a means of analysing issues in contention, nor as a set of guidelines to structure negotiations. This unwillingness to see the ILC rules as useful is much less to do with their imprecision than with the intractable state of MENA international relations.

When international relations reach a stage of sufficient strength, reciprocal regard and mutually understood dependence between two or more riparians it will be possible to develop institutions that will establish rules, agreed sanctions, implementing bodies, courts and all the formal instruments required for the administration of international water law (Wouters 1999a).

On the 'international watercourse': international water law and the United Nations chosen definition

> International law has to be devised that copes with water as a public good within national economies and as a form of common property at the international level.

International water law is the concern of professional and academic lawyers. The development of international water law is, however, a political process. International water lawyers are just one group of players in the game developing international water law. They contribute principles of water law which evolve from ideas formed in diverse legal regimes. They contribute ideas to the political process of interacting water using entities. There are laws which apply to resources including land, property and the natural environment. Other laws regulate commercial transactions including employment, finance, international finance and international trade. Law also deals with circumstances where the transactions go wrong causing injury and requiring rules of compensation. In addition there is a huge range of social conventions to do with family and the community which have become embedded in culturally specific legal regimes. At the national level constitutional and common law define the obligations of the state and its citizens with respect to national identity, and to the entitlements and obligations of individuals and communities. Isolated communities develop their own customary law which operates effectively in pre-modern communities.

The status of international water law in the anarchic circumstances of international relations is sometimes captured in the term customary international water law. What happens in the international river basins is not subject to recognised codified practice. Riparians do what is in their own interests to the limits that their political antennae tell them is possible. If they cannot access enough water as a consequence of the actions of another riparian, such as Egypt on the Nile, Jordan and the Palestinian Authority on the Jordan, or Turkey on the Orontes, they do not resort to international law or even armed conflict. Syria and Iraq on the Tigris-Euphrates are not yet in water deficit. Water short economies in practice solve the problem through the economic device of importing water intensive commodities such as grain. Turkey is not yet short of water as an economy.

To date international water law has not been effective in the MENA region in establishing rights, entitlements or obligations. Legal instruments do not yet secure water for economies enduring water deficits. Many, including water lawyers, do not understand why the apparent water crises do not focus the attention of legislators on the potential of legal remedies. The reason is that it is politically expedient for them to de-emphasise the water issue, including the proposed legal remedies. The solution provided by virtual water in the imported grain is effective, politically invisible and therefore politically stress free. A legal process would be contentious, politically fraught and problematic.

Legal principles that derive from other areas of law which deal with static property and transactions related thereto are of limited value in the realm of water. Water is not the same as land. It is very dynamic; much more like finance than land. Financial transactions function naturally in markets. Water does not. Laws relating to water at all levels, and especially at the international level, need to be flexible (Du Bois 1995). They need to deal with both regular and unpredictable fluctuations in flow. In these circumstances market instruments would seem to be ideal. But the universal perception in the MENA region is that water should not be commodified at any level in the big water using agricultural sector. The consequence is that well argued outsider cases recommending an economic approach to water (Fisher 1994 and 1995) are very strongly resisted. In addition water is generally low in economic value and does not merit expensive administrative, regulatory and legal overheads, at least in weak economies.

We have something of an impasse which economists see as perverse. If water could be commodified, they argue, then water could be traded by market instruments and regulated in a legal framework. In theory this would also be possible at the international level. Both the market system and legal frameworks would make possible the achievement of significant transaction cost savings. In real international political economies the market is not an option. International law has to be devised that copes with water as a public good within national economies and as a form of common property right at the international level. Customary law at the politically remote local level can still successfully regulate the use of water in dynamic environmental and social circumstances in the MENA region (Trottier 1999, Wolf 2000b, Lichtenthaler 2000).

It is a paradox that customary law operates effectively with respect to water in remote pre-modern communities. Such successful operation depends on comprehensive awareness across the community and regulation by social rather than legal sanctions (Wolf 2000b, Lichtenthaler 2000). By contrast international river basins are characterised by the absence of transparency. Hydrological data are not shared. Where water data do reach professionals in another riparian state they are likely to be biased to serve the interests of the data providing entity. There is no equivalent of the effective sanctioning social system which is possible in the intimate and transparent societies of remote communities.

Problems of deconstruction

The extent to which it is possible to deconstruct complex political reality to enable the deployment of effective legal principle to guide governments on international water issues is worth analysis. It would be useful to know what features have to be considered if governments are to reach agreements on, and manage considerately (with 'due diligence' (ILC Article 7 1994) or 'with best efforts under the circumstances' (ILC Article 7 1994, McCaffrey 1995:395), shared water resources. If such deconstruction is possible it would then be possible to determine whether the international law professionals have elected to address that deconstructed complex which is susceptible to legal principle.

The waters of concern to economies and communities are shown in Table 7.3.

The waters that are susceptible to analysis, negotiation and agreement according to international water law are the waters

Table 7.3 National waters; legal principles on transnational waters are of NO concern in the national domain

Water types	Characteristics
• waters which fall as rainfall and remain naturally within the borders of the state and are available for use from	
• surface flows	variable, monitoring easy
• soil profiles [non-evident, never accounted]	variable, **not monitored,** silent, not accounted, can be 90% of national water
• groundwater	Non-variable, monitoring easy
• ancient groundwater: fossil groundwater	Non-variable, often deep and poor quality, monitoring easy
• 'virtual water'; water embedded in water intensive and strategically important commodities such as grain	Delivered by effective and operational systems; cheap, subsidised, silent, politically benign, can be **50–90 % [non-evident, never accounted] of the national water budget in MENA economies. All the regions economies depend on 'virtual water'**
Transnational waters: the target for legal principles* vis-à-vis 'international watercourses' [* equitable utilisation, 'no harm', prior notification]	
• **waters that flow into the state on the surface and underground**	Variable if no upstream engineering, They benefit, downstream riparians Normally developed first. **Can be the major waters** in an economy—e.g. **Egypt** from the Nile; **Israel** from West Bank groundwater

Table 7.3 National waters; legal principles on transnational waters are of NO concern in the national domain *cont'd*

Water types	Characteristics
• **waters that flow out of the state on the surface and underground**	Very variable without upstream regulation Can be the major water in a water 'exporting' economy—e.g. Turkey, Ethiopia.

Negotiations focus on:
- volume and to some extent on quality

Other negotiable issues:
- obligation to use efficiently
- provide securely
- obligation to inform other riparians about new uses and new engineering

susceptible to measurement. The underlined entries in Figure 7.3 are the measurable waters. Variation in the volumes and quality of such waters can be monitored. Figure 7.3 is useful in that it shows that waters that are allocated and managed come from diverse, sometimes hard to measure, and always quantified by biased observers. Even if monitoring of the variable phenomena were possible there is a tendency to bias data about these measurable variables to serve the interests of the provider. Unfortunately attempts to create river basin databases have not been successful. The Hydromet project in the Nile basin was based in Uganda. The vision of a comprehensive hydrometeorological basin-wide was lost in the destruction of public records in the civil wars in Uganda in the late 1970s and the 1980s (Kabanda and Kahanghire 1995:221). Elsewhere there have been no sustained exchanges of hydrometeorological data in the MENA river basins.

International waters can be subject to analysis, negotiation and agreement on three other issues. First, the obligation to use efficiently; secondly the assurance of secure flows; thirdly the obligation to inform other riparians about new proposed uses and engineering activity. The first issue defies definition. The second has not yet achieved prominence as a negotiation issue. It should have figured in the Israel-Palestine peace treaty of 1994, but did not. In the Palestine-Israel negotiations in the year 2000 the subject of secure and insecure water was not on the agenda. The principle of the obligation to inform other riparians about new uses is only observed by the World Bank through its operational procedure on international waters—OP 7.5 (World Bank 1992).

The role of the international agencies though indirect can be significant. On the Euphrates the World Bank was prepared to lend money for the first two hydro-power dams, the Keban and the Karakoya Dams completed in the 1970s and the 1980s. The Ataturk Dam, a multi-purpose dam, including substantial irrigation projects, could not be funded by the World Bank. Not all international bodies were so scrupulous. MENA governments could assemble funds from other international sources. Private banks were not tied by World Bank type regulations. Even some Northern government ministries with sister overseas development ministries regulated by World Bank type regulations were prepared to underwrite the risk of national civil contractors. The principles of international law could be successfully brought to bear by legal opinion requested by NGOs such as Greenpeace. The following from a 'legal opinion' sought by Greenpeace in May 2000 over the Ilusu Dam on the Tigris is an example of how pressure can be exerted indirectly on riparians contemplating a major civil project.

'The obligation to notify, consult and negotiate under general international law

Five States and other members of the international community, including international organisations, have long recognised the need to ensure the availability of information on activities and other circumstances which could affect the interests of states in relation to shared natural resources. This arises from the principle of "good neighbourliness", reflected in Article 74 of the UN Charter and the dicta of the International Court of Justice ("ICJ") that the principle of sovereignty embodies "the obligation of every state not to allow its territory to be used for acts contrary to the rights of other states". Typically, this is provided for in international agreements by related commitments:

- the requirement to provide information to potentially affected states on particular activities, and
- the requirement to engage in consultation and, if necessary, negotiation. (Crawford et al. 2000:3)

The legal opinion pressed on the Department of Trade and Industry of the United Kingdom by Greenpeace was successful in that DTI withdrew its guarantee of support to the UK contractor. This is not the same as saying that the Ilusu Dam on the Tigris River would not go ahead. The Turkish Government is expert in viring funds from acceptable international projects to unacceptable ones. And other private sector banks would come forward to meet the project costs.

Transnational waters and why international water law is so difficult to develop and install?

> 'People have really got to want law and order before they do anything about it. Deep down around here they don't really care.'
>
> > Elderly former sheriff to Gary Cooper as Cooper went around the town in the Hollywood film 'High Noon' looking for support in facing down the vengeful 'lawless' men arriving imminently by train.

> 'But in actual international disputes, it seems probable that the facts and circumstances of each case, rather than any *a priori* rule, will ultimately be the key determinants of the rights and obligations of the parties. Difficult cases, of which there are bound to more in the future, will be solved by cooperation and compromise, not by rigid insistence on the rules of law. This is one of the lessons of the World Court's judgment in the Gabikovo/Nagymaros case.
>
> (McCaffrey 1998:7)

> 'As is its practice, the ILC did not indicate which of the provisions codify, and which progressively develop the law. But it seems clear that the most important elements of the [May 1997 ILC] Convention—equitable utilisation, "no harm", prior notification—are in large measure, codifications of existing norms.'
>
> (McCaffrey 1998:12)

In 1996, the UN's Sixth Committee of the International Law Commission, convened as the 'Working Group of the Whole', commenced meetings on the Watercourses Draft Articles addressing the whole proposed Convention. The Commission had been engaged on the preparing for the drafting of the Convention since the mid-1970s. The first two weeks of meetings revealed the extent of controversy that existed on key issues. At the end of this first session (November 1996), it was questionable whether States could agree on a text and some believed that agreement would never be reached. At the second two-week session March/April 1997, following much debate, and many

Table 7.4 Voting Record of the MENA countries of the Group of the Sixth Committee (of the ILC of the UN) on the Text as a Whole of the Watercourses Draft Articles—March/April 1997

For (5 **of 38**)	Against (1 **of 4**)	Abstained (3 **of 22**)
Algeria	Turkey	Egypt
Iran		Israel
Jordan		Lebanon
Sudan	*(Figures in brackets MENA votes of all votes of those present)*	
Tunisia	*(130 states did not attend/vote out of c191)*	

Did not vote of MENA countries: (12)
Cyprus, Iran, Iraq, Kuwait, Libya, Morocco, Oman, Qatar, Saudi Arabia, Syria, UAE, Yemen

Did not vote of countries relevant to MENA: (6)
Burundi, Eritrea, Kenya, Tanzania, Uganda, Zaire

Did vote and relevant to MENA countries: (2)
Ethiopia—for, Rwanda—abstained

inevitable compromises, the Working Group of the Whole took the unusual step of voting on a revised, albeit highly contested, draft text. By a vote of 42 States for, 3 against and 18 abstentions, a final text was adopted by the Working Group of the Whole.

Figure 7.4 shows the way the MENA countries voted on the whole draft. The voting reflects the interests of the participants. Those with serous reservations voted against or abstained including the three MENA hegemons, Turkey, Egypt and Israel.

Figure 7.5 reveals the positions taken by the participants on the vexed issues of *equitable and reasonable utilisation* and *no harm*. The drafting challenge for Articles 5–7 was considerable. The inevitably vague and sometimes awkward outcome of the contention that preceded the vote are the Articles shown below, being the text from the May 1997 UN Convention accepted for circulation to member states by the UN General Assembly.

PART II. GENERAL PRINCIPLES

Article 5
Equitable and reasonable utilization and participation

1. Watercourse States shall in their respective territories utilize an international watercourse in an equitable and reasonable manner. In particular, an international watercourse shall be used and developed by watercourse States with a view to attaining optimal and sustainable utilization thereof and benefits therefrom, taking into account the interests of the watercourse States concerned, consistent with adequate protection of the watercourse.
2. Watercourse States shall participate in the use, development and protection of an international watercourse in an equitable and reasonable manner. Such participation includes both the right to utilize the watercourse and the duty to cooperate in the protection and development thereof, as provided in the present Convention.

Article 6
Factors relevant to equitable and reasonable utilization

1. Utilization of an international watercourse in an equitable and reasonable manner within the meaning of article 5 requires taking into account all relevant factors and circumstances, including:

   (a) Geographic, hydrographic, hydrological, climatic, ecological and other factors of a natural character;
   (b) The social and economic needs of the watercourse States concerned;
   (c) The population dependent on the watercourse in each watercourse State;
   (d) The effects of the use or uses of the watercourses in one watercourse State on other watercourse States;
   (e) Existing and potential uses of the watercourse;
   (f) Conservation, protection, development and economy of use of the water resources of the watercourse and the costs of measures taken to that effect;
   (g) The availability of alternatives, of comparable value, to a particular planned or existing use.

2. In the application of article 5 or paragraph 1 of this article, watercourse States concerned shall, when the need arises, enter into consultations in a spirit of cooperation.
3. The weight to be given to each factor is to be determined by its importance in comparison with that of other relevant factors. In determining what is a reasonable and equitable use, all relevant

factors are to be considered together and a conclusion reached on the basis of the whole.

Article 7
Obligation not to cause significant harm

1. Watercourse States shall, in utilizing an international watercourse in their territories, take all appropriate measures to prevent the causing of significant harm to other watercourse States.
2. Where significant harm nevertheless is caused to another watercourse State, the States whose use causes such harm shall, in the absence of agreement to such use, take all appropriate measures, having due regard for the provisions of articles 5 and 6, in consultation with the affected State, to eliminate or mitigate such harm and, where appropriate, to discuss the question of compensation.

The decision to vote on the two issues *of equitable utilisation* and *no*

Table 7.5 Voting Record of the MENA countries of the Working Group of the Sixth Committee (of the ILC of the UN) on the Convention on the Law of Non-navigational Uses of International Watercourses on the Revised Article 5–7 [i.e. equitable utilisation and no harm] of the Watercourses Draft Articles—March/April 1997

For (6 **of 38**)	Against (1 **of 4**)	Abstained(3**of 22**)
Algeria,	Turkey	Egypt,
Iran,		Sudan,
Israel,		Lebanon
Jordan,	*(Figures in brackets MENA votes of votes of those*	
Syria,	*present in bold)*	
Tunisia,	*(129 States did not attend/vote out of c194)*	

Did not vote of MENA countries: (11)
Cyprus, Iran, Iraq, Kuwait, Morocco, Oman, Qatar, Saudi Arabia, Syria, UAE, Yemen

Did not vote of countries relevant to MENA: (6)
Burundi, Eritrea, Kenya, Tanzania, Uganda, Zaire

Did vote and relevant to MENA countries: (2)
Ethiopia—abstained, Rwanda—abstained

harm together faced participants with a difficult dilemma. The notion of *equitable utilisation* contradicts that of *no harm* in a number of respects. *Equitable utilisation* also calls on diverse principles which contradict the notion of *sovereignty*. In a rough and ready political world upstreamers cleave to *sovereignty* and downstreamers cleave to the principle of *no harm*. Equitable utilisation is not seen as a compromise position by either group. By linking *equitable utilisation* to *no harm* rather than voting on *equitable utilisation* alone keen observers of the ILC process were denied a key insight, namely which nations supported or opposed the operationalisation of principles of *equitable utilisation.* It is argued in this chapter that the political contention provides the prime explanatory logic. The political process which linked equitable utilisation and no harm at the 1997 ILC meetings can be seen as fitting the interests of both upstreamers and downstreamers. Both hydropolitical ideologies could vote or abstain without revealing which of the two ideas, equitable utilisation or no harm they were favouring or disfavouring. The political reality is that upstreamers and downstreamers remain unwilling to come out in favour of the non-concrete principles of equitable utilisation.

The voting on these contentious articles caused Israel to move from a position of abstention on the whole draft to voting in favour on Articles 5–7. Was Israel signifying that it was determined that the no harm rule should have a place? The answer to this question is yes. By voting for Articles 5–7 was Israel endorsing the principles of equitable utilisation? The answer to this second question is almost certainly no. In the years since the general endorsement of the whole Convention, during which the Palestine-Israel Peace Process has included the highest profile international water dispute, Israel has shown no inclination to recognise equitable utilisation principles.

Sudan moved from voting in favour of the whole draft to abstention on Articles 5–7. Here again the link between equitable utilisation and no harm in the vote has made it difficult to detect how the parties prioritised their positions. Egypt and Turkey remained resolute in giving no support.

When the vote took place two months later at the General Assembly, when many more MENA countries were represented than at the earlier Sixth Committee meeting, the outcome was surprisingly clear. Bearing in mind that no permanent commitment was made by voting at the General Assembly, as there had to be follow-up stages of ratification and signature, the outcome of the vote was remarkable. Four states did not approve the Convention. The group included the three major powers—Turkey, Egypt and Israel.

Table 7.6 Voting Record of the MENA countries of the UN General Assembly on the Convention on the Law of Non-navigational Uses of International Watercourses 21 May 1997

For (15 **of 104**)	Against (1 **of 3**)	Abstained (3 **of 27**)
Algeria	Turkey	Egypt
Bahrain		Israel
Cyprus		Lebanon
Iran		
Jordan		
Kuwait		
Morocco		
Oman		
Qatar		
Saudi Arabia		
Sudan		
Syria		
Tunisia		
UAE	*(Figures in brackets MENA votes of all votes in bold)*	
Yemen	*(33 States were absent out of c191.)*	

Did not vote of MENA countries: (2)
Iraq, Libya

Did not vote of countries relevant to MENA: (6)
Eritrea, Uganda, Zaire

Did vote and relevant to MENA countries: (4)
Burundi —against; Ethiopia, Rwanda, Tanzania—abstained

Lebanon was the fourth member of the anti-Convention group reflecting its anxiety about any external pressure that might steer it towards discussing water issues with Israel. However, by the time voting took place at the UN General Assembly on the whole draft Convention, Lebanon had aligned with Syria which has taken a leading position in supporting the Convention. Fifteen MENA countries, a remarkable 75 per cent of the total, voted in favour of initiating the referral of the Convention to the UN membership for ratification and signature. Iraq and Libya were absent and amongst the 25 per cent of MENA members which did not vote.

Few of the countries outside the MENA region which 'export' large volumes of water to the region voted. Ethiopia and Tanzania

Table 7.7 Ratification of the UN General Assembly on the Convention on the Law of Non-navigational Uses of International Watercourses 21 May 1997

Signatures by June 2000
Cote d'Ivoire, Finland, Germany, Hungary, **Jordan**, Luxembourg, Namibia, Netherlands, Norway, Paraguay, Portugal, South Africa, **Syria**, **Tunisia**, Venezuela, and **Yemen**.
Ratifications by June 2000
Finland, Hungary, **Jordan**, **Lebanon**, Norway, South Africa, and **Syria**.

abstained. That Burundi voted against and Rwanda abstained suggests that reaching basin-wide agreements on the Nile will continue to be difficult.

Signatures on and ratifications of the Convention (Table 7.7) have come in slowly. Syria was the first nation to notify its support in August 1997, closely followed by South Africa. MENA governments are over represented amongst the signatories with 25 per cent of those signing while comprising only five per cent of the UN membership. Signatories from the MENA region and other regions are almost all downstream states. Most of them are minor riparians in terms of power within their respective river basins or relatively isolated in hydrological terms and therefore anticipating little contention. Some signatories have been influenced by strong individuals in the field of law in their governments. Dr Kader Asmal in South Africa has been particularly influential in getting the South African Government to support the Convention and has taken an equally strong lead in the SADEC region over cooperation over water. Prominent lawyers in Jordan have played a similar role in Jordan. The rate at which ratification takes place will be determined by a wide variety of factors. The leadership of informed and enthusiastic individuals will be important, even more important will be the cold evaluation by politicians and professional civil and military officials of current and future individual riparian interests.

Conclusion: which is prime—international relations or international law?

'The law should only present vague standards and avoid strict rules, as the former induce negotiation while the latter induce litigation.'

> Eyal Benvenisti in a paper delivered at the Workshop on the joint management of aquifers held in Istanbul in May 1997 convened by the Truman Institute for the Advancement of Peace and the Palestinian Consultancy Group. (Haddad and Feitelson 1997)

The arguments in the chapters which addressed the economics, the politics and the international relations of water in the MENA region showed that explanations from politics were always the most powerful. Economic development was shown to be a very important source of explanation on for example when water policy reform could be initiated. But when the question was asked what enabled the economic development to take place the answer was found in the quality of the political institution building. Political development has been fundamental to economic development. When economies have become strong and diverse they have had the capacity to reform their water allocation and management policies.

This pattern is also true in the field of water law at the national level. At the international level, however, such is the attachment to the advantages of low levels of regulation to the hegemon, whether a super-power or a major regional power such as Turkey, Egypt and Israel, progress towards the adoption of an international legal regime for water has been and will continue to be very slow.

In these circumstances the dominance of high politics over the relatively low politics of water contention and conflict has and will prevail. Where, in addition, political institution building has been sub-optimum, as some have argued it has been in many MENA countries, then political economies have become 'trapped in the institutional cages they have built' (Waldner 1999:1). Institutions created to bring political stability and economic development have frequently brought the political stability but failed to facilitate economic development. The state building attempted in the MENA region has tended to produce institutions that impede rather than facilitate economic development. The relevance to international law is that states that lack the institutional capacity to build economic development will also lack the capacity and the will to develop international negotiating capacity to play a part in river basin relations to identify constructive and convergent solutions to water disputes.

These inauspicious circumstances have not deterred outsiders and international agencies from attempting to create institutions to develop cooperation amongst riparians (Nile 2002 1992–2000, World Bank et al. 1998). Informal insider processes have also been initiated by scientists and activists for peace (Isaac and Shuval 1995, Feitelson and Haddad 1992–1998). The establishment of international law is, however, much more at the stage of contended principle than agreed practice. International lawyers claim that customary international law provides guidelines for those contending over shared waters (Wouters 1997, 1998). Based on general principles of consideration for neighbours, and not causing them harm, customary international water law provides a context which does little as yet to deter water using entities from pursuing their own interests. The incipient international regulation proposed by the ILC Convention (1997) does not have effective compliance procedures. (Wouters 1999) In a MENA region where water issues are so peripheral to or deeply subordinate to the major security sub-complexes of the Levant and the Gulf the development of international legal processes will remain very much in the shadow of international relations. Unfortunately international law comes a very poor second to virtual water as a stress free and adoptable solution to coping with the regional water deficit for weak economies.

PART 4: THE FUTURE

CHAPTER 8

The future role of water in Middle East and North African society

> Not all waters are equal: some waters are more evident, more accessible, more manageable, more costly and more economically valued, more integral to society, more political, more multi-functional, more conflictual, more negotiable, and more prone to litigation than others.

It is easy to repeat water clichés—water is essential for life and economic security, and to coin water banalities such as each of us comprises 60 per cent water. Humanity's relationship with water has been as long as human history. Communities have moved to where there was rainfall or accessible surface water or groundwater. With the final settlement of North America and of the 'virgin lands' in Central Asia (Breznehev 1978) in the twentieth century there are few remaining undeveloped environments suitable for agriculture. And some of those upon which hands were laid in the pioneering endeavours of the nineteenth and twentieth centuries have been found to be inadequately watered when tested over a century rather than a few seasons.

Some misconceptions: defining the water deficit

The reason it is so difficult to make sense of the water situation in the Middle East and North Africa is that the indicators of shortage are hard to read and those arguing that there are shortages are often describing different aspects of water shortage. For example a potential shortage of drinking and domestic water is politically significant because the problem has to be solved locally. In the event this problem has everywhere in the MENA region been solved locally.

A shortage of water for food production is also politically significant. But this problem is impossible to solve locally and the political consequences have to be accommodated very differently.

The main reason that MENA water is difficult to discuss or evaluate in the region is that water for food is a very political matter. People do not want to admit that water supplies are insecure. What can be said about water is fixed within a MENA 'sanctioned discourse'. That discourse insists that water is sufficient and secure.

To understand the water budget of an economy it is necessary to recognise that water is available from a number of different sources world-wide. The richest water endowments are in those regions where naturally occurring soil water is secure from one year to the next.

All economies in the Middle East, most economies in Africa south of the Sahara and many economies in Asia, including the giant economies of China and India, are already coping with serious water deficits or will face water deficits imminently because parts of these regions endure unreliable rainfall.

Meanwhile, populations world-wide are still rising rapidly and every individual added to a population not balanced by the death of another member of the community requires a future supply of at least one thousand tonnes (cubic metres) of new water every year to meet their food needs. Scientists, agency professionals and government officials are having to adjust their perspectives on water availability world-wide as more and more economies cross into water deficit. The deficits affect the agricultural sector. Domestic and industrial water needs rarely pose unmanageable challenges, except occasionally in the poorest economies.

The traditional remedy to water shortage has been the deployment of engineering measures such as storage dams and networks for water distribution to average seasonal and annual fluctuations in supply and move water from surplus to deficit regions. These measures are known as supply management. The language of supply management engineering is well diffused and has been deeply institutionalised during the twentieth century. The institutions are well established in both the South and the North.

In water deficit circumstances these supply management measures have been augmented by demand management policies. Demand management is more concerned with the efficiency of water use than with the volumes of water available. Demand management measures are only partly technological. Just as important are economic instruments and regulatory systems. Economic instruments and regulation are not rooted in engineering science or easily assimilated by the monolithic bureaucracies which have allocated and managed the water resources for the past century or more. Economic instruments such as water tariffs, water markets, and regulation are advocated by outsider social theorists and economists.

Environmental principles discovered recently in the North and the regulatory instruments by which they can be established are also not naturally rooted in communities in the South. They are especially alien to the institutions which the hydraulic mission of Southern governments and professionals has spawned.

The water predicament of the MENA region can be defined by insiders and by outsiders. One of the main reasons for writing these chapters has been to draw attention to these parallel and unconnected Southern and Northern discourses. When the respective discourses connect the outcome tends to be contentious, sometimes very contentious, for entirely predictable reasons. At the turn of the twenty-first century the MENA South had very different assumptions from those held by Northern professionals and agencies. They differ on the volume, nature and significance of deficits. They especially differ on how to accommodate or remedy the deficits.

The rest of this concluding chapter will attempt to integrate the somewhat deconstructed explanations of the preceding chapters. Emphasis will be given to social theory in achieving an understandable synthesis. It is intended that the synthesis will help those engaged in the separate Southern and Northern discourses to relate to each other.

A theory of the hydropolitical contention of social, economic and environmental water demand. Water for whom in the MENA region: competing demands and responsive networks of consensus.

> This analysis is being made at the height of the contention between the MENA insider discourse and the new outsider Northern discourse.

> The capacity to construct new knowledge about the economic and environmental role of water in a political economy comes relatively late in the process of socio-economic development.

Water is necessary to meet three basic demands. Water supports communities, economies and the environment. It has social, economic and environmental dimensions. Until communities placed demands on surface and groundwaters the environment was green, with vegetation, or brown, without vegetation, permanently and/or seasonally, according to the availability of water. The environment and its ecosystems were shaped by natural hydrological and biological processes operating on naturally occurring water. Ecosystems needed no voice to exert their entitlement to water in order to continue in their, albeit, non-equilibriated natural state. It is assumed here that natural environments are in constant flux reflected in variable rainfall, variable soil moisture, variable river flows and global climate change. They are not in equilibrium. (De Angelis 1987, Benke et al. 1993, Stott et al. 2000) For communities to survive in simple or complex political economies they have to be able to adapt to short term and long term, non-equilibriated, environmental change.

The first demands on naturally occurring water by human communities were made to meet their social needs. They needed water for drinking and to meet their food needs. The raising of food was integral to society. Food was raised to ensure subsistence. It was not an economic activity in a twentieth century sense. The demands placed by subsistence users are by definition associated with low technology and were environmentally benign. The important idea in this narrative about the use and allocation of water in the MENA region is that water in agriculture was initially regarded as a social resource. Water was still regarded as a social resource in the MENA region at the turn of the twenty-first century. Where the volumes of water involved were limited it was usual to regard such water as a free entitlement provided by a generous deity from the natural bounty associated with generous deities.

By the late nineteenth century the demands on naturally occurring MENA waters were made by very much larger farming communities. These communities were armed with higher technology than those available to subsistence farmers. Water was an input to the major rural activity, agriculture. Increasingly through the twentieth century water was also needed by industry. Water became an economic input, albeit unrecognised in agriculture, in both agriculture and in industry. Population growth and diversifying economies imposed unprecedented demands on the naturally occurring water. The new levels of demand and their impact on water in the environment were not, however, associated with a shift in perception on the part of users. The heavy water users in agriculture, using 70 to 90 per cent of the water, were locked into their comfortable and enduring belief that water was a social resource. Water for agriculture was an entitlement legitimised by the tenets of their religion and ancient social convention. Without a voice the need of water to maintain the health of the environment was ignored.

The role of politics in integrating the contention over the use and allocation of water is illustrated in Figures 8.1 and 8.2. Water is conceptualised as playing a role in the environment, in providing society's needs and in providing an input to the economy. Each of these uses of water is at a point of a triangle. The relative and changing demands through time for water by society, the economy and the environment are illustrated on the diagrams. As the volumes of water required by the respective users increase the areas taken up in the triangle by water for social, economic and environmental purposes proportionately increase. The impact of the three demands differs according to the effectiveness with which the ideological and political arguments of the three contending demands can be communicated and legitimised. Only socially acceptable demands can be legitimised. Claims from those constructing knowledge on the basis of economics or environmental theory have not to date been able to achieve a prominent role in MENA discourses.

Those demanding water for social, for economic and for environmental sustainable purposes are differently equipped in political terms when contending over water. They differ in their ideological associations and in the extent that those advocating them can reach for politically feasible arguments. And these differences vary with time. However, past ideology is preferred to new concepts. As a result water is perceived as a social resource rather than an economic and environmental one in the MENA region, because water was first perceived as a social resource.

The extent to which water policies are politically feasible and ideologically compatible with the beliefs of a community are fundamental both for the continuation of existing policies and the adoption of new ones. If the adoption of new outsider ideas, such as the economic and environmental value of water, requires the transformation of beliefs and ideology, then the problem of water reform should be seen as the political process which it is. The transformation of beliefs and ideologies is not a technical challenge, except insofar as communications technology is deployed. Dams, pipes and pumps do not change ideas and beliefs. Ideas and communication which are part of the social and political milieu do.

Figures 8.1 and 8.2 also emphasise the ways that perceptions of MENA insiders and the perspectives of outsiders differ. In the MENA region the strong voice of numerous agricultural users of

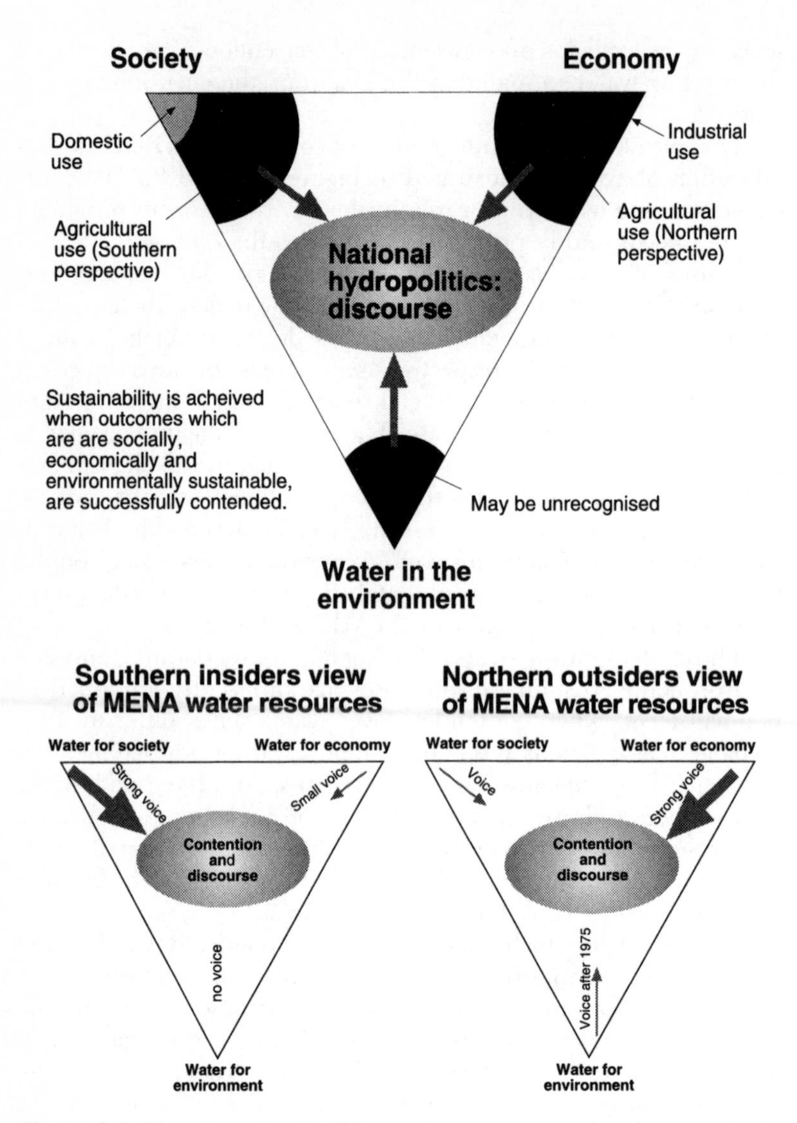

Figure 8.1 Water in national politics: society, economy and environment in the South and the North

water is the determining force in the national discourse. In the South the environment has a small voice or no voice at all. In the North, by contrast, the environment had by the end of the twentieth century a strong voice in hydropolitical contention. Water for the economy

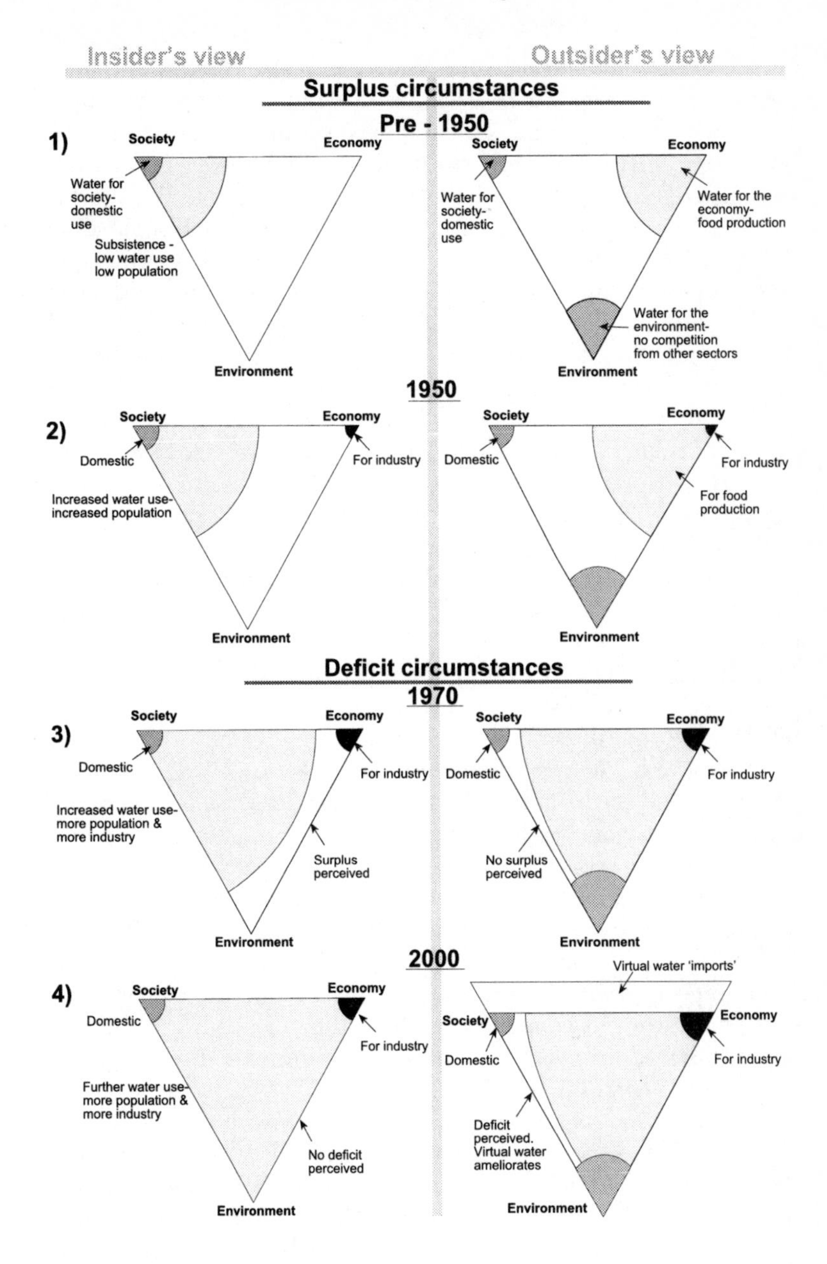

Figure 8.2 Sectoral growth in water demand and perceptions of the consequences in the South and the North

also had a strong voice in the discourse in the North. The forces determining hydropolitical discourses are different in the South and the North. When water short circumstances are recognised in these different Northern and Southern social and professional circumstances the politics of innovation to ameliorate the shortage will be different in these separate circumstances. The discourses are different. Those attempting to implement water policy reform will face different political challenges. In the media saturated, risk aware North, communities are assaulted daily by news of constructed risks to which they should be reflexive (Beck 1999). In the South the local risks are well understood, but instead of being amplified by the media, governments and media conspire to de-emphasise them.

There is little point in looking for model answers for the current water deficit problems of the scarcely industrialised economies of the semi-arid MENA region in the current water policies of the industrialised semi-arid political economies of the North. The water policy reform narratives of Northern political economies can be useful in one single respect. Their past experience might provide helpful insights on what came before the reformed present. In the MENA region, Israeli experience has demonstrated that it is the process of industrialisation that enables water policy reform. Such reforms become possible as economic strength and diversity have been achieved. The capacity to construct new knowledge about the economic and environmental role of water in a political economy comes relatively late in the process of socio-economic development. Developing the capacity to achieve water policy reform is one of a vast range of options available to industrialised political economies with high social adaptive capacity (Ohlsson and Turton 1999).

Without an awareness of the determining role of the socio-economic development neither the insider nor the outsider can understand the processes upon which they are trying to have an impact. In addition they cannot communicate their separate discourses to each other.

A comparison of the MENA insider's view with that of outsider professionals and scientists is shown in Figure 8.2. The water demand situations before 1950, in about 1950, in about 1970 and at the end of the twentieth century are shown. The different perspectives of insider MENA water users, water professionals and governments are shown alongside the perspectives of Northern professionals and scientists. Evidence of the difference in perspective of insiders and outsiders can be found in the way they relate to the idea of the region, and especially of its individual economies, being short of water. By the end of the twentieth century water interests in the MENA region perceived that some economies were approaching water deficit circumstances. In the North it had been concluded that the MENA region had in practice run out of water much earlier—in the 1970s. The Northern narrative argued that the MENA region solved its water deficit problem after 1970 by importing virtual water. The MENA insider narrative still did not include this insight by the turn of the century. Virtual water was still economically, and more importantly politically and socially, silent in the MENA narrative despite its important role in solving the regional water deficit.

The essential idea highlighted in Figure 8.2 is the different way that MENA insiders and outsiders from the North conceptualise the role of water in MENA political economies. For the MENA insider water used in agriculture is a social issue. For the outsider water is an economic input. For the outsider water also plays an important role in sustaining the environment. It was not always thus for the outsider. Outsiders until very recently, until the early 1980s, used to recommend the good news of the hydraulic mission. The hydraulic mission was not considerate of the environment and only selectively inspired by economic factors. Consideration for the economic and environmental importance of water has only very recently been learned in the North (Carter 1982). The analysis in this section is being made at the height of the contention at the turn of the twenty-first century between the MENA insider discourse and the new outsider Northern discourse. A version of this contention took place in the United States a quarter of a century ago.

The above theory about the hydropolitical landscape and the hydropolitical contention of social, economic and environmental water demand is a very helpful point of departure in bringing together other very useful concepts relevant to understanding the way water is allocated and managed in the MENA region. We shall look next at who plays this hydropolitical game.

Who is playing the game: competing inspirations

A number of social theorists have pointed out that individuals and communities take different approaches to the way they perceive and utilise resources (Douglas 1970, Douglas et al 1982, Thompson et al., Hoekstra 1998). Douglas developed her group/grid theory, shown in Figure 8.3, to provide insights into the attitudes and expectations that determined the approach of major groups of players. First, there are individuals and political and bureaucratic elites which can exert influence on affairs through holding high positions in political hierarchies. These are the hierarchists with power, democratically or otherwise founded, in positions in government departments, in religious bodies and in the institutions that regulate and provide public services. Secondly, Douglas termed the mass of the people, that is civil society, as fatalists. Their role was to be responsive to the influences of individuals and institutions in the other three categories. Thirdly, she identified individualists or entrepreneurs, who sought opportunities to use and combine resources to provide goods and services. Finally, there were egalitarians or ethicists who are inspired by principles of equity and in the case of water, also by principles of environmentalism and economics. They want to improve the way that water is allocated and managed.

The theory is not predictive. But it is very helpful in highlighting the political relationships between those inspired by the different approaches to allocating and managing water. It is also helpful in identifying those contributing to the water policy discourses and especially in identifying the sources of new ideas that might influence regional hydropolitical processes.

Another very important role of the theory is that it helps to gain insights into the insider and outsider thinking on water policy reform and the lack of fit between insider and outsider approaches. Figure 8.3c shows the insider situation. By the end of the twentieth century the MENA approach to water allocation and management was still based on a public sector model. The preference for public sector water management prevailed in most of the rest of the world where perhaps only six per cent of even domestic water supplies were provided by the private sector.

Figure 8.3d summarises the approach of outsiders to addressing the problem of allocating and managing water in the MENA region. Outsiders, led by the World Bank, advocate that the private sector should play a major role in rehabilitating and extending water delivery systems in both the urban and the rural sectors. The private sector contributions would be made by international water companies based in France, the United Kingdom and the United States. Such projects were already being put in place by the year 2000 in cities in Turkey, Jordan and in Gaza.

The significance of an influence of an alliance between hierarchists and entrepreneurs are known to be potentially very powerful. The military-industrial complexes of the United States and western European economies are examples of the form. They have become 'essential' revenue earning and job providing elements in the political economies of these economies. They are fundamental,

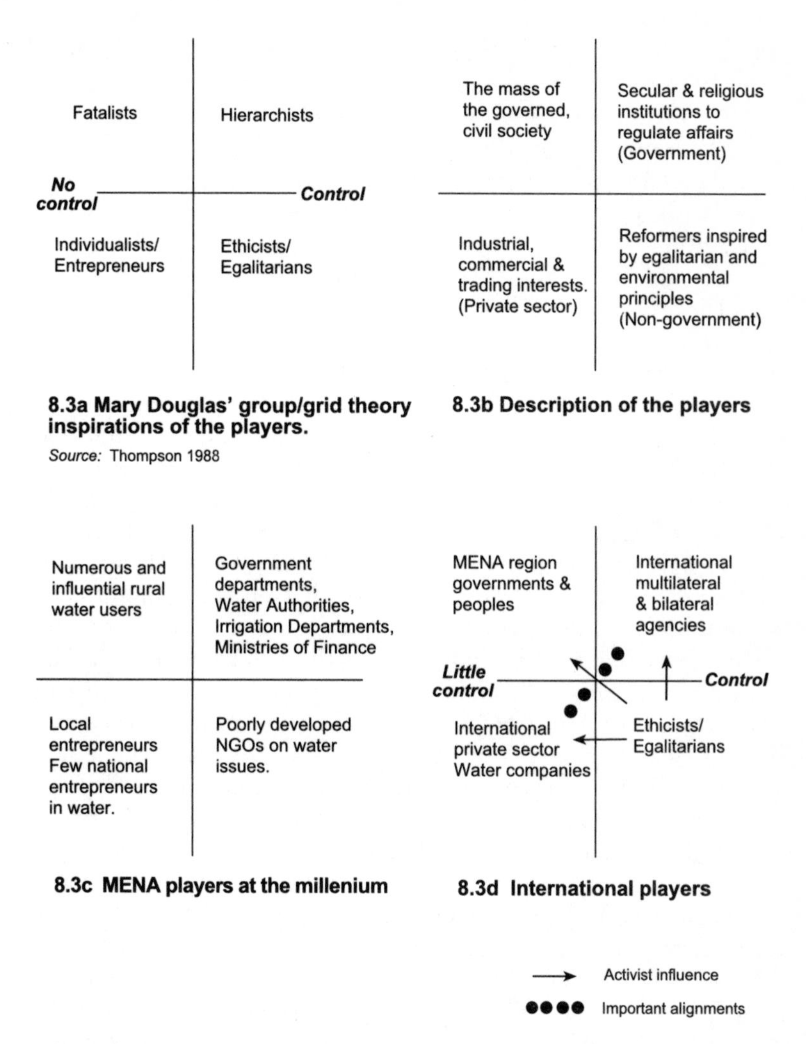

Figure 8.3 The Douglas/Thompson theory and the MENA water sector

frightening and very distorting features of the respective political systems. There is immense intuitive resistance world-wide to such an innovation in the water sectors of national economies. The resistance is very unyielding indeed in the MENA countries. Even in Israel where the social-adaptive capacity has been sufficient to privatise the national water agency, Mekorot, the process of privatisation was resisted and slow after its initiation in 1998 and thereafter. The intuitive resistance is equally strong to a global hydro-industrial complex allying Northern hierarchies and Northern Trans-National Company (TNC) entrepreneurs.

The outsider international community is also populated by activist agencies and scientists. Bodies such as Greenpeace, International Union for the Conservation of Nature and the World Wide Fund for Nature are active and influence the multi-lateral international funding bodies. They can have serious influence on international funding as shown in the Ilisu Dam project in south-east Turkey where Greenpeace sought legal opinion on the legitimacy of UK Government financial support for a Turkish project which had not been approved by downstream riparians. (Cameron et al. 2000) The NGOs can have significant indirect influence by pressing Northern governments and institutions to observe the new ethical principles for which the activist NGOs have gained currency. Their direct influence in the region is small. The influence of MENA NGO activist groups is as yet limited.

The MENA region is characterised by states which aspire to regulate their affairs from the centre despite some denials, for example by Libya's leader. This aspiration is found in both the remaining monarchical regimes of the MENA region and especially in the socialist and/or Islam inspired governance systems of Egypt, Israel, Iraq, Iran, Libya, Syria and Tunisia. Privatising reforms were in progress at the turn of the twenty-first century in almost all these economies. But the appetite for such reform in the water sector was as poor as it was in the very different circumstances facing Northern governments, such as the Netherlands government in the year 2000. Privatisation is not just a fiscal, legal and regulatory issue. It is mainly an ideological and political issue affecting communities and society. The innovation of water privatisation will only be achieved by influencing the beliefs and expectations of water users. It remains to be seen how the radical privatised model of the UK water industry finally shapes. The extraordinary growth in the size and cost of the regulatory public superstructure set up to ensure first socially acceptable water prices for consumers and especially for the poor and secondly the quality of the water provided and thirdly environmental standards. Regulatory bodies will be needed and will have to be funded for both the public and the private sector models. But experience shows that there are significant budgetary burdens on the public purse when the water sector is privatised.

The group/grid theoretical framework of Douglas does not predict but it does help explain policy and operational outcomes. Deployed as a temporally dynamic analytical framework it is useful in showing how an individual in a government hierarchy would be influenced by activist NGO scientists. Such a pattern of influence and contention explains the position taken up by Jimmy Carter in 1976 when in the office of President of the United States. He was persuaded by the NGO environmentalist arguments and took on, at great political cost to himself, the vested interests of senators, the Bureau of Reclamation and the Corps of Engineers (Carter 1982). The confrontation of the World Bank by NGO conservationist and sustainability arguments is an example of such political processes which led to the 'greening of the Bank' (Werksman 1996, Williams 1997). The progressive cultivation of the World Bank by some of the major transnational private sector water companies during and since the 1990s is another global level example of the phenomenon illuminated by the framework. The incentives to influence hierarchies are very high.

The theory is also helpful in highlighting the different social and political circumstances that exist in civil society. Civil society is influential in different ways in the South than in the North. In the North, democratic institutions and globalised communication systems, albeit manipulable by elites and the media, enable very diverse and high levels of interaction between those in civil society and the ideas and norms being advocated by governments, NGOs and the private sector. The daily media overflow with stories about economic and technical inefficiency, wasted resources and environmental pollution. There are also mechanisms in the North by which civil society can feedback its preferences once it has been alarmed by the constructed knowledge of the activists.

The governance systems in the MENA region, insofar as they affect water policy and management, are locked in sanctioned discourses which align government and civil society perspectives on water and on norms of management. So strong is this alignment that it is impenetrable to outsider recommendations. Meanwhile, the ethicist approach as advocated by insider NGOs and the market inspired approaches of the insider private sector entrepreneurs have very weak voices and no political status in the MENA region. The alignment of the government hierarchies and civil society on how water should be perceived and allocated in MENA economies has generally proved strong enough to withstand the energetic advocacy of the Northern agencies [hierarchy], of Northern private sector companies [entrepreneurs] and of Northern NGOs [egalitarians and ethicists]. The government/civil society alignments in the MENA political economies have also been proof against their own weak NGOs and the poorly developed private sector.

The major driving force that will lead to the transformation and reform of water policy and management in the MENA region is progressive socio-economic development. The water policy narrative will shift in a number of ways. First, greater economic diversity and strength will enable civil society to be differently employed with livelihoods less dependent on local water. The basis of the economic security of civil society will be transformed. Secondly, those in hierarchies in government will have the option to escape from the sanctioned discourse determined by a majority of civil society being dependent on irrigated farming. Thirdly, in the private sector local entrepreneurial projects will play a role in a much higher proportion of national economies. These new actors will have a louder voice in influencing policies, including water policies, in the progressively more diverse and stronger MENA economies. Fourthly, MENA NGOs will gain in strength and in their capacity to mobilise opinion within civil society, to lobby directly private sector interests and influence national and local water policies.

First order and second order scarcity: different responses to scarcity and social adaptive capacity

> Devoting resources to providing engineering and institutional remedies in the narrow water sector will not solve the water problem for the whole political economy. Strengthening and diversifying the political economy will, on the other hand, create changed circumstances that will enable water problems to be solved.

> An improvement in social adaptive capacity can compensate for a physical water shortage. An improvement in the volume and quality of water cannot compensate for a shortage in social adaptive capacity in the same measure.

In the preceding sections social theory has shown itself to be helpful in analysing the hydropolitical contexts in which water policy is made and reformed. Water users and professionals in the MENA region view their water policy options very differently from the way Northern professionals and scientists perceive them. The latter tend to discount politics. The MENA water users and professionals are, however, driven by social priorities expressed in the politics of sanctioned discourses from which it is impossible to escape. Douglas and Thompson have provided useful approaches for the analysis of the players involved in such water policy making games. Another body of theory developed by Ohlsson and Turton (1999) is useful in explaining why water policy reform is difficult. The theory is also useful in gaining an understanding of the social and political dynamics of water policy reform.

Ohlsson (1999) drew attention to a very important feature of water scarcity. He pointed out that water scarcity has two dimensions. He identified first order water scarcity. This is the physical shortage of water. Second order scarcity he argues is the lack of capacity to ameliorate the shortage. Ohlsson called second order scarcity the lack of 'social adaptive capacity'. Both first order and second order scarcity can change over time. First order scarcity can occur and worsen when demand rises to outstrip supply. Second order scarcity can vary according to the pace at which social adaptive capacity can be strengthened. An improvement in social adaptive capacity can compensate for a physical water shortage. An improvement in the volume and quality of water cannot compensate for a shortage in social adaptive capacity in the same measure.

Figure 8.4 illustrates how much more important it is to be rich in socio-adaptive capacity than in water availability. Only when low social adaptive capacity is combined with water deficits is there an intractable situation. An economy need not be significantly hampered by a water deficit. It is possible to have a strong economy with high levels of social infrastructure in circumstances of both well endowed or poorly endowed water resources. Figure 8.4 also shows that levels of social adaptive capacity achieved have not been related to the levels of water resources enjoyed by the respective economies. The achievement of a high level of social adaptive capacity is not determined by water availability. The corollary of this absent relationship is that the solution to water scarcity for the MENA economies will be found in the overall political economy and not in the water sector itself. Devoting resources to providing engineering and institutional remedies in the narrow water sector will not solve the water problem for the whole political economy. Strengthening and diversifying the political economy will, on the other hand, create changed circumstances that will enable water problems to be solved.

An economy with social adaptive capacity can cope with the first order physical water scarcity. An economy without this capacity is vulnerable to the severe economic and social impacts of water shortages. Regions such as the Middle East and North Africa have demonstrated that it is possible to cope with periodic and permanent water deficits. A variety of strategies have enabled amelioration. The oil rich have been able to access the big water required to provide food via international trade in food. Even the economies without major oil resources, such as Egypt and Syria, have resorted to this

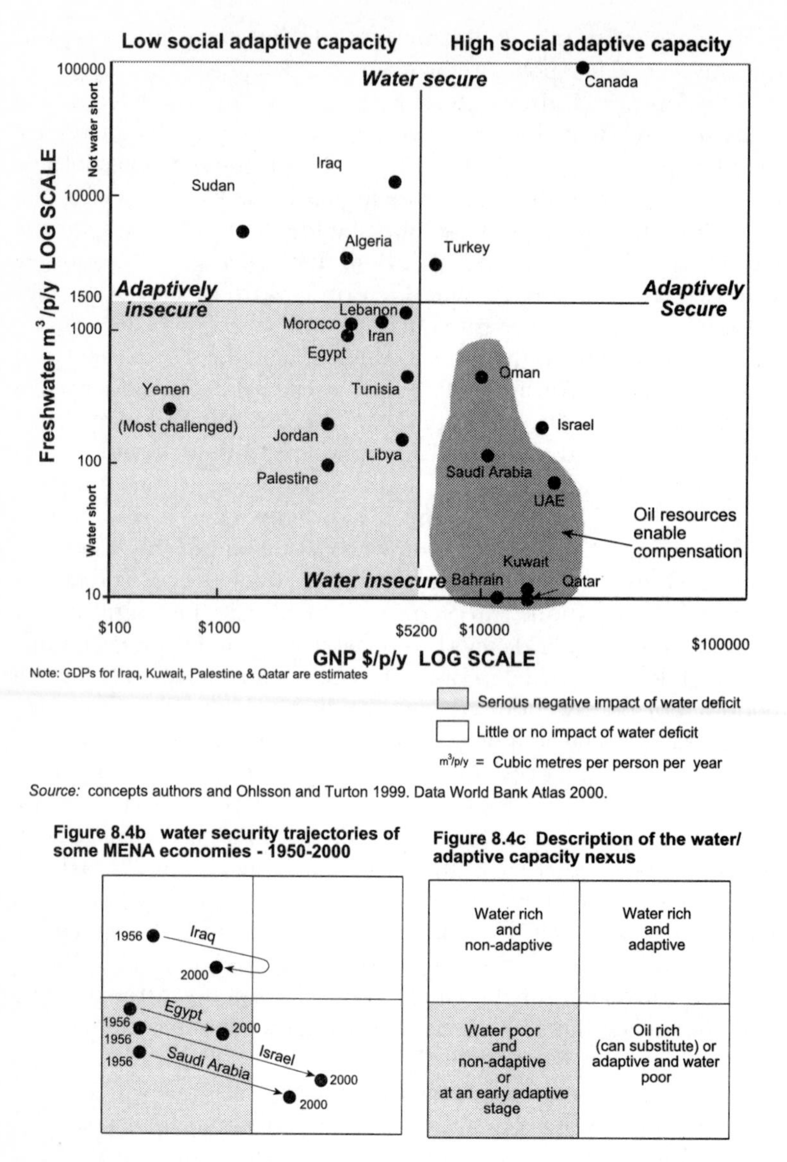

Figure 8.4 An analysis MENA economies in relation to water deficits and shortages of social adaptive capacity

form of amelioration. Only Israel has demonstrated how easy it is to cope with a water deficit with the social adaptive capacity of an economy which is diverse and strong.

The Ohlsson theory is important because it emphasises that it is not new water or technical/productive approaches to managing old and new water that are the major sources of a solution to a water deficit. It is the social and political processes that enable industrialisation that in turn enable water policy reform. Lest economics, however, appear to be prime it is necessary to recall the message of chapter 4 in which it was argued that socio-economic development can only take place when political and social institutions have been put in place.

Outsiders and insiders—parallel discourses

> Many now in senior positions in the MENA region who studied in the major engineering schools in US universities recall that even as late as 1984 the good news about the economic and environmental values of water were not part of the curricula. (Unver 2000)

Social theory is also useful in helping to explain the existence of different discourses on water of MENA insiders and professionals and scientists from the North. The concepts of industrial modernity and reflexive modernity were shown in chapter 5 to have powerful explanatory power for Northern political economies. Figure 8.5 illustrates the trajectory of the hydraulic mission (Swyngedouw 1999) which inspired policy makers and engineers. In the North from the second half of the nineteenth century governments, entrepreneurs, engineers and other professionals inspired by the achievements of the Enlightenment, science, engineering and capitalism turned their attention to providing water for agriculture. Figure 8.5 illustrates the resulting trajectory of the hydraulic mission as expressed in increasing levels of water use mainly for irrigation.

A significant shift took place in the North in the mid-1970s by which time perceptions of the value of the environment had been significantly impacted by the green movement in North America. Water was one of the principal issues on the green agenda in the 1960s and the 1970s. The green campaigns succeeded and by the mid-1980s water policies began to change. Water was one of many areas of environmental policy that conformed to the notion of reflexive modernity (Beck 1992, 1999). In the water sector an indicator of the adoption of a reflexive approach is revealed in the levels of water use. Figure 3.9 (p144) confirmed how the grand narrative of sustainable water use was observed in some industrialised semi-arid regions. Australia, California and Israel all conformed to the reflexive model.

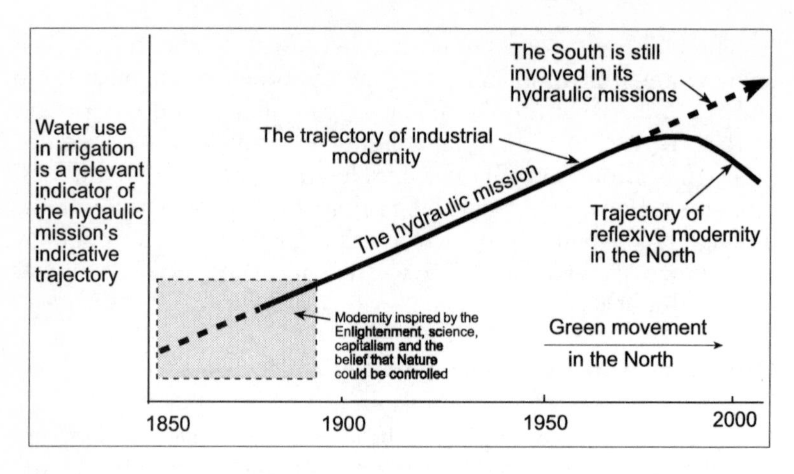

Figure 8.5 The hydraulic mission, industrial modernity, reflexive modernity—explaining parallel discourses in the North and the South

The reflexive pattern has not been adopted in the MENA region except in Israel. Even in Israel the reflexive principles were not proof against new international political circumstances after the commencement of the Peace Talks in 1992. There is significant contention on the part of Northern NGOs to encourage reflexiveness in the region. For example the campaign by Greenpeace (Cameron et al. 2000) and the *Guardian* newspaper against the construction of the Ilisu Dam on environmental and human rights grounds. But the inspiration of governments and professionals in the MENA region remained at the turn of the twenty-first century firmly on the hydraulic mission track.

The outsider discourse and the MENA insider discourses are parallel and when they connect misunderstandings and conflict occur. The outsiders are perversely unaware of how their newly won wisdom is regarded by MENA insiders. The new messages from the North are regarded as self-serving, ill-informed and ingenuous by MENA professionals and politicians. The way that the message is pressed upon institutions and communities that are not understood by the outsiders is regarded as crass. Many of the professionals have first hand experience on how recent the conversion in the North has been. Many now in senior positions in the MENA region who studied in the major engineering schools in US universities recall that even as late as 1984 the good news about the economic and environmental values of water were not part of the curriculum.

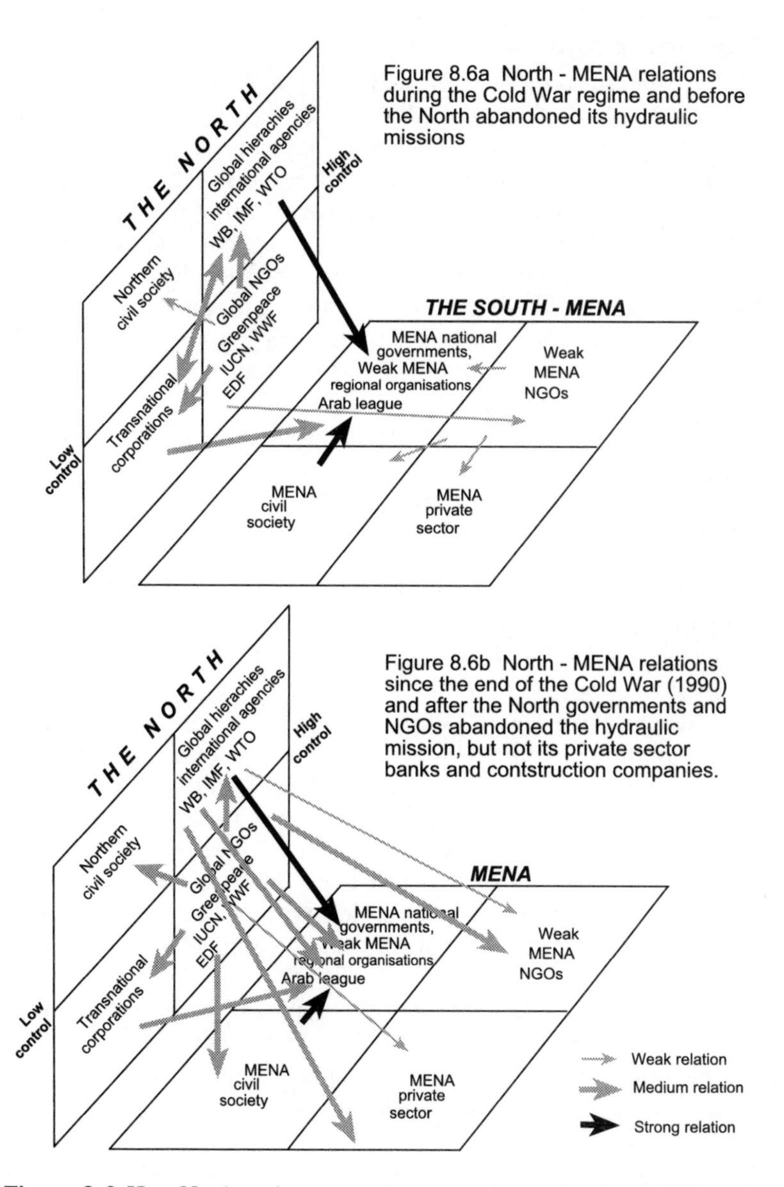

Figure 8.6 How Northern interests and messages have related to MENA socio-political entities over water during and after the Cold War, and during and after the North abandoned its hydraulic mission

Discovering common theory

One of the consequences of deploying such a diverse range of theory is the discovery that the different analysts are often saying the same thing. They have somewhat laboriously come up with explanations in different theoretical language which reach similar conclusions about how and why perceptions of water can change and how and when water policy reform can occur. The theories all have in common, however, the primacy of political factors. They all also argue that it will be via an understanding of water societies and their social norms that the water problems of communities will be successfully addressed. Influencing societies and their norms will be achieved by either changing the awareness of the water problem or improving the socio-economic status of the water using political economies via socio-economic transformation outside the water sector. Using both these approaches is likely to be even more effective than using them independently.

Specialists in politics have provided important insights. On awareness and focusing government and society on the water problem the notion of Kingdon (1984) of the 'window of opportunity' is useful. In the water sector droughts provide brief windows of opportunity to enable activists, hydrologists, hydraulic engineers, government agencies, the media and water using communities to examine their assumptions about water availability. Sometimes the perceptions converge for a sufficient period for water policy reform to be advanced. The experience of Israel in the 1986 and 1991 droughts is an example of such changes in a political economy with high social adaptive capacity. Neighbouring Jordan did not achieve similar advances despite experiencing the same drought because it did not have the same social adaptive capacity (Allan 1996a).

Environmental scientists have also contributed the very similar notion of 'emblematic events' (Hajer 1996). They also argue that extreme environmental events can capture the attention of water users, governments and the media to enable shifts in the political agenda. As a consequence water gains prominence and bringing the allocation of resources to improve institutional and physical infrastructures in the water sector.

In the field of international relations a similar idea has been promoted by Buzan et al. (1999 and 2000). Normal politics are insufficient to bring about changes in perception on the security significance of water. Securitisation requires that the discourse on water and security has to be made more prominent. Buzan has termed this different and much more publicly prominent mode of politics, security politics.

Political ecology provides a useful framework for all these different disciplinary explanations. Those inspired by principles of political ecology start with the assumption that ecological circumstances are in constant flux and that human systems have to strive to be sufficiently dynamic to cope with such non-equilibrial change. Political ecologists presume that the game of managing environmental resources will be played by players with different levels of political influence. Environmental management outcomes will be related to these unequal power relations. Powerful elites are more likely to steer environmental policy to advance their elite interests than those of the less powerful majority. The significance of windows of opportunity, emblematic events and security politics will be affected by the political ecological circumstances that obtain.

Economic theory also indicates that there has been change in the way water has been allocated and managed over the last two decades of the twentieth century. The environmental-Kuznets curve and the Karshenas curve (1994) both suggest that socio-economic development, and especially the achievement of higher standards of living through economic diversification enable communities and nations to become considerate of the environment and its secure future. They see such change as economic rather than political. They conclude that socio-economic development enables policy change.

An even more overarching theory which contextualises the above social, political and environmental theory is that of industrial modernity and reflexive modernity (Beck 1999), elsewhere termed modernity and post-modernity. All the change points identified by the above theories—the windows of opportunity of Kingdon (1984), the emblematic events of Hajer (1996), the prosperity enabled changes in environmental policy of Karshenas (1994) and the awareness of water as a national security issue (Buzan et al. 1999) occurred in the North after the conceptually narrow hydraulic mission of industrial modernity had been replaced by awareness of the complexity of environmental and water problems and their management.

Reflexiveness is a Northern condition. It is not a strongly established notion in the MENA region and the dissonance when new Northern ideas from the North encounter old Northern ideas still strong in the MENA region has been the main focus of the chapters of this study.

Conclusion

> By ignoring the mediating role of politics,
> Through which their advocacy
> Could bring reform,
> Outsider water professionals
> Denied themselves
> The very capacity to persuade
> Which they most craved.

Many faces and personalities have been in mind during the writing of this analysis of Middle East and North African water. One has recalled farmers and engineers struggling with the not very fertile sand or the awkward clay of 'new lands' in Egypt. And the dilemmas facing farmers drilling ever deeper in pursuit of over-pumped aquifers in Libya and Oman. The daily toil and the livelihood impacts of challenging water resource circumstances were plain to see. The water resource is often locally inadequate in the region, and the soil circumstances difficult. One has also recalled meetings with ministers of water and of agriculture and with water commissioners in a number of MENA countries. They impressed by their skill in making comments which showed their understanding not only of a question but also of the need to shape an answer that would be safe in diverse political circumstances. They have to cope with formidable political challenges rather than with unyielding soils and scarce water. One has also observed that though shifts in perception take place very slowly in the inflexible water politics that prevail perceptible changes have taken place even during the two years of writing the book. The harmony that exists between farmers and the politicians running agriculture and water ministries in the MENA region has been commented upon throughout the study. The reasons for this harmony on the part of the heavy water users and officials in the ministries responsible for water is a very important feature of the insider hydropolitics of the region. That outsiders perversely ignore the political logic of the political systems they encounter is disappointing. This perverseness immensely attenuates the possibility of their having a useful impact on them.

One has discovered over the years that people do not like being written about by outsiders. This attitude exists in substantial measure in the MENA region. The analysis of orientalism (Said 1986) has sensitised social scientists and many professional commentators relating to the MENA region to the offensive, even obscene tendency, of using language that speaks ill of the region. In the event there is a reciprocal tendency in the region itself. It is not necessary to contend, quench nor even to acknowledge outsider comment on the region. Insiders turn a very effective blind eye and deaf ear. Insider readers presume that an outsider author who wants to say anything of importance will do so in Arabic. Material in English can be sublimely ignored. Reaction, critical or not, is unnecessary on something that does not exist. As a result the embarrassing consequences of publicly commenting on biased published outsider analyses are comprehensively avoided.

A number of individuals from the international outsider water community have also been constantly in mind during the writing. Many of them are hyper-busy professionals in international agencies. They enjoy privileged exposure to the challenges facing officials and farmers in the MENA region. They have more chance to relate to opinion formers in the region than any other group of outsiders. They have detailed experience which they have often shared but which is difficult to make public. They have more insights into the perceptions and expectations of insiders than any other community. One has had the will to complete the study not because one has better information than this well informed group. The study has been written because many from this privileged group have often said that so intense is their professional life that they do not have time to think as deeply as they would like never mind write an analysis informed by experience.

The analysis has a theoretical emphasis for a number of reasons, some of which were not evident at the outset. First the appearance of numerous books on water on the MENA region made it unnecessary to write another one describing the water resources of the region and commenting on the significance of water scarcity. Secondly, the main reason for the theoretical slant has been that the more one analyses water in the MENA region the more one realises that outsiders do not understand the perceptions of insiders. Outsiders are especially blind to, and therefore uninformed about, the societies which use water, and the political processes which determine how water is allocated and managed in the region. For this reason there is in the analysis a strong emphasis on politics and the socio-political circumstances which are the contexts in which water uses are prioritised and in which water shortages are coped with.

The main and non-intuitive conclusion of the study is that despite the evidence that the MENA water resources are being used heavily the over-riding insider perception is that the water resources will be sufficient for the needs of the peoples of the region. One has heard ministers of water, agriculture and planning follow each other closely on a conference platform to assure their professionals and scientists, as well as outsider visitors, that this is the case as late as the mid-1990s. The reason that people and politicians protest so much that water will be adequate is political. Adequacy of water is not primarily a scientific or professional issue, although professionals and scientists have useful things to say about water deficits. Water security and water adequacy, however defined, are political issues. Peoples and their leaders need to assure each other that they are secure. Water is an important dimension of communal and national security in arid and semi-arid countries.

Other potential readers have also been in mind throughout the writing. Outsiders are certain that they have useful things to say about the allocation and management of MENA water. For example senior economists are funded to visit Egypt constantly to discuss principles of water use efficiency and the pricing of water. The good ones rediscover the basics of political economy and the impossibility of introducing prices that would change behaviour. There remains an army of professionals, unedified by such deep first hand analysis. They remain convinced that technical and economic instruments can be deployed because they make sense. They do not realise that their technical and fiscal solutions are in practice political problems for those who would implement them. This group is a target for this analysis that has been constructed in these chapters.

The early sections of this chapter have explained why MENA insiders and outsiders from the North are differently inspired when identifying solutions to the region's water problem. Most of us in the North who lived through the third quarter of the twentieth century were unaware of the success of the green movement in constructing a new perspective on the environment and its diverse values. That success moved the North and its governments, water professionals and scientists into a new trajectory in the fourth quarter of the century. The green activists stimulated economists to invent environmental economics; stimulated ecologists to concoct the slippery concept of sustainability; and sociologists to insist that sustainability was a complex notion in which sustainability had a social dimension. This trajectory was not underpinned by the old assumption that nature could be controlled. On the contrary there was widespread and politically legitimised opinion that the environment could be destroyed by too much technical intervention and too much ill-conceived getting and spending. It has to be emphasised that the new principles did not spring from the minds of clearly argued economic and environmental science. That came later. They were a response to the construction of knowledge of environmental activists and their colourful and contentious campaigns.

The professionals and governments of the MENA region have not yet been seized by this Northern ideology. The last quarter of the twentieth century, and especially its last decade, were periods when the outsider professionals performed, albeit unwittingly, in the MENA region an advocacy role which emphasised the environmental and economic value of water. It was unwitting because the Northern professionals did not understand the political process which had transformed their own science and professional norms. In the North this advocacy role had been performed by environmental NGOs and other activists. A political process resulted in the conversion of Northern professionals and scientists to environmental and economic wisdom in the third quarter of the century in the North. The conversion process had been politically stressful. But apparently the complicated and very political process of conversion, with all the usual and unavoidable construction of knowledge did not leave the Northern professional with an awareness of the role of politics. And certainly not with a respect for politics. As a result the encounters between the recently converted politically blind Northern outsiders and the MENA insiders totally immersed in the politics of water have been contentious and unfruitful. The contention has been carried out in two mutually incomprehensible language registers. The Northern professionals have been speaking the languages of ecology and economics. The MENA professionals and politicians have been speaking the language of society and politics. Figure 8.1 conceptualised the role of politics in mediating the concerns of society, the economy and the environment with respect to water. By ignoring the mediating role of politics, through which sound advocacy could bring reform, the outsider Northern water professionals denied themselves the very capacity to persuade which they most craved.

The MENA region is in the midst of a number of very important discourses over water. First, there is a MENA insider discourse which takes its shape from the entrenched expectations of farming communities that have been involved for centuries in developing water to meet what they regard as social needs, namely water for the household and for livelihoods. This centuries long discourse is so familiar that it is difficult to gainsay. MENA farmers, politicians and bureaucratic and engineering institutions are the stakeholders and they have developed an enduring consensus on how water should be perceived and allocated. In this discourse social values take precedence over economic and environmental values. Secondly, there is an insider/outsider discourse over water policy with the insider and outsider stakeholders contending over the retention of the socially inspired *status quo* or its extensive reform. Outsiders shamelessly argue for economic and environmental principles without an idea on how the political process of implementing them could be achieved by unwilling governments and water users. This study has been about this second discourse. Through discourse analysis such as that attempted in this and earlier chapters it is possible first, to identify the social logic of the existing water policies in the region, secondly, to illuminate the significance of the alien provenance of the outsider economic and ecological principles for insiders and thirdly, to understand the inevitable attenuation of the discourses through the resort of MENA governments to expedient, operationally and politically invisible, virtual water imports.

Glossary

Aquifer	An underground geological stratum that is saturated with water and through which water can be transmitted.
Blue water	The freshwater of high quality surface flows—streams, rivers, and in surface storage—ponds and lakes and non-saline inland seas.
Brown water	The water present in soil profiles and available for use by vegetation and crops. Soil water is either on its way down in response to gravity, on its way upwards through capillary forces, or held in place by the surface tension of the soil particles. Soil water should not be regarded as a 'water bank'. The useful water available is that water which the roots of plants can intercept. The downward flow of water may be in excess of that which can be intercepted. In this case the water continues downwards to the aquifer (groundwater). [See soil water]
Consumptive use	The water consumed during domestic, municipal, industrial and agricultural uses. Crop production requires about 90 per cent of water mobilised by an economy and is associated with high consumptive use—between 30 and 60 per cent of water infiltrated or applied; in comparison water used for industrial and domestic needs is mainly returned to the hydrological system, usually with decreased quality. It is usual to assume that 70 per cent of domestic and industrial water is available, after treatment, for re-use.
Demand management	The use of technology to reduce the levels of use of water and the use of financial and economic instruments (subsidies and taxes), and regulation to reduce the demand for water.
Discourse	*Discourse* in the text is used after the manner of the French historian-philosopher Foucault (1969, 1971, 1980). He distinguished between coercive

	power and discursive power. He argued that power should be regarded as the outcome of contention. Even those with apparent substantial coercive power find that they have to contend with other players and other interests. He showed that knowledge and especially its construction by some groups was integral to the outcomes of contention. Power is the outcome of contention which accommodates as a 'network of consensus'. Confusingly both the process of contentious discourse and its contended outcome, power, are both called discourse in English. The notion is very helpful in understanding the way the players play the games of water politics in the individual MENA national entities. The idea is especially useful in helping to understand the contention between insider water users and managers and outsider professionals and scientists.
Externality	The unintended real (non-monetary) side effect of one party's actions on another party that is ignored in decisions made by the party causing the effects.
Freshwater	Freshwater for drinking should have less than 500 parts per million of total dissolved solids. 'Freshwater' of poorer quality can be used for other domestic and many industrial uses. Water for agriculture can be as poor as 1500 parts per million and for some crops such as rice and date palms over 2000 parts per million.
Green water	Green water is the water in the biosphere.
Groundwater	The water in the ground beneath the soil layer. It is stored in aquifers. Groundwater flows in response to gravity.
Invisible water or non-evident water	Invisible or non-evident water is soil water and water embedded in commodities which require water for their production. For example a tonne of grain requires 1000 tonnes (m3) of water to

	produce it. A community or economy can balance its water needs by accessing invisible water outside its national boundaries.
New water	Engineers mobilise new water. New water can result from projects to regulate the flow of rivers with natural variable regimes. Pumping water from rivers and surface storage and from groundwater can make new water available. New water can be manufactured by desalination processes. New water can also be recovered by water treatment processes.
Non-evident water	See invisible water.
Problemshed	The term captures the operational context in which decision makers and their problems exist. The term watershed defines a tract with limited and variable water resources. When the water resources of a watershed become insufficient for the needs of the resident population and their livelihoods, a political economy has to reach outside the watershed for solutions. It is in the 'problemshed' that solutions can be found such as virtual water.
Riparian	An owned tract, a sub-national entity, a nation state which lies across, or has a river bank on, a flowing river. Also refers to groundwater bodies.
Sanctioned discourse	Any group of people operates and communicates on the basis of shared assumptions and shared interests. It is difficult to change deeply embedded shared beliefs. New knowledge from outside contends with inflexible beliefs. Such belief systems create what has usefully been described as a sanctioned discourse (Tripp 1996) which is a term which is readily accepted by non-social theorists. [see also 'discourse']

Secure water	Water can be hydrologically secure and/or socio-politically secure (Nicol 2000). Year on year and seasonal variations in rainfall and other physical circumstances in relation to the territorialisation of water using communities (eg international borders) determine hydrological security. Socio-political securitisation of water is determined by social adaptive capacity (Ohlsson and Turton 1999) including economic diversity and strength (Allan 1996a & Allan and Karshenas 1996). Engineers can remedy hydrological insecurity. Socio-political security is achieved by institutions including regulatory activity, water policy reform and a range of interventions which are achievable when an economy becomes strong and diverse.
Sewage	Liquid refuse or waste matter carried by sewers.
Soil water	Water in the soil profile. Soil water is very difficult to monitor as it is either moving down through the profile after rainfall or it might be held in the profile through the surface tension of soil particles. Soil water can move back upwards first as a result of capillary forces characteristic of soil structures or secondly after becoming biological water in the roots of plants. Soil water can also be drawn up into the soil profile from groundwater through capillary action. Soil water also evaporates from the surface of wet soil.
Supply management	Supply management is the approach to managing water that provides new water to meet new water demands. Engineering solutions are very important in supply management strategies.
Wastewater	The water that emerges from domestic, urban, municipal and industrial activity. Wastewater can be treated by sewage systems and re-used.
Water re-use	Water processing systems which treat the biological and other contamination caused during

	water using activities in domestic, municipal and industrial systems.
	Water is also re-used in agriculture where water drains into drainage channels and groundwater. Eygpt uses its irrigation water 2.3 times.
Watershed	The edge of a natural river basin. Sometimes means the same as the river basin being the area drained by a river system.
Well	
Dug well	An excavated pit from which water can be drawn. Dug wells can be deepened and given further productive life when the pit can be excavated no further by drilling a tube well.
Observation well	A borehole drilled with a drilling machine and then developed as an observation well with a pump.
Tubewell	A borehole drilled with a drilling machine and then developed as a production well with a pump.
Wetlands	Areas of marsh, fen, peat land, or water that include natural, artificial, permanent, and temporary areas with static or flowing water that is fresh, brackish or marine.
Willingness to pay	Willingness to pay is a monitoring and analytical approach which seeks to determine by observation and asking how much an individual or community is willing to pay for water services provision. The method can also be deployed to determine how people and communities value the environment and environmental amenity. The World Bank has cultivated this methodology.

Abbreviations and acronyms

AOAD	Arab Organization for Agricultural Development—based in Khartoum
AQUASTAT	Water resources database compiled by FAO
FAO	Food and Agricultural Organization of the United Nations
GWP	Global Water Partnership
ILC	International Law Commission of the United Nations. In the area of international freshwater the ILC was responsible from the early 1970s for the development of the ILC Convention on the Law of non-navigational uses of international water courses. (UN General Assembly 21 May 1997)
IWMI	International Water Management Institute, Colombo. Formerly IIMI.
MENA	Middle East and North Africa—in this study it includes the countries from Morocco in the west to Iran in the east and from Turkey in the North to Yemen in the south of Arabia and the Sudan in north-east Africa. This definition does not coincide with the World Bank MENA definition.
TNC	Trans-National Corporation
UNCED	United Nations Conference on Environment and Development—Rio de Janeiro 1992
UNDP	United Nations Development Program—based in New York
UNEP	United Nations Environment Program—based in Nairobi
WWC	World Water Council

Abbreviations/measurements units

acre	one acre = 220 x 22 yards = 4840 square yards = 43560 square feet. Is equal to c 4046 square metres = 0.4046 hectare
acre foot	43560 cubic feet = 1234 m3
bn m3	billion cubic metres = c810373 acre feet
dunum	dunum = 0.1 hectare
feddan	feddan = one acre very closely = 0.42 hectare
ha	hectare = 2.4 acres/feddans
km (1000 m)	kilometre = 0.62 miles
km2	100 hectares
km3	cubic kilometres = on billion cubic metres = c810373 acre feet
m	metre
m3	cubic metres
mm	millimetres
mm3	cubic millimetres
Mm3	million cubic metres
ppy	per person per year

References

ABARE (Australian Bureau of Agricultural and Resource Economics), 1989, *US grain policies and the world market,* ABARE Policy Monograph No. 4, Canberra: Australian Government Publishing Service.

ABARE, 1995, US Farm Bill 1995: *US agricultural policies on the eve of the 1995 farm bill,* ABARE Policy Monograph No. 5, Canberra: Australian Government Publishing Service, Canberra. ISSN 1037–8286, ISBN 0 642 22755 1, pp 195.

Abdel Haleem, Muhammad, 1989, Water in the *Qur'an*, *Islamic Quarterly,* London, XXX111, 1, pp 33–49.

Abu Zeid, M., 1990, Environmental impacts of the High Aswan Dam: a case study, in Thanh, N. C. and Biswas, A. K., *Environmentally sound water management,* Oxford: Oxford University Press, ISBN 0 19 562744, pp 247–270.

Abu Zeid, M, 1997, Intervention, stating that irrigation water re-use in Egypt was approximately 2.3 times, at the Nile 2002 meeting in Addis Ababa, February 1997.

Abu Zeid, M. and Abdel-Dayem, S., 1993, Agricultural drainage water reuse in Egyptian Biswas, A. et al., *Water for sustainable development in the 21st century,* Oxford: Oxford University Press.

Al-Kloub, B. and Allan, J. A., 1997, *Virtual water as a remedy to water deficits,* Amman.

Al-Kloub, B. and Shemmeri, T. A. T., 1995, Application of multicriteria decision-aid to rank the Jordan-Yarmouk Co-riparians according to the Helsinki and ILC Rules, in Allan, J. A. and Court, J. H .O., *Water, peace and the Middle East: negotiating resources in the Jordan Basin,* London: Tauris Academic Studies, pp 185–210.

Al-Hayat, 1996, 14 February 1996, London.

Allan, J. A., 1971, *Agriculture and economic change in the Tripoli area*, Libya, PhD thesis, University of London.

Allan, J. A., 1981a, *Libya: the experience of oil,* London: Croom Helm.
Allan, J. A., 1981b, The High Dam is a success story, *The Geographical Magazine*, LII, 6, March 1981, pp 393–396.
Allan, J. A., 1983a, Some phases in extending the cultivated area in the nineteenth and twentieth centuries, *Middle Eastern Studies,* Vol. 19, No. 2, pp 470–481.
Allan, J. A., 1983b, Natural resources as national fantasies, *Geoforum*, pp 243–247.
Allan, J. A., 1987, The development of irrigated agriculture in Syria, in Allan, J. A., *Politics and the economy in Syria*, London: School of Oriental and African Studies, pp 35–47.
Allan, J. A., 1989, Water resources evaluation and development in Libya, 1969–1989, *Libyan Studies,* Vol. 20, pp 235–242.
Allan, J. A., 1993, Fortunately there are substitutes for water otherwise our hydropolitical futures would be impossible, in ODA, *Priorities for water resources allocation and management,* London: ODA, London, pp 13–26.
Allan, J. A., 1994, Overall perspectives on countries and regions, in Rogers, P. and Lydon, P. (editors) *Water in the Arab World: perspectives and prognoses,* Cambridge, Massachusetts: Harvard University Press, pp 65–100.
Allan, J. A., 1995, Water in the Middle East and in Israel-Palestine: some local and global resource issues, in Haddad, M. and Feitelson, E. (editors) *Joint management of shared aquifers: second workshop,* Jerusalem: Palestine Consultancy Group, Jerusalem and the Harry S. Truman Institute for the Advancement of Peace, pp 31–42.
Allan, J. A., 1996a, The political economy of water [in the Jordan basin]: reasons for optimism but long term caution, in Allan, J. A. and Court, J. H .O., *Water, peace and the Middle East: negotiating resources in the Jordan Basin,* London: Tauris Academic Studies, pp 75–120.
Allan, J.A., 1996b, The Jordan Israel Peace Agreement—September 1994, Appendixes 1 and 2, in Allan, J. A. and Court, J. H. O., *Water, peace and the Middle East: negotiating resources in the Jordan Basin,* London: Tauris Academic Studies, pp 207–221.
Allan, J. A., 1996c, The Israel–PLO Interim Agreement—September 1995, Appendixes 3 and 4, in Allan, J. A. and Court, J. H. O., *Water, peace and the Middle East: negotiating resources in the Jordan Basin,* London: Tauris Academic Studies, pp 223–240.
Allan, J. A., 1997a, Drought as a concept, drought as an instrument of policy, in Feitelson, E. and Haddad, M., *Implementation issues of joint management—specifically, monitoring issues, verification, data*

management and requirements, and implications of droughts, Jerusalem: Palestinian Consultancy Group and the Truman Institute for the Advancement of Peace, Hebrew University.

Allan, J. A., 1997b, *'Virtual water': a long term solution for water short Middle Eastern economies?* Proceedings of the 1997 Leeds Conference, Leeds: the British Association.

Allan, J. A., 1998, The challenges, imperatives and current responses for sustainable water resources management, paper presented at the MENA/MED Conference on *Water for sustainable growth*, Washington: World Bank, with EC, EIB, 1–3 June 1998.

Allan, J. A., 1999, Israel and water in the framework of the Arab-Israeli conflict, *Occasional Paper 15*, www.soas.ac.uk/geography/waterissues/ London: SOAS.

Allan, J. A., 2000, Contending environmental knowledge on water in the Middle East: global, regional and national contexts, in Stott, P. A. and Sullivan, S., *Political ecology*, London: Edward Arnold.

Allan, J. A. and Chambers, D., 1997, *Agricultural production and trade trends: based on FAO data*, London: SOAS Water Issues Group.

Allan, J. A. and Karshenas, M., 1996, in Allan, J. A., *Water, peace and the Middle East*, London: Tauris Academic Studies, pp 117–130.

Allan, J. A. and Mallat, C., *Water in the Middle East: legal and commercial issues*, London: Tauris Academic Publications.

Allan, J. A. and Radwan, L., 1997, Perceptions of the value of water and water environments, paper at the Erasmus Conference convened in the Department of Geography and Environmental Management, University of Middlesex on the *Value of value of water and water environments.* London: University of Middlesex.

Altaf, M. and Hughes, M., 1991, Measuring the demand for improved urban sanitation services: results of a contingent valuation study in Ouagadougou, Burkina Faso, *Development Studies*, 31, 10, pp 1763–1776

Amar Appeal, 1994, *An environmental and ecological study of the marshlands of Mesopotamia*, London: Amar Appeal Trust, 2 Vincent Street, London SW1P 4LD.

Ambric, 1980–85, Reports of the American-British Company on the Cairo Waste Water Project, Cairo: Ambric.

Amery, H. A. and Wolf, A. T., 2000, *Water in the Middle East: a geography of peace*, Austin, Texas: University of Texas Press.

Amery, H. A., 2000, A popular theory of water diversion from Lebanon: toward public participation for peace, in Allan, J. A., *Water in the Middle East: a geography of peace*, Austin, Texas: University of Texas Press, pp 121–149.

AOAD, 1997a, *The Tunis Statement*, Khartoum: Arab Organization for Agriculture and Development.
AOAD, 1997b, Arab food security in the light of water constraints and changes in international trade, paper at the *Middle East Economic Forum* convened by the World Bank and the Arab Organization for Agricultural Development, Marrakech: AOAD, PO Box 474, Khartoum.
Arab Research Centre, 1993, Tigris-Euphrates issues, Proceedings of a conference in London convened by the Arab Research Centre, London: Arab Research Centre.
Arab Research Centre, 1995, Water, food and trade in the Middle East, Proceedings of a conference in Cairo convened by the Arab Research Centre and the Ministry of Agriculture and Land Reclamation, London: Arab Research Centre.
Arlosoroff, S., 1995. Personal communication.
Arlosoroff, S., 1996, Managing scarce water: recent Israeli experience, in Allan, J. A., *Water, peace and the Middle East*, London: Tauris Academic Studies, pp 21–48.
Arlosoroff, S., 1997, *Report to the Water Commissioner of Israel on the Israeli water sector*, Tel Aviv: Office of the Water Commissioner
Attiga, A., 1971, The impact of oil on Libyan agriculture, in Allan, J. A., Mclachlan, K. S. and Penrose, E. T., *Libya: agriculture and economic development*, London: Frank Cass, pp 9–17.
Bagis, A. I., 1989, *GAP (South Eastern Anatolia Project): the cradle of civilisation regenerated*, Istanbul: Interbank.
Bagwati, M., 1993, Trade and environment, *Scientific American*, November 1993.
Bakker, K., 1998, Water crises and water policy, Unpublished PhD, Oxford: University of Oxford Department of Geography.
Baudrillard, J., 1987, *Forget Foucault and forget Baudrillard: an interview with Sylvere Lotringer*, New York: Semiotexte (originally published in 1977 as Oublier Foucault).
Bazzaz, F. A., 1994, Global climate change and its consequences for water availability in the Arab World, in Rogers, P. and Lydon, P. (editors) *Water in the Arab World: perspectives and prognoses.* Cambridge, Massachusetts: Harvard University Press, pp 243–252.
Beaumont, P., 1994, The myth of water wars and the future of irrigated agriculture in the Middle East, *International Journal of Water Resources Research*, Vol. 10.1, pp 9–21.
Beaumont, P., 2000, Conflict, coexistence and cooperation: a study of water use in the Jordan basin, in Amery, H. A. and Wolf, A.

T., 2000, *Water in the Middle East: a geography of peace*, Austin, Texas: University of Texas Press.

Beck, U., 1992, From industrial to risk society, *Theory, Culture and Society*, Vol. 9, pp 97–123.

Beck, U., 1995, *Ecological politics in an age of risk*, Cambridge: Polity Press.

Beck, U., 1996a, *The reinvention of politics*, Cambridge: Polity Press.

Beck, U., 1996b, The sociology of risk, in Goldblatt, D., *Social theory and environment*, Cambridge: Polity Press.

Beck, U., 1999, What is a 'risk (society)'?, *Prometheus*, Vol. 1.1, Winter 1999, pp 75–79.

Benke, R. H., Scoones, R. H. and Kerven, C. (editors), 1993, *Range ecology at disequilibrium: new models of natural variability and pastoral adaptation in African savannas*, London: IEED and ODI.

Ben Meier, Meier, 1992, personal communication.

Ben Meier, M., 1994, Water management policy in Israel: a comprehensive approach, in Isaac, J. and Shuval, H. *Water and peace in the Middle East*, Amsterdam: Elsevier, pp 33–40.

Ben-Shahar, H., Fishelson, G. and Hirsch, S., 1. (Howell, *Middle East peace*, London: Weidenfeld and Nicolson.

Benvenisti, Eyal, 1997, Standards or rules? The definition of water rights, in Haddad, M. and Feitelson, E. (editors) *Water rights and allocations*, in Proceedings of the Fourth Workshop on the management of shared aquifers, Jerusalem: Truman Institute for the Advancement of Peace and the Palestinian Consultancy Group, pp 61–78.

Bethemont, J., 1999, *Les grands fleuves*, Paris: Armand Colin.

Bilen, O. and Uksay, S., 1991, *Background report on comprehensive water resources management policies – an analysis of Turkish experience*, Washington: The World Bank, 24–28 June 1991.

Bilen, O., 1997, *Turkey and water issues in the Middle East*, Ankara: Southeastern Anatolia Project (GAP), Regional Development Administration.

Birkett, C., Murtugudde, R. and Allan, J. A., 1999, Indian Ocean climate event brings floods to East Africa's lakes and the Sudd Marsh, *Geophysics Research Letters*, Vol. 26 No. 8 pp. 1031–1034, 15 April 1999.

Biswas, A., 2000, The Dublin principles revisited, *Water International*, Vol. 24.4.

Bourdieu, P., 1977, *Outline of a theory of practice*, Cambridge Studies of Social Anthropology, Cambridge: Cambridge University Press.

Brans, H. P., de Haan, Esther J., Nollkaemper, André, Ronzema, Jan, 1997, *The scarcity of water: emerging legal and policy issues*,

London-The Hague-Boston: Kluwer International, International Environmental Law and Policy Issues, 299 pp ISBN 90-411-0657-X.

Brechin, S. R., 1997, Beyond postmaterialist values: national versus individual explnations of global environmentalism, *Social Science Quarterly,* Vol. 78.1, pp 16–20.

Brechin, S. R., 1999, Objective problems, subjective values, and global environmentalism: evaluating the post-materialist argument and challenging a new explanation, *Social Science Quarterly,* Vol. 80:4, December 1999.

Breznehev, L., 1978, *The virgin lands,* Moscow: Progress Publishers.

Briscoe, J. 1997, Managing water as an economic good, in Kay, M., Franks, T., Smith, L., (editors) *Water: economics, management and demand,* London: E & FN Spon

Brooks, D., 1996, Economics, ecology and equity: lessons from the energy crisis, in Isaac, J. and Shuval, H., *Water and peace in the Middle East,* Amsterdam: Elsevier, pp 441–450.

Brooks, D., 1997, Water demand management: conceptual framework, scope and definition, Paper presented at a workshop on *Water demand management research network,* Cairo, Egypt (12 May 1997). Convened by IDRC. Available on http://www.idrc.ca/books/focus.html, IDRC, Ottawa, Canada.

Brown, L., 1996a, *Tough choices: facing the challenge of food scarcity,* Washington DC: World Watch Institute, Environmental Alert Series.

Brown, L., 1996b, *Who will feed China: wake-up call for a small planet,* Washington DC: World Watch Institute, Environmental Alert Series.

Brown, L. R. and Kane, H., 1995, *Full house: reassessing the Earth's population carrying capacity,* Washington DC: World Watch Institute, Environmental Alert Series.

Bulloch, J. and Darwish, A., 1993, *Water wars,* London: Gollancz.

Buros, O. K., 1997, *ABC's of desalting,* Topsfield, Maryland: International Desalination Association.

Burrill, A., 1996, Intangible economic values of water and water environments, in Allan, J. A. and Radwan, L., *Perceptions of the economic value of water and water environments,* Proceedings of the Erasmus Workshop held at Middlesex University, Department of Geography and Environmental Management.

Buzan, B., Waever, O. and de Wilde, J., 1998, *Security: a new framework for analysis,* London and Boulder: Lynne Rienner.

Buzan, B. and Waever, O., 2000, *Security complexes and sub-complexes,* London and Boulder: Lynne Rienner. [In press]

Cairncross, F., 1992, The freedom to be dirtier than the rest, *The Economist*, 30 May 1992.

Cameron, J., Sands, P., and Boisson de Chazournes, L., 2000, The matter of an application pursuant to the export and investment guarantees act and in the matter of the proposed Ilisu Dam, Opinion provided for Greenpeace, London, 3 April 2000.

Caponera, D. A., 1990, *Principles of water law and administration,* Rotterdam: Balkema.

Caponera, D. A., 1996, Conflicts over international river basins in Africa, the Middle East and Asia. *RECIEL,* 5.2, pp 97–106.

Carter, J., 1982, *Keeping faith: memories of a president,* New York: Bantam Books, ISBN 0 553 05023 0, 622 pp.

Castels, M., 1996, *The information age: economy, society and culture,* Oxford: Blackwells.

Chambers, R., 1997, *Whose reality counts? putting the first last,* Intermediate Technology Publications, UK.

Chalabi, H. and Majzoub, T., 1995, Turkey, the waters of the Euphrates and public international law, in Allan, J. A. and Mallat, C., (editors) *Water in the Middle East, legal, political and commercial implications,* London: Tauris Acadmic Press, pp 189–238.

Chatterton, B. and Chatterton, L., 1996, Closing a water resource: some policy considerations, in Howsam, P. and Carter, R. C., *Water policy: allocation and management in practice,* London: E & FN Spon, pp 355–361.

Cheret, I., 1998, French experience implementing water policy reform, paper presented at the MENA/MED Conference on *Water for sustainable growth,* Washington: World Bank, with EC, EIB. 1–3 June 1998.

Chesworth, P., 1995, History of water use in the Sudan and Egypt, in Howell, P. P. and Allan, J. A., *The Nile: sharing a scarce resource,* Cambridge: CUP.

CIA, 1994, *Satellite images showing the drainage of the Iraq marshes of southern Iraq,* NASA Eros Data Center TM Imagery, Washington DC: Central Intelligence Agency.

CNNI, 1997, *Insight* programme on *Water in Palestine*, CNN International by Jonathan Mann, 15 June 1997.

Coase, R., 1960, The problem of social cost, *Journal of Law and Economics,* October, pp 1–44.

Cohen, A., 1999, A dry Israel must cut water flow to Jordan, in *Ha'aretz,* Jerusalem: Ha'aretz, 15 March 1999.

Cohen, Amiram, 2000, Government to create desalting facility, in *Ha'a'retz [newspaper]*, Jerusalem: Ha'aretz, 18 April 2000.

Cohen, J. E., 1991, *International law and the water politics of the Euphrates, New York University Journal of International Law and Politics* 24, pp 503–556.

Collins, R. O., 1990, *The waters of the Nile: hydropolitics and the Jonglei Canal 1900–1988,* Oxford: Clarendon Press, ISBN 0 19 821784 6

Collins, R. O., 1995, History, hydropolitics and the Nile: Nile control – myth or reality?, in Howell, P. P. and Allan, J. A., *The Nile: sharing a scarce resource,* Cambridge: Cambridge University Press, pp 109–135.

Conway, D., 1993, *The development of a grid-based hydrologic model of the Blue Nile and the sensitivity of Nile discharge to climate change,* PhD thesis, Climate Research Unit, University of East Anglia, UK.

Conway, D. and Hulme, M, 1993, Recent fluctuations in precipitation and run-off over the Nile sub-basins and their impact on Mile discharge, *Climatic Change,* Vol. X.**X.**

Conway, D. and Hulme, M, 1996, The impacts of cliamte variability and future climate change in the Nile basin on water resources in Egypt, *Water Resources Development,* Vol. 12.3, pp 277–296.

Conway, D., Krol, M., Alvamo, J. and Hulme, J., 1996, Future availability of water in 1996: the interaction of global, regional and Basin scale driving forces in the Nile Basin, *Ambio*, Vol. 25, No. 5, August 1996.

Crawford, J., Sands, P. and Chazournes, L. B. de, 2000, *Opinion: in the matter of an application pursuant to the [UK] Export and Investment Guarantees Act and in the matter of the proposed Ilisu Dam [Turkey*], London; Friends of the Earth [NGO], 3 April 2000.

Dabbagh, T., Sadler, P., Al–Saqabi, A. and Sadeqi, M., 1994, Desalination and emergent option, in Rogers, P. and Lydon, P., *Water in the Arab World: perspectives and progress,* Cambridge: Harvard University Press, pp 203–241.

De Angelis, D. L. and Waterhouse, J. C., 1987, Equilibrium and non-equilibrium concepts in ecological models, *Ecological Monographs*, Vol. 57, pp 1–21.

de Haan, Esther J., 1997, Balancing free trade in water and the protection of water resources in GATT, in Brans, H. J. et al. , *The scarcity of water: emerging legal and policy issues,* London-The Hague-Boston: Kluwer International, International Environmental Law and Policy Issues, pp 245–258.

Dellapenna, J., 1995, Why are true markets in water rare? Or why should water be treated as public property, in Haddad, M. and Feitelson, E., *Joint management of shared aquifers,* Jerusalem: The

Palestinian Consultancy Group and the Truman Research Institute for the Advancement of Peace, Hebrew University of Jerusalem, pp 148–168.

Dellapenna, J., 1997, The Nile as a legal and political structure, in Brans, H. J. et al., *The scarcity of water: emerging legal and policy issues*, London-The Hague-Boston: Kluwer International, International Environmental Law and Policy Issues, pp 121–133.

Del Moral, 1996, The debate on the financial and economci regulation of water in contemporary hydrological planning in Spain, in Proceedings of the Erasmus Conference on the *Value of water and water environments*, Department of Geography and Environmental Management, London: University of Middlesex, pp 37–48.

Dinar, A. and Subramanian, A., (editors), 1997, Water pricing experiences: an international perspective, *World Bank Technical Paper No. 386*, The International Bank for Reconstruction and Development, Washington DC, USA.

Dinar, A., (editor), 2000, *The political economy of water pricing reforms*, Oxford: Oxford University Press for the World Bank, Stock No. 61594, ISBN 0 19 521594 X.

Douglas, M., 1970, *Natural symbols: explorations in cosmology*, London: Barrie and Rockcliffe.

Douglas, M. and Wildavsky, A., 1982, *Risk and culture: an essay on the selection of technical environmental dangers*, Berkeley: University of California Press.

du Bois, F., 1995, Water law in the economy of nature, in Allan,, J. A. and Mallat, C., *Water in the Middle East: legal and commercial issues*, London: Tauris Academic Publications, pp 111–126.

Dutton, R. W., 1999, *Changing rural systems in Oman: the Khabura Project*, London and New York: Kegan Paul International, ISBN 0 7103 0607 5, 243 pp.

Dyson, T., 1994, Population growth and food production: recent global and regional trends, *Population and Development Review*, Vol. 20.2, 403.

Eco, U., 1987, *Travels in hyperreality: essays.* Translated from the Italian by William Weaver, London: Pan Books (Translation of: Semiologio quotidiano – English edition originally published: London: Secker & Warburg, 1986).

Editor's Foreword to *International Water Law. Selected Writings of Professor Charles B. Bourne*, London: Kluwer Law International. http://www.dundee.ac.uk/cepmlp/water/html/documents.html

El-Erian, Mohammed A. and Fischer, Stanley, 1996, *Is MENA a region?*, IMF Working Paper, Washington: International Monetary

Fund, Direction of Trade Statistics. (Quoted in Israeli Ministry of Foreign Affairs, 1996, *Programs for regional cooperation,* Document presented to the Middle East/North Africa Economic Summit held in Cairo, November 1996.)

Elmusa, Sharif S., 1996, *Negotiating water: Israel and the Palestinians*, Washington, DC: Institute of Palestinian Studies.

Ergil, D., 1991, The water of Turkey and international problems, *Dis Politika Bulteni,* No. 1, April 1991, pp 40–62.

Evans, T., 1995, History of Nile flows, in Howell, P. P. and Allan, J. A., *The Nile: sharing a scarce resource,* Cambridge: CUP, pp 27–64.

EXACT, 1998, *Overview of Middle East water resources*, Executive Action Team (EXACT), Middle East Water Data Banks Project, a Multilateral Working Group of the Middle East Peace Process, published in Washington: United States Geological Survey.

Falkenmark, M., 1977, Water and mankind—a complex system of mutual interaction, *Ambio*, Vol. 6.1, pp 3–9.

Falkenmark, M., 1979, Main problems of water use and transfer of technology, *Geo Journal,* , Vol. 3, pp 435–443.

Falkenmark, M., 1984, New ecological approach to the water cycle: ticket for the future, *Ambio*, Vol. 13.3, pp 152–160..

Falkenmark, M., 1986a, Fresh waters as a factor in strategic policy and action, in Wasting, A. A. (editor), *Global resources and international conflict: environmental factors in strategic policy and action,* New York: Oxford University Press.

Falkenmark, M., 1986b, Fresh water—time for a modified approach, *Ambio*, Vol. 15.4, pp 194–200.

Falkenmark, M., 1987, Water related limitations to local development, *Ambio,* Vol. 16.4, pp 191–200.

Falkenmark, M., 1989, The massive water scarcity now threatening Africa: why isn't it being addressed, *Ambio*, Vol. 18.2, pp 112–118.

Falkenmark, M., 1990, Global water issues confronting humanity, *Journal of Peace Research*, Vol. 27.2, pp 177–190.

Falkenmark, M., 1997, Water scarcity-challenges for the future, in Brans, H. J. et al., *The scarcity of water: emerging legal and policy issues,* London-The Hague-Boston: Kluwer International, International Environmental Law and Policy Issues, pp 21–39.

Falkenmark. M., 2000, Plants: the real water consumers, *GWP Newsflow*, Stockholm: Global Water Partnership. (see www.gwp.sida.se)

FAO, 1995a, *The water resources of Africa,* Rome: FAO, Land and Water Development Division, Aquistat Programme.

FAO, 1995b, *Irrigation in Africa,* Rome: FAO, FAO, Land and Water Development Division, Aquistat Programme, XXX pp. ISBN 92–5–003727–9.

FAO, 1997a, *Water resources of the Near East region: a review,* Rome: FAO, Land and Water Development Division, Aquistat Programme. 38 pp.

FAO, 1997b, *Irrigation in the Near East region in figures,* Water Reports No. 9, Rome: FAO, Land and Water Development Division, Aquistat Programme. 281 pp.

FAO, 1997c, *FAO production and trade statistics,* most accessible on http:/ /fao.org.

Feitelson, E., 1996, Economic and political dimensions in changing perceptions of water in the Middle East, paper presented at the EU Erasmus Conference on *Valuing water and water environments,* London: Middlesex University Department of Geography and Environmental Management.

Feitelson, E. and Haddad, M. (editors), 1995, *Joint management of shared aquifers,* Harry S. Truman Research Institute, Hebrew University and the Palestinian Consultancy Group, Jerusalem, November 1994.

Feitelson, E. and Haddad, M. (editors), 1996, *Joint management of shared aquifers,* Harry S. Truman Research Institute, Hebrew University and the Palestinian Consultancy Group, Jerusalem, November 1995.

Feitelson, E. and Haddad, M. (editors), 1997, *Joint management of shared aquifers,* Harry S. Truman Research Institute, Hebrew University and the Palestinian Consultancy Group, Bethlehem, May 1996.

Feitelson, E. and Haddad, M. (editors), 1998, *Joint management of shared aquifers,* Harry S. Truman Research Institute, Hebrew University and the Palestinian Consultancy Group, Istanbul, May 1997.

Ferguson, D., Haas, C., Raynard, P. and Zadek, S., 1996, *Dangerous curves,* Report for the New Economics Foundation for the WWF International, Gland (Switzerland): WWF.

Ferguson, J., 1990, *The anti-politics machine*, Cambridge: Cambridge University Press. Also 1999 & 1996 Minneapolis: Minnesota University Press.

Fishelson, G., 1992, The allocation of marginal value product of water in Israeli agriculture, Israeli Agriculture, Reference Number WP/028.

Fisher, F. M., 1994, *The economic framework for water negotiation and management.* The Harvard Middle East water project. Cambridge: Kennedy School of Government, Harvard University.

Fisher, F. M., 1995a, Removing water as a source of conflict. *Middle East International,* 6 January 1995, 20–21.

Fisher, F. M., 1995b, The economics of water dispute resolution, project evaluation and management: an application to the Middle East, Cambridge, Massachusetts: Harvard University, Kennedy School of Government, Institute of Social and Economic Policy in the Middle East. Paper presented at the Stockholm Water Symposium, 14 August 1995.

Foucault, M., 1969, *The order of things: an archaeology of knowledge,* translated by Sheridan-Smith, A. M., London: Tavistock, 1974.

Foucault, M., 1971, Orders of discourse, *Social Science Information,* Vol. 10.2.

Foucault, M., 1980, *Power/knowledge: selected interviews and other writings, 1972–1977*, edited by Colin Gordon; translated by Colin Gordon ... [et al.], Brighton: Harvester Press.

Garang, J. de M., 1981, *Identifying and selecting and implementing rural development in the Jonglei Projects Area*, Southern Sudan, Unpublished PhD thesis at Iowa State University.

Gardaño, H., 1998, Water reform in Mexico: the legal framework, in World Bank, Economic Community, European Investment Bank, *Water for sustainable growth: the MENA/MED water initiative,* Washington: World Bank. Proceedings of a World Bank, EC, EIB Regional seminar on water policy reform, 1–3 June 1998 in Cairo.

Garrfinkle, A., 1994, *War, water and negotiation in the Middle East: the case of the Palestine-Syrian border*, Tel Aviv: Tel Aviv University Press.

Giddens, A., 1984, *The constitution of society*, Cambridge: Polity Press.

Giddens, A., 1990, *The consequences of modernity*, Cambridge: Polity Press.

Gleick, Peter H. (editor), 1993, *Water in crisis: a guide to the world's freshwater resources*, Oxford: Oxford University Press.

Godana, B. A., 1985, *Africa's shared water resources: legal and institutional aspects of the Nile, Niger and Sengal rivers systems,* London: Frances Pinter and Lynne Rienner; Boulder, Colorado.

Gokçe, B., 1998, *Assistance for urban-rural integration and community programmes in Halfeti-Sanliurfa. Projet scope and first year progress report,* Ankara: Prime Minsistry of the GAP Project Regional Development Administration.

Goldblatt, D., 1996, *Social theory and the environment,* Cambridge: Polity Press.

Goldblatt, M., 1996, unpublished MSc, *The provision, pricing and procurement of water: a survey of WTP in two informal settlements in Greater Johannesburg,* Department of Economics, University of the Witwatersrand, Johannesburg.

Goleman, D., 1997, *Vital lies, simple truths,* London: Bloomsbury.
Gray, B. E., 1994, The role of laws and institutions in California's 1991 Water Bank, in Carter, H. O. et al., *Sharing scarcity: gainers and losers in water marketing,* San Francisco, p 133.
Green, C., 1998, Perceptions of the economic value of water and water environments, in Allan, J. A. and Radwan, L., *Perceptions of the economic value of water and water environments,* Proceedings of the Erasmus Workshop held at Middlesex University, Department of Geography and Environmental Management.
Groner, Eli, 2000, Green light for desalination, in *Ha'aretz [newspaper]*, Jerusalem: Ha'a'retz, 18 April 2000.
Güner, Serdar, 1996, The Turkish-Syrian war of attrition: the water dispute, *Occasional Paper,* Ankara: Bilkent University Department of International Relations.
Güner, Serdar, 1997, *Water alliances in the Euphrates-Tigris basin*, Paper at the NATO Advanced Research Workshop, 9–12 October 1996, Budapest, Hungary. Available Department of International Relations, Bilkent University, Ankara.
Gupta, J., Junne, G. and Wurff, Richard van der, 1993, *Determinants of regime formation,* Working Paper 1, Department of International Relations and Public International Law, Institute of Environmental Studies, Amsterdam: Frie Universiteit.
GWP-MENA, 1999, *Report of the water for food group of the MENA Technical Advisory Group*, December 1999, Stockholm: Global Water Partnership
GWP, 2000, *Towards water security: a framework for action*, Report of the Global Water Partnership to the Hague Water Forum, March 2000, Wallingford (UK): HR Wallingford.
Ha'aretz, 2000, Water shortage is a dire threat, *Ha'aretz newspaper,* Jerusalem, Editorial Wednesday, 14 June 2000.
Haas, E. B., 1980, Why collaborate? Issue-linkage and international regimes, *World Politics,* Vol. 32, pp 357–405.
Haas, P. M., 1989, Do regimes matter? epistemic communities and Mediterranean pollution control, *International Organization,* 43.3, summer 1989, pp 377–403.
Haas, P. M., 1992, Introduction: epistemic communities and international policy coordination, *International Organization,* Vol. 46.1., pp 1–35.
Haddad, M. and Feitelson, E., 1995, *Joint management of shared aquifers*, Proceedings of the Workshop held in Bethlehem in February 1995, Jerusalem: Palestinian Consultancy Group and The Truman Institute for the Advancement of Peace of the Hebrew University.

Haddad, M. and Feitelson, E., 1997, Joint management of shared aquifers, Proceedings of the Cyprus meeting in May 1997, Jerusalem: Palestinian Consultancy Group and The Truman Institute for the Advancement of Peace of the Hebrew University.
Haddad, M. and Mizyed, N., 1996, Water resources in the Middle East: conflict and solutions, in Allan, J. A. (editor), *Water and the Middle East Peace Process: negotiating resources in the Jordan Basin,* London: Tauris Academic Studies, pp 1–16.
Haddedin, M., 1995, personal communication.
Haddadin, M., 1996, Water management: a Jordanian viewpoint, in Allan, J. A. (editor), *Water and the Middle East Peace Process: negotiating resources in the Jordan Basin,* London: Tauris Academic Studies, pp 55–70.
Haddedin, M., 2000, personal communication.
Hajer, M., 1996, *The politics of environmental discourse: ecological modernization and the policy process,* Oxford: Clarendon Press.
Hamilton, K., and O'Connor, J., 1994, *Genuine saving and the financing of development,* Washington: World Bank.
Handley, C., 1996, *Baseline socio-economic survey in the Ta'iz Governorate, Yemen,* Sana'a: UNDP, UNDP/DDSMS Project YEM/93/101, January 1996, 52 pp plus annexes.
Handley, C., 1999, *Water resources and water resource development in the Taiz region of Yemen,* Unpublished PhD, London: SOAS in the University of London.
Handley, C. and Dottridge, J., 1997, Causes and consequences of extreme water shortage in Ta'iz, Yemen, *International Association of Hydrologists,* Nottingham Conference, September 1997, 6 pp.
Hardin, G., 1968, The tragedy of the Commons, *Science*, 162, pp 1243–1248.
Hartman, L. and Seastone, D., 1970, Water transfers, in National Water Commission, *Water policies for the future,* Washington: National Water Commission.
Higazi, Abd al-Aziz, 1995, Lecture in London at the Arab Research Centre. Professor Higazi was Minister of Finance under President Nasser from 1968 and became the main economic strategist under President Sadat, Deputy Prime Minister in 1973, Prime Minister from after the October War (1973) until April 1995.
Hillel, D., 1994, *Rivers of Eden: the struggle for water and the quest for peace in the Middle East,* Oxford: Oxford University Press.
Hoekstra. A., 1998, *Perspectives on water: an integrated model-based exploration of the future,* Utrecht: International Books.
Homer-Dixon, T. F., 1991, On the threshold: environmental changes as causes of acute conflict, *International Security*, Vol. 16.1, pp 76–116.

Homer-Dixon, Thomas F., 1994a, Environmental scarcities and violent conflict: evidence from cases, *International Security*, Vol. 19.1 (Summer), pp 5–40.

Homer-Dixon, T. F., 1994b, Environmental scarcities and violent conflict: evidence from cases, *International Security*, Vol. 19.2, pp 5–40.

Homer-Dixon, T. F. and Percival, V., 1996, *Environmental scarcity and violent conflict: briefing book,* American Association for the Advancement of Science.

Howell, P. P. and Lock, M., 1995, The control of the swamps of the southern Sudan: drainage schemes, local effects and environmental constraints on remedial development in the flood region, in Howell, P. P. and Allan, J. A., *The Nile: sharing a scarce resource,* Cambridge: Cambridge University Press, pp 243–280.

Howell, P. P., Lock, M. and Cobb, S., 1988, *The Jonglei Canal: impact and opportunity,* Cambridge: Cambridge University Press.

Howsam, P. and Carter, R. C., 1996, *Water policy: allocation and management in practice,* London: E & FN Spon.

Hudson, H., 1996, Resource-based conflict: water (in)security and its strategic implications, in Solomon, H. (editor), *Sink or swim? Water, resource security an state co-operation.* IDP Monograph Series.

Hunting Technical Services, 1979, *Suez Canal Region Integrated Agrcltural Development Study, Report No. 2,* Hemel Hempstead: Hunting Technical Services.

Hurst, H. H., Black, R. P. and Simaika, Y. M., 1966, *The Nile Basin,* Cairo: Ministry of Public Works.

Ibn Manzur, 1959, *Lisan al-'Arab*, Beirut, Vol. 3, pp 175ff.

IFPRI (International Food Production Research Institute), 1995, *Global food security,* IFPRI Report, October 1995, Washington DC: IFPRI, see IFPRI@www.cgiar.org/ifpri/

IFPRI (International Food Production Research Institute), 1997, *China will remain a grain importer,* IFPRI Report, Feb. 1997, Washington DC: IFPRI, see IFPRI@www.cgiar.org/ifpri/

IIMI, 1997, The mission of the International Institute for Irrigation Management (IIMI)—now known as the International Water Management for Institute, Colombo, Sri Lanka: IIMI.

ILC—International Law Commission, 1994, *Report of the International Law Commission on the work of its forty-sixth session,* New York: UN, GAOR, 49th Session, Supp. No. 10, p 197, UN Doc A/49/10 (19914).

ILC—International Law Commission, 1997, *Convention on the law of the non-navigational uses of international watercourses,* Annex to the United Nations General Assembly Resolution 51/229, 21 May 1997, New York: United Nations.

Inan, K., 1990, The South-East Anatolia Project and Turkey's relations with Middle Eastern countries, *Middle East Business and Banking,* Vol. 9.3, March 1990, pp 4–6.

International Atomic Energy Agency, 1992, *Report: Technical and economic evaluation of potable water production through desalination of sea water,* Vienna: IAEA.

Ionides, M. G., 1953, *The disputed waters of the Jordan,* Middle East Tributary, Vol. 2, pp 153–164.

Izaac, J., and Kuttab, J., 1994, Approaches to the legal aspects of conflict on water rights in Palestine?Israel, in Izaac, J. and Shuval, H., *Water and peace in the Middle East,* Amsterdam: Elsevier, pp 239–250.

Isaac, J., 1999, personal communication.

Isaac, J., and Shuval, H., 1994, *Water and peace in the Middle East,* Amsterdam: Elsevier.

Islam, N., 1995, *Population and food in the early 21st century: meeting future food demand of a rising population,* Washington DC: IFPRI, 239 pp.

Israel Bureau of Statistics, 1994, *Statistical Abstract 1993,* Jerusalem: Central Bureau of Statistics.

Israel-PLO Interim Agreement, 1995, The text is available in Appendices 2 and 3 in Allan, J. A. (editor), *Water and the Middle East Peace Process: negotiating resources in the Jordan Basin,* London: Tauris Academic Studies, pp 223–240.

Jellali, M. and Jebali, A., 1994, Water resource development in the Maghreb countries, in Rogers, P. and Lydon, P. (editors), *Water in the Arab World: perspectives and prognoses.* Cambridge, Massachusetts: Harvard University Press. pp. 147–170.

Jonathan Mann, Atlanta: Cable Network News International, 15 June 1997.

Jordan-Israel Peace Treaty, 1995, Annex II—Water Related Matters, The text is available in Appendices 1 and 2 in Allan, J. A. (editor), *Water and the Middle East Peace Process: negotiating resources in the Jordan Basin,* London: Tauris Academic Studies, pp 207–221.

Junne, G., 1991, Beyond regime theory, *Acta Politica,* Vol. 27.1, January 1992, pp 9–28.

Kabanda, B. and Kahanghire, P., 1995, Irrigation and hydro-power potential and water needs in Uganda, in Howell, P. P. and Allan, J. A., *The Nile: sharing a scarce resource,* Cambridge: Cambridge University Press, pp 217–226.

Karshenas, M., 1994, Environment, technology and employment: towards a new definition of sustainable development, *Development and Change,* Vol. 25.2, pp 723–757.

Katz, D., 1994, A model for evaluating conflict over shared water resources: the Arab-Israeli conflict as a case study, Unpublished thesis, University of Washington, 140 pp.

Katz-Oz, A., 1994, Former Minister of Agriculture and head of Israeli water delegation at the Peace Talks, 1994, quoted in Katz, 1994, p 132.

Keller, A. A. and Keller, J., 1995. *Effective efficiency: a water use efficiency concept for allocating freshwater resources*. Center for Economic Policy Studies, Winrock International, Arlington, Virginia.

Kessler, P., 1997, Economic instruments in water management, in Kay, M., Franks, T. and Smith, L. (editors) *Water: economics, management and demand,* London: E & FN Spon.

Khassawneh, A., 1995, The International Law Commission and Middle East Waters, in Allan, J. A. and Mallat, C., *Water in the Middle East: legal, political and commercial implications,* London: Tauris Academic Studies, pp 21–28.

Khatib, Ahmad, 1999, Jordan 'strongly' rejects Israeli plan to reduce water supplies, *Jordan Times*, Amman: Jordan Times, 16 March 1999.

Kibaroglu, A., 1998, *Management and allocation of the waters of the Euphrates-Tigris basin: lessons drawn from global experiences,* Unpublished PhD dissertation at Bilkent University in the Department of International Relations, Ankara.

Kingdon J., 1984, *Agendas, alternatives and public policies,* New York: Harper-Collins.

Kinnersley, D., 1994, *Coming clean*, London: Penguin.

Kliot, N., 1994, *Water resources and conflict in the Middle East,* London: Routledge.

Kolars, J., 1990, The course of water in the Arab Middle East, *American-Arab Affairs,* Vol. 33, Summer 1990, pp 56–68.

Kolars, J., 1991, *The future of the Euphrates River*, Prepared for the World Bank Conference, Washington DC.

Kolars, J., 1992a, *Fine tuning the future Euphrates-Tigris system*, Presentation to the Council on Foreign Relations, US Government, Washington DC.

Kolars, J. F., 1992b, The future of the Euphtrates River, in Le Moigne, G. (editor), *Country experiences with water resources management: economic, institutional and environmental issues*, World Bank Technical Publication, No 175. Washington: World Bank.

Kolars, J. F. and Mitchell, W. A, 1991, *The Euphrates River and the Southeast Anatolia Development Project,* Carbondale: Southern Illinois University Press.

Kuznets, S., 1973, Modern economic growth: findings and reflections, *American Economic Review,* Vol. 63.

L'vovich, M. I., 1969, *Water resources in the future,* Moscow: Prosveshcheniye. (In Russian)

L'vovich, M. I., 1974, *Water resources and their future,* Moscow: Mysl' Publishing. (In Russian) And 1979 US translation available from the American Geophysical Union, Washington DC.

Lancaster, William and Fidelity, 1999, *People, land and water in the Arab Middle East: environments and landscapes in the Bilad ash-Sham*, Amsterdam: Harwood Academic Publishers, 458 pp.

Langford, J., 1998, The Australian experiences in water resources management, in World Bank, Economic Community, European Investment Bank, *Water for sustainable growth: the MENA/MED water initiative,* Washington: World Bank. Proceedings of a World Bank, EC, EIB Regional seminar on water policy reform, 1–3 June 1998 in Cairo.

Latham, J., *Cropping and water use on the Gefara Plain of Libya,* Study commissioned by the Libyan Government and carried out by FAO, Tripoli: FAO.

Le Heron, R. and Roche, M., 1995, A 'fresh' place in food's space, *Area,* Vol. 27.1, pp 23–33.

Le Heron, R., 1993, *Globalization of agriculture,* Oxford: Pergamon Press.

Libiszewski, S. C., 1991, *Water disputes in the Jordam Basin region and their role in the resolution of the Arab-Israel conflict,* Occasional Paper 13, Zurich: Center for Security Studies and Conflict Research.

Lichtenthaler, G., 1999, *Water management and community participation in the Sa'adah Basin of Yemen,* Study for a World Bank Pilot Project, London: SOAS.

Lichtenthaler, G., 2000, *Water resources in the Sa'adah area*, Yemen, Unpublished PhD thesis, London: SOAS-University of London.

Lipsey, R. G. and Lancaster, K. J., 1956–57, The general theory of the second best, *Review of Economic Statistics*, Vol. 24, pp. 11–32.

List, M., and Rittberger, V., 1992, Regime theory and international environmental management, in Hurrell A., and Kingsbury, B., (editors), *The international politics of the environment,* Oxford: Clarendon Press, pp 85–109.

Littlefair, K., 1998, *Water use and water pricing in southern Kerala,* London: SOAS-University of London, Unpublished dissertation in the Department of Geography.

Lonergan, S. C., 1995, Use of economic instruments for efficient water use, in Haddad, M. and Feitelson, E., *Joint management of shared aquifers,* Jerusalem: The Palestinian Consultancy Group and the Truman reserach Institute for the Advancement of Peace, Hebrew University of Jerusalem, pp 114–127.

Lonergan, S. and Brooks, D., 1993, *The economic, ecological and geo-political dimensions of water in Israel,* Centre for Sustainable Regional Development, Victoria, British Colombia: University of Victoria.

Lonergan, S. C. and Brooks, D., 1994, *Watershed: the role of freshwater in Israeli-Palestinian relations,* Ottawa: IDRC Press.

Lopez-Gunn, E., 1997, The estimation and management of minimum flow in rivers and the role of such information in ecosystem management, Paper given at the SOAS Water Issues Group seminar.

Lowdermilk, W., 1944, *Palestine: the land of promise,* New York: Harper Row.

Lowi, M. R., 1990, *The politics of water under conditions of scarcity and conflict: the Jordan River and riparian states*, Unpublished PhD, Politics Department, Princeton University.

Lowi, M. R., 1994, *Water and power: the politics of a scarce resource in the Jordan River Basin,* Cambridge: Cambridge University Press.

Lundquist, J., 1996, The triple squeeze on water, Paper given at the Erasmus Workshop held at the University of Middlesex. London: University of Middlesex.

Lutz, E. and Munasinghe, M., 1991, Accounting for the environment, *Finance and Development,* World Bank, Washington D. C. 19–21.

Maktari, A. M. A., 1971, *Water rights and irrigation practices in Lahj,* Cambridge: Cambridge University Press.

Mallat, C., 1993, *The renewal of Islamic law,* Cambridge: Cambridge University Press.

Mallat, C., 1995a, Law and the River Nile: emerging international rules and the Sharia'h, in Howell, P. P. and Allan, J. A., *The Nile: sharing a scarce resource,* Cambridge: Cambridge University Press, pp 356–384.

Mallat, C., 1995b, The quest for water use principles: reflections on Shari'a and custom in the Middle East, in Allan, J. A. and Mallat, C. (editors), *Water in the Middle East: legal, political and commercial implications,* London: Tauris Academic Studies, pp 127–150.

Manley, R. E. and Robson, J., 1994, *Hydrological study of the marshes of southern Iraq,* Exeter: Wetlands Ecosystem Research Group, University of Exeter, UK.

Mann, M. E. and Bradley, R. S., 1998, Global climate variations over the past 250 years: relationships with the Middle East, in Albert, J., Berhardsson. M. and Kenna, R., *Transformations of Middle Eastern natural environments: legacies and lessons,* Yale School of Forestry and Environmental Studies, New Haven, Connecticut, Yale University.

Martinez-Alier, J., 1995, Commentary: the environment as luxury good or 'Too poor to be green', *Ecological Economics*, Vol. 13.1, pp 1–10.

Mayer, F. W., 1992, Managing domestic differences in international negotiations: the strategic use of internal side-payments, I*nternational Organization*, Vol. 46.4, Autumn 1992, pp 793–818.

McCaffrey, S., 1995, The International Law Commission adopts draft articles on on international watercourses, *American Journal of International Law,* Vol. 89, pp 395ff.

McCaffrey, S., 1997a, Water scarcity: institutional and legal responses, in Brans, H. J. et al., *The scarcity of water: emerging legal and policy issues,* London-The Hague-Boston: Kluwer International, International Environmental Law and Policy Issues, pp 43–56.

McCaffrey, S., 1997b, Middle East water problems: the Jordan River, in Brans, H. J. et al., *The scarcity of water: emerging legal and policy issues,* London-The Hague-Boston: Kluwer International, International Environmental Law and Policy Issues, pp 158–165.

McCaffrey, S., 1998, Legal issues in the United Nations Convention on International Watercourses: prospects and pitfalls, Paper delivered at World Bank Seminar on international watercourses, Washington: World Bank, November 1997.

McCaffrey, S. and Sinjela, M., 1998, The United Nations Convention on International Watercourses, *American Journal of International Law,* Vol. 92, pp 97ff.

McCalla, A., 1997, The water, food and trade nexus, Paper delivered at the MEN-MED conference convened by the World Bank in Marrakesh, May 1997.

McCulley, P., 1996, *Silenced rivers: the ecology and politics of large dams,* London: Zed Books.

McKenzie, H. S. and Elsaleh, B. O., 1995, The Libyan Great Man-Made River project: project overview, *Proceedings of the Institute of Civil Engineers: Water, Maritime and Energy,* Vol. 106, Water Board Panel Paper 10340, pp 103–122.

McLachlan, K. S., 1988, *The neglected garden: the politics and ecology of agriculture in Iran,* London: I.B. Tauris.

McNicholl, G., 1984, Consequences of rapid population growth: an overview and assessment, *Population and Development Review,* Vol. 10.2, pp 177–240.

Meade, J. E., 1955, *Trade and welfare,* London: Oxford University Press.

Medzini, A., 2000a, *The Euphrates River: an analysis of a shared river system in the Middle East,* London: SOAS, University of London.

Medzini, A., 2000b, *The River Jordan: the struggle for frontiers and water:* 1920–*1967,* London: SOAS Water Issues Group

Meinzen-Dick, R. and Ro Segrant, M. W., 1997, Water as an economic good: incentives, institutions and infrastructure, in Kay, M., Franks, T. and Smith, L. (editors) *Water: economics, management and demand,* London: E & FN Spon.

Meir, Meir Ben, 1994, Water management policy in Israel: a comprehensive approach, in Isaac, J. and Shuval, H. (editors), *Water and peace in the Middle East,* Amsterdam: Elsevier, pp 35–39.

Mendeluce, J. M. M. and del Rio, J. P., 1998, Water challenges and responses: the Spanish experience, World Bank, Economic Community, European Investment Bank, *Water for sustainable growth: the MENA/MED water initiative,* Washington: World Bank. Proceedings of a World Bank, EC, EIB Regional seminar on water policy reform, 1–3 June 1998 in Cairo.

Merrett, S., 1996, *An introduction to the economics of water,* London: UCL Press.

Migdal, J., 1988, *Strong societies and weak states: state-society relations and state capabilities in the Third World,* Princeton: Princeton University Press.

Ministry of Energy and Infrastructure, 1995, *Economic reassessment of the Dead Sea Hydro Project,* Ministry of Energy and Infrastructure, State of Israel, Jerusalem, January 1995.

Ministry of Oil and Mineral Resources, 1995, *The water resources of Yemen: a summary and digest of available information*, San'a and Delft: Ministry of Oil and Mineral Resources, General Department of Hydrology and Institute of Applied Geoscience/TNO, Delft.

Ministry of Water and Irrigation, 1995, *Water resources data for Jordan,* Amman: Ministry of Water and Irrigation.

Mitchell, R. E., 1993, Management for efficient and equitable use. In Kay, M., Franks, T. and Smith, L., (editors) *Water: economics, management and demand,* London: E & FN Spon.

Mohieldeen, Y., 1999, *Responses to water scarcity: social adaptive capacity and the role of environmental information. A case study from Ta'iz, Yemen,* Unpublished MA thesis completed at SOAS in the University of London. See www.soas.ac.uk/geography/waterissues

Morris, J., 1996, Water policy: economic theory and political reality, in Howsam, P. and Carter, R. C., *Water policy: allocation and management in practice,* London: E & FN Spon, pp 228–234.

Morris, J., Weatherhead, E. K., Dunderdale, J. A. L., Green, C. and Tunstall, S., 1997, Feasibility of tradeable permits for water abstraction in England and Wales, in Kay, M., Franks, T. and Smith, L. (editors) *Water: economics, management and demand,* London: E & FN Spon.

Mubarak, Ali. (administrator and engineer in Egyptian governments between 1848 and 1892), 1891, in *al-Azhar*, Vol. 4, No. 10, May 1891, pp 309–315.

Mubarak, Ali, 1891, Sharah al hadith al-nabwi—a saying of the Prophet in the *Hadith*, as quoted by 'Ali Mubarak (administrator and engineer in Egyptian governments between 1848 and 1892) in *al-Azhar*, Vol. 4, No. 10, May 1891, pp 309–315.

Murugan, G., Pushpangadan, K. and Navaneetham, K., 1996, *Travel time, user rate and cost of supply, drinking water in rural Kerala, India,* Working Paper No. 266, Centre for Development Studies, Kerala, India.

Myers, N., 1989, Population growth, environmental decline and security issues in Sub-Saharan Africa, in Hjort af Ornäs, A. and Mohamed Salih, M. A., (editors), *Ecology and politics: environmental stress and security in Africa,* Stockholm: Scandinavian Institute of African Studies.

Naff, T., 1992, The Jordan basin: political, economic and institutional issues, in LeMoigne, G., Barghouti, S., Feder, G., Garbus, L. and Xie, M. (editors), *Country experiences with water resources management: technical and environmental issues.* World Bank technical paper No, 175, Washington DC: The World Bank.

Naff, T. and Matson, R., 1984, *Water in the Middle East: conflict or cooperation,* Boulder, Colorado: Westview.

Nasser, Y., 1996, Palestinian management options and challenges within an environment of scarcity and power imbalance, in Allan, J. A. (editor), *Water, peace and the Middle East,* London: Tauris Academic Press. pp. 45–56.

Nasser, Y., 1996, Management options and challenges within an environment of scarcity and power imbalance, in Allan, J. A. (editor),*Water and the Middle East Peace Process: negotiating resources in the Jordan Basin,* London: Tauris Academic Studies, pp 55–70.

National Water Commission, 1973, *Water policies for the future,* Washington DC: National Water Commission.

Newspot, 13 August 1992, Ankara.

Nicol, A, 2000a, A wider definition of water security: socio-political security, personal communication.

Nicol, A., 2000b, *Decentralisation and river basin management in Ethopia* Unpublished PhD in the University of London.

Nile 2002, 1993–97, Annual meetings held in Nile basin countries, Aswan 1993, Khartoum 1994, Arusha 1995, Entebbe 1996, Addis Ababa 1997, Kigali 1998, Tanzania 1999, Addis Ababa 2000.

Nile Waters Agreement, 1960, Agreement between the United Arab Republic and the Republic of the Sudan for the full utilisation of the Nile Waters, signed at Cairo, 8 November 1959, and the Protocol Concerning the Permanent Joint Technical Committee, signed in Cairo, 17 January 1960, in *Revue égyptienne de droit international,* Vol. 15, 1960.

North, D., 1990, *Institutions, institutional change and economic performance,* Cambridge: Cambridge University Press.

Ochet, A., 1999, *Fantasy economics,* Unesco Courier, Paris: Unesco, March 1999, pp 31–32.

Ohlsson, Leif, 1999, Environment, scarcity, and conflict—a study of Malthusian concerns, PhD dissertation (12 February), Dept. of Peace and Development Research, University of Göteborg.

Ohlsson, L. and Turton, A. R., 1999, The turning of a screw: social resource scarcity as a bottle-neck in adaptation to water scarcity, *Occasional Paper No. 19*, SOAS Water Issues Group. www.soas.ac.uk/geography/waterissues/

Oman WRA, 1998, *National master water plan for Oman,* Water Resources Authority of Oman: Ruwa, Muscat.

Oman, 1995, Proceedings of the Muscat Seminar on Water Resources in the Middle East, 1995, Muscat: Ministry of Water Resources.

Oslo Accord, 1996, available as Appendix 2—The Oslo Accord, 1995, in Allan, J. A., 1996, *Water and peace in the Middle East*, London: Tauris Academic Studies. pp 227–242.

Pacific Consultants, 1980, *New Lands productivity in Egypt: technical and economic feasibility,* Contract No. NE-C-1645, Project No. 263–0042, Cairo: USAID.

Palestinian National Authority, 1999, *Draft Water Law (Arabic),* Ramallah: Palestinian National Authority.

Pallas, P. P., 1976, Water resources of Libya, in Salem, M. J. and Busrewil, M. T., (editors), *The Geology of Libya,* Vol. II, London: Academic Press, pp 539–594.

Palmer, S. E., 1963, Comments of Stephen E. Palmer, Jr., at the American Embassy, Tel Aviv, on the Jordan Waters Contingency Planning sent to the Department of State, 23 October 1963. Ref POL 33-1 ISA-Jordan.

Pearce, D., Markandya and Barbier, E., 1989, *Blueprint for a green economy*, Earthscan: London.

Pearce, D. and Turner, K. T., 1990, *Economics of natural resources and the environment.* New York and Hemel Hempstead: Harvester Wheatsheaf.

Pearce, F., 1992, *The dammed: rivers, dams, and the coming world water crisis,* London: The Bodley Head.

Penrose, E.T., 1971, The economic setting, in Allan, J. A., Mclachlan, K. S. and Penrose, E. T., *Libya: agriculture and economic development,* London: Frank Cass, pp 1–8.

Perry, C., 1996, *Alternative approaches to cost sharing for water services for agriculture in Egypt,* IIMI Research Report 2, Colombo: IWMI.

Perry, C., 2000, personal communication.

Perry, C., Rock, A. and Seckler, 1997, *Water as an economic good: a solution or a problem,* IIMI Research Report 14, Colombo: IWMI.

Pigram, J. J., 1997, The value of water in competing uses in water, in Kay, M., Franks, T. and Smith, L, (editors), *Water: economics, management and demand,* London: E & FN Spon.

Postel, S., 1992 and 1997, *Last oasis: facing water scarcity,* New York: Norton, ISBN 0 393 31744 7. See www.worldwatch.org

Postel, S., 1999, Pillar of sand: can the irrigation miracle last, New York: Norton—see www.worldwatch.org

Prakash, B. A., 1994, *Kerala's economy, performance, problems, prospects,* New Delhi: Sage Publications India Pvt Ltd.

Radwan, L., 1994, Water management in the Nile Delta, Unpublished Oxford University DPhil thesis.

Radwan, L., 1996, Irrigation in delta Egypt, *Geographical Journal,* Vol. XX, pp 256–267.

Redclift, M., 1987 [1995 reprint], Sustainable development: exploring the contradictions, London: Routledge.

Reisner, M., 1993, *Cadillac desert,* New York: Penguin Books. Reprinted 1993.

Renner, M., Pianta, M. and Franchi, C., 1991, International conflict and environment degradation, in Väryan, R. (editor), *New directions in conflict theory: conflict resolution and conflict transformation,* Washington DC: International Social Science Council.

Renner, M., Pianta, M. and Franchi, C., International conflict and environment degradation, in Väryan, R. (editor), *New directions in conflict theory: conflict resolution and conflict transformation,* Washington DC: International Social Science Council.

Reuters (Cairo), 1997, Report of speech by Egypt's Minister of Agriculture, Cairo: Reuters, 12 May 1997.

Reuters (Dubai), 1997a, *Feature on water policy in Saudi Arabia,* Dubai: Reuters, April 1997.

Reuters (Dubai), 1997b, *Saudis battle to save water,* Dubai: Reuters, 28 July 1997.

Roberts, D. G. M. and Flaxman, E., 1985, Greater Cairo wastewater project: history, development and management, *Institute of Civil Engineers Proceedings*, Vol. 78.1, August 1985.
Roberts, D. G. M. and Fowler, D., 1995, *Built by oil,* Reading: Ithaca.
Rodda, J., 1996, Estimates of the global availability of freshwater, Displayed at a lecture at the Water Issues Group seminar at SOAS, University of London, March 1996.
Rogers, P., 1995, 1994, The agenda for the next thirty years, in Rogers, P. and Lydon, P. (editors), *Water in the Arab World: perspectives and prognoses,* Cambridge, Massachusetts: Harvard University Press., pp 285–316.
Rogerson, C., 1996, Willingness to pay for water: the international debates, *Water Studies Association*, 22.4, pp373–80.
Rosengrat, M. and Schleyer, R. G, 1995, *Vision for food, agriculture and environment,* A 2020 Vision IFPRI Paper, June 1995, Washington DC: IFPRI (International Food Pcgiar.org/ifpri/
Roy, A., 1999, *The cost of living,* London: London: Harper Collins (Flamingo).
Sabry, Ismail, 1995, The Arab League agricultural projects in Sudan, in Arab Research Centre, *Water, food and trade in the Middle East,* Proceedings of a conference in Cairo convened by the Arab Research Centre and the Ministry of Agriculture and Land Reclamation, London: Arab Research Centre.
Saeijis, H. L. F. and van Berkel, M. J., 1997, The global water crisis: the major issue of the twenty-first century, a growing explosive programme, in Brans, H. J. et al., *The scarcity of water: emerging legal and policy issues,* London-The Hague-Boston: Kluwer International, International Environmental Law and Policy Issues, pp 3–20.
Said, E., 1986, *Orientalism*, London:
Said, R., 1993, *The river Nile: geology, hydrology and utilization,* Oxford: Pergamon Press, ISBN 008 041886 4.
Said, R., 1995, Origin and evolution of the Nile, in Howell, P. P. and Allan, J. A., *The Nile: sharing a scarce resource,* Cambridge: CUP, pp 17–26.
Salameh, E., 1992, *Water resources of Jordan, Present status and future potentials,* Amman: Fridrich Ebert Stiftung/Royal Society for the Conservation of Nature.
Sax, J. I., 1994, Understanding transfers: community rights in the privatisation of water, *West-Northwest,* Vol. 1, p 13.
Schiffler, M. et al., 1994, *Water demand management in an arid country: the case of Jordan with reference to industry.* Berlin: German Development Institute.

Schur, M., 1994, The need to pay for services in the rural water sector, *South African Journal of Economics*, 62.4, pp 419–431.
Seckler, D. and de Silva, R. 1996, *The IIMI Indicator of International Water Scarcity*, Research Report, International Irrigation Management Institute, Colombo, Sri Lanka.
Sen, A., 1981, *Poverty and famines: an essay on entitlement and deprivation*, Oxford: Oxford University Press.
Serageldin, I., 1993, *Water supply, sanitation, and environmental sustainability: the financing challenge*, Washington: World Bank Directions in Development Paper.
Serageldin, Ismail, 1994, *Water supply, sanitation, and environmental sustainability*, Washington DC: The World Bank, Directions in Development series, ISBN 0 8213 3022 5.
Serageldin, Ismail, 1995, *Toward sustainable management of water resources*, Washington DC: The World Bank, Directions in Development series, ISBN 0 8213 3413 1.
Shafik, N., 1994, Economic development and environmental quality: an econometric analysis, *Oxford Economic Papers*, No. 46 (October Supp), pp 757–773.
Shapland, G., 1997a, *Rivers of discord: international water disputes in the Middle East*, London: Hurst & Company.
Shapland, G., 1997b, *Rivers of conflict*, London: Hurst Publishers.
Sharett, M., 1952, Speech to the Knesset of Prime Minister Moshe Sharett, quoted in Feitelson and Haddad 1994, p 73.
Sharon, Ariel, 1999, Speech of the Israeli Infrastructure Minister Ariel Sharon to a German delegation quoted in *Ha'aretz [newspaper]*, Jerusalem, 3 February 1999.
Sherk, W., Wouters, P. and Rochford, S., 1999, Water wars in the near future? Reconciling competing claims for the world's diminishing freshwater resources—the challenge of the next millennium, Paper of the CEMLP at Dundee University. http://www.dundee.ac.uk/cepmlp/water/html/documents.html
Sherman, M., 1999, *The politics of water in the Middle East: an Israeli perspective on the hydro-political aspects of the conflict*, London: Macmillan Press.
Shiffler, M., 1995, *The economcis of water in industry of Jordan*, Berlin: GTZ.
Shiklamanov, I. A., 1985, Large-scale water transfer, in Rodda, J. (editor), *Facets of hydrology*, Vol. II, London: John Wiley, pp 345–387.
Shiklamanov, I. A., 1986, Water consumption, water availability and large–scale water projects in the world, in *Proceedings of the international symposium on the impact of large water projects on the environment*, UNESCO, Paris, 21–31 October 1986.

Shiklamanov, I. A., 2000, Appraisal and assessment of world water resources, *Water International*, Vol. 25.1, pp 11–32.
Shiklamanov, I. A. and Markova, O. A., 1987, *Specific water availability of runoff transfers in the world,* Leningrad: Gidrometeoizdat.
Shiklamanov, I. A. and Sokolov, A. A., 1983, Methodological basis of world water balance investigation and computation, in Proceedings of the Hamburg Workshop, August 1981, IAHS Publication Bo. 148, pp 77–91.
Shuval, Hillel, 1996, Towards resolving conflicts over water between Israel and its neighbours: the Israeli-Palestinian shared use of the Mountain Aquifer as a case study, in Allan, J. A., *Water, peace and the Middle East,* London: Tauris Academic Studies, pp 137–168.
Shuval, Hillel, 1997, Contribution to *Insight* with Jonathan Mann, Atlanta: Cable Network News International, 15 June 1997.
Slaughter Burley, A.-M., 1993, International law and international relations theory: a dual agenda, *The American Journal of International Law*, Vol. 87, No. 205, pp 205–239.
Smith, C. G., 1966, The disputed water of the Jordan, *Transactions of the Institute of British Geographers*, 40.1, pp 111–119.
Soffer, A., 1999, *Rivers of fire,* New York: Rowman & Littlefield.
Southern Development Investigation Team, 1955, *Natural resources and development potential in the Southern Provinces of the Sudan,* Khartoum: Sudan Government.
Stein, A. A., 1990, *Why nations cooperate,* Ithaca: Cornell University Press.
Stevens, G., 1965, *The Jordan River partition,* Stanford, California: Stanford University Press.
Stoner, R., 1995, Future irrigation planning in Egypt, in Howell, P. P. and Allan, J. A., *The Nile: sharing a scarce resource,* Cambridge: Cambridge University Press, pp 281–298.
Storer, D., 1994, The role of privatisation in water management, in Allan, J. A. and Mallat, C. (editors), *Water in the Middle East: legal, political and commercial implications,* London: Tauris Academic Studies, pp 261–269.
Stott, P. A. and Sullivan, S., *Political ecology*, London: Edward Arnold.
Sutcliffe, J., and Parks, Y. P., 1987, Hydrological modelling of the Sudd and Jonglei Canal, *Hydrological Science Journal,* Vol 32, pp 143–159.
Sutcliffe, J., and Parks, Y. P., 1989, Comparative water balances of selected African wetlands, *Hydrological Science Journal,* Vol. 34, pp 49–62.

Sutcliffe, J., and Parks, Y. P., 1995, The water balance of the Bahr el Ghazal swamps, in Howell, P. P. and Allan, J. A., *The Nile: sharing a scarce resource,* Cambridge: Cambridge University Press, pp 281–298.

Sutcliffe, J. and Lazenby, J., 1995, Hydrological data requirements for planning Nile management, in Howell, P. P. and Allan, J. A., *The Nile: sharing a scarce resource,* Cambridge: Cambridge University Press, pp 163–192.

Swyngedouw, E., 1999a, *Modernity and hybridity – the production of nature: water and modernisation in Spain,* Paper presented to the SOAS Water Issues Study Group, University of London, 25 January 1999.

Swyngedouw, E., 1999b, *Sustainability, risk and nature: the political ecology of water in advanced countries,* Proceedings of a workshop convened in the University of Oxford, 15–17 April 1999. Available from the Geography Department, University of Oxford.

Tarlock, A. D., 1997, Current trends in United States water law and policy: private property rights, public interest limitations and the creation of markets, in Brans, Edward H. P., J. de Haan, Esther J., Nollkaemper, André, and Rinzema, Jan, *The scarcity of water: emerging legal and policy responses,* London-The Hague-Boston: Kluwer International, ISBN 90 411 0657 X, 1997, pp 183–195.

TeccoNile, 1996, *Nile River Basin Action Plan,* Cairo: TeccoNile Secretariat.

Thesiger, W., 1967, *The marsh Arabs,* London: Harmondsworth Press/ Penguin.

Thompson, M., 1988, Socially viable ideas of nature: a cultural hypothesis, in Baark, E., and Svedin, U. (editors*), Man, nature and technology: essays on the role of ideological perceptions,* London: Macmillan Press.

Thompson, M. and Warburton, M., 1988, Uncertainty on a Himalayan scale, in Ives, J. and Pitt, D. C., *Deforestation: social dynamics in watersheds and mountain ecosystems,* London: Routledge. pp 1–53, ISBN 0 415 00456 X.

Todini, E. and O'Connell, P. E., 1979, *Hydrological simulation of Lake Nasser,* IBM/IH UK (also WMP technical Report 15, Cairo.)

Toussoun, Prince Omar, 1925, *Memoire sur l'histoire du Nil,* MIE, Vol. IX.

Tripp, C. H., 1996, personal communication.

Turton, A., 1997, The hydropolitics of Southern Africa: the case of the Zambesi River Basin as an area of potential co-operation based on Allan's concept of 'virtual water', Unpublished masters thesis in international relations, University of South Africa.

Turton, A., 1999, Impediments to inter-state co-operation in international river basin commissions within arid regions: can existing theory allow for predictability?, *Occasional Paper 7*, London: SOAS Water Issues Group, see www.soas.ac.uk/geography/waterissues/

Turton, A., 1999, personal communication.

UNCED, 1992, *The UN Conference on Environment and Development*, United Nations: Rio de Janeiro.

Unver, I. H. O, 1999, *Speeches and presentation on the Souteastern Anatolia Project (GAP) delivered by Olcay Unver*, President of the GAP Regional Development Administration.

USAID (US Agency for International Development), 1996, *Environmental Health Project, coping with intermittent water supply: problems and prospects,* Dehra Dun, Uttar Pradesh, India, Activity Report, 26.

US Bureau of Reclamation, 1958, *Studies of the Abay River water resources*, Denver: US Bureau of Reclamation.

US Corps of Engineers, 1957, *The Ethiopian valleys*, Study by The US Corps of Engineers, Washington DC: US Corps of Engineers.

USGS, 2000, *Report on the marshes on the lower Tigris-Euphrates,* Washington: USGS. website www.usgs.gov

Vesilind, P. J., 1993, Water: the Middle East's critical resource. *National Geographic,* May 1993.

Vidal, J., 1995, Ready to fight to the last drop, *Guardian Weekly*, 13, 20 August 1995.

Voice of Israel, 1991, Agriculture Ministry announces reduction in water quotas, *Voice of Israel*, Jerusalem and IDF Radio, Tel Aviv, 2200 gmt, 27 January 1991.

Waldner, D., 1999, *State building and late development,* Ithaca: Cornell University Press.

Wali, Y., 1997, Egypt wants to withdraw from its key position in global grain markets, Reuters, Cairo: Reuters, 12 May 1997.

Ward, C., 1997, *Reflected in water,* London: Cassell.

Ward, C., 1998, *Yemen: towards a water strategy, an agenda for action,* Sana'a: The World Bank, Report No. 15718-YEM.

Ward, R. C. and Robinson, M., 1990, *Principles of hydrology,* Maidenhead (UK): McGraw-Hill.

Waterbury, J., 1978 , *Egypt: burden of the past, options for the future,* Bloomington and London: Indiana University Press.

Waterbury, J., 1979, *The hydropolitics of the Nile,* Syracuse: University of Syracuse Press, ISBN 0 8156 2192 2.

Waterbury, J., 1994, Transboundary water and the challenge of international cooperation in the Middle East, in Rogers, P. and Lydon, P. (editors), *Water in the Arab World: perspectives and prognoses.* Cambridge, Massachusetts: Harvard University Press, pp 39–64.

Whittington, D., Lauria, D. T., Okun, D. A. and Mu, X., 1989, Water vending activities in developing countries: a case study of Ukunda, Kenya, *Water Resources Development,* Vol. 5:3, September 1989.

Whittington, D., Mu, X. and Roche, R., 1990, Calculating the value of time spent collecting water: some estimates for Ukunda, Kenya, *World Development,* Vol. 19:2.

Whittington, D. and McClelland, E., 1992, Opportunities for regional and international co-operation in the Nile Basin, *Water International,* 17, pp 144–154.

Whittington, D., Lauria, D. T., Okun, D. A. and Mu, X., 1991, A study of water vending and willingness to pay for water in Onitsha, Nigeria, *Water Development,* Vol. 19, pp 2–3.

Wilcox, W., and Craig, J. I., 1913, *Egyptian irrigation,* 3rd edition, London: Spon.

Wilkinson, J., 1977, *Water and tribal settlement in South-East Arabia: a study of the aflaj of Oman*, Oxford Research Studies in Geography, Oxford: Clarendon Press.

Winpenny, J., 1991, *Values for the environment: a guide to economic appraisal,* London: Her Majesty's Stationery Office.

Winpenny, J., 1994, *Managing water as an Economic resource*, London: Routledge, ISBN 0–415–10378–9.

Winpenny, J. T., 1996, The value of water valuation, in Howsam, P. and Carter, R. C., *Water policy: allocation and management in practice,* London: E & FN Spon, pp 197–204.

Winpenny, J. T., 1997, *Demand management: an international perspective,* World Bank Technical Paper No. 386, The International Bank for Reconstruction and Development, Washington DC, USA

WMO—World Meteorological Organization, 1995, *African conference on water resources: policy and assessment. Report of the Conference,* Addis Ababa: WMO and UN Economic Commission for Africa.

WMO—World Meteorological Organization, 1998, *Global estimates of water,* Geneva: World Meteorological Organization.

Wolf, A. T., 1995a, *Hydropolitics along the Jordan River: scarce water and its impact on the Arab-Israeli conflict*, Tokyo: United Nations University Press.

Wolf, A. T., 1995b, International dispute resolution: the Middle East Multilateral Working Group on Water Resources, *Water International,* Vol. 20.2, pp 141–150.

Wolf, A., 1996a, *'Hydrostrategic 'territory in the Jordan Basin: water, war and Arab-Israeli peace negotiations,* Tuscaloosa, Alabama: University of Alabama, Department of Geography.

Wolf, A. T., 1996b, Middle East water conflicts and directions for conflict resolution, 2020, *Vision Initiative Monograph* No. 12, Washington DC: International Food Policy Research Institute.

Wolf, A. T., 2000a, Hydrostrategic territory in the Jordan basin: water, war and Arab-Israeli peace negotiations, in Amery, A. A. and Wolf, A. T.,*Water in the Middle East: a geography of peace,* Austin, Texas: University of Texas Press.

Wolf, A. T., 2000b, Indigenous approaches to water conflict negotiations and implications for international waters, *International Negotiation,* Special issue *Negotiating in International Watercourses: Water Diplomacy, Conflict and Cooperation*, Vol X.X, pp xxx–xxx. [in press]

Wolf, A. T., 2000c ongoing, *International Freshwater Treaties Database and Interstate Freshwater Compacts Database,* Corvallis: Oregon State University, Department of Geography.

World Bank, 1992, *Operational Procedure on international waters OP 7.5*, Washington: World Bank.

World Bank, 1993, *Water resources management,* Washington DC: The World Bank, A World Bank Policy Paper, 140 pp. ISBN 0 8213 2638 8.

World Bank, 1997a, *World development report: the state in a changing world,* Washington: The World Bank, 265 pp. ISBN 0 19 521114 6.

World Bank, 1997b, *Water, food and trade,* Washington: World Bank. Proceedings of a World Bank, EC, conference on Mediterranean Development, May 1997 in Marrakech.

World Bank, 1998a, *National water plan of the Water Resources Authority of the Palestinian Authority*, Jerusalem: World Bank/WRA of the Palestinian Authority.

World Bank, 1998b, *MENA Water Initiative,* Papers of the meeting held in Cairo, Washington: World Bank.

World Grain Council, 1997, *Monthly cereal price statistics,* London: World Grain Council.

World Water Council, 1999, *Water for food in the MENA region,* Paris: World Water Council Working Group on Water for Food in the MENA region.

World Water Council, 2000, *A Vision for World Water at the Millennium*, prepared for the World Water Forum Marc 2000, The Hague: World Water Council.

Wouters, P. K., 1998, *International water law*, Unpublished doctoral thesis, Geneva: Graduate Institute of International Studies & University of Geneva.

Wouters, P. K., 1999a, *Geneva strategy and framework for monitoring compliance with international watercourse agreements: a proposed compliance review procedure,* Dundee: Dundee University, Water Law and Policy Programme, University of Dundee, Scotland. http://www.dundee.ac.uk/cepmlp/water/html/documents.html

Wouters, P. K., 1999b, The relevance and role of water law in the sustainable development of freshwater. Replacing 'hydro-sovereignty' and vertical proposals with 'hydro-solidarity' and horizontal solutions, Paper at the Stockholm International Water Institute Annual Conference 1999.

Wouters, P. K., 2000, personal communication.

WWC, 2000, see World Water Council, 2000.

Young, O. R., 1989, *International cooperation: building regimes for natural resources and the environment,* Ithaca: Cornell University Press.

Young, R. A., 1986, Why are there so few transactions among water users?, *American Journal of Economics,* Vol. xx.

Yusf 'Ali, A., 1971, The glorious Qur'an, translation and commentary, note number 1044 to 7:73; see also the commentary by Al-Jamal, Sulayman b. 'Umar, Al-Futuhat al-ilahiyya, on 54:28; and M. B al-sadr, Iqtisadund, Beirut, 1981, pp 519–523 and passim.

INDEX